4c

3

Introduction to Computer Methods for Microwave Circuit Analysis and Design

For a complete listing of the *Artech House Microwave Library*, turn to the back of this book . . .

Introduction to Computer Methods for Microwave Circuit Analysis and Design

Janusz A. Dobrowolski

Warsaw University of Technology

Artech House
Boston • London

© 1991 Artech House, Inc.
685 Canton Street
Norwood, MA 02062

International Standard Book Number: 0-89006-505-5
Library of Congress Catalog Card Number: 91-4553

10   9   8   7   6   5   4   3   2   1

To my son, Mateusz

Contents

Preface

This book presents appropriate parts of microwave circuit theory, and the mathematical and numerical methods required for computer-aided analysis and design of linear microwave circuits.

The book has eleven chapters. Chapter 1 is an introduction. Chapter 2 presents matrix representations of microwave circuits used in computer-aided analysis and design. The next four chapters concentrate on the theory and computational methods associated with frequency-domain analysis, noise analysis, and sensitivity analysis of linear microwave circuits. Chapters 6, 7, and 8 deal with matrix equation solution techniques used in computer-aided circuit analysis. The sparse matrix techniques are discussed.

Computer-aided tolerance analysis methods are discussed in Chapter 9. Chapter 10 deals with tolerance design methods. The final chapter of the book presents optimization methods and their application to computer-aided design (CAD) of microwave circuits.

The book is intended for graduate courses or as a self-study handbook for microwave engineers and researchers who are interested in CAD of microwave circuits. The reader is assumed to have completed courses in undergraduate network theory, electronics, and microwaves, or their equivalents.

The text of this book has been used for several years at the Warsaw University of Technology to teach a fourth-year undergraduate-level course and graduate courses in electrical engineering.

The author wishes to thank his colleagues and students for their helpful comments, especially the invaluable suggestions and constructive criticisms of the text from Dr. Jacek Wojciechowski, Dr. Andrzej Ruszczynski, and Dr. Zbigniew Nosal of the Warsaw University of Technology.

JANUSZ DOBROWOLSKI
WARSAW UNIVERSITY OF TECHNOLOGY
SEPTEMBER 1990

Chapter 1

Introduction

Since the 1970s, significant progress has been made in the computer-aided design of microwave circuits. A computer with appropriate software has become an indispensable tool in all phases of microwave circuit design procedures. Proper use of the computer in the design procedure allows better understanding of the particular design problem, which, as a result, leads to a significant decrease in the time and cost of experimental investigations. This fact is important for the design and manufacture of modern microwave networks because of the limitations for incorporating any modifications in circuits fabricated by microwave integrated circuit (MIC) technology and monolithic microwave integrated circuit (MMIC) technology.

Figure 1.1 schematically presents the sequence of steps for the CAD procedure that have to be performed to design a microwave circuit. In the initial design stage, the microwave engineer mostly uses his or her knowledge of microwaves, experience, and intuition to determine the configuration and topology of the circuit. The performance of the initial circuit design is evaluated by computer-aided analysis methods. Computer-aided analysis allows application of accurate and sophisticated models of passive and active circuit elements. Thanks to the computer, we may quickly and efficiently investigate alternative designs.

In the next stage of the CAD procedure, characteristics of a designed circuit computed by circuit analysis subroutines are compared with the given specifications. If the results do not satisfy the desired specifications, the circuit design parameters must be appropriately changed. The sequence of *circuit analysis, comparison of circuit characteristics with design specifications*, and then *parameter modification* is performed until acceptable performance goals for the circuit are met. This kind of CAD of microwave circuits is performed by using optimization methods. The optimization methods also may be used for parameter estimation of passive and active devices on the basis of experimental data.

Information on the influence of small changes in design parameters on circuit characteristics provides *sensitivity analysis*. Sensitivity analysis may be used in the

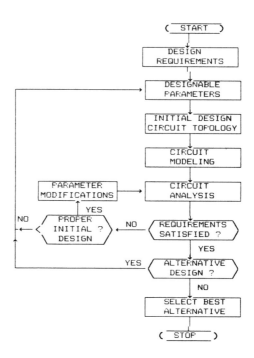

Figure 1.1 Computer-aided design procedure.

optimization subroutines for efficient computation of the derivatives (gradient vector) of performance functions.

Because design should correspond to practical circuits that are built in a manufacturing environment, parameter tolerance analysis should be used to determine *worst-case design*. Random variations in the parameter values are simulated by Monte Carlo techniques. By using statistical circuit analysis, the random variation expected during manufacturing may be taken into account by the designer. Significant increases in manufacturing yield can be obtained by using statistical rather than nominal design. Worst-case circuit design and statistical circuit design are very important for the manufacture of high-yield MMICs. In the case of the MMICs, circuit parameter values are especially difficult to control.

In the last few years, minicomputers and microcomputers have become relatively inexpensive so that small companies and individual engineers can buy and use them in their microwave work. Obviously, computational methods will become increasingly important because the means of applying them in microwave circuit design will continue to grow. In parallel with the progress in computer hardware technology, four major innovations in numerical methods have influenced all aspects of computer-aided microwave network analysis and design. These milestone CAD methods of electronic and microwave circuits are *sparse matrix techniques* applied for solution of systems of linear equations, *adjoint circuit and system methods* for sensitivity computation, application of the efficient, constrained *gradient optimization methods*, and the use of *statistical methods* for tolerance analysis and design.

Since the mid-1970s, a number of computer programs for microwave and RF network analysis and design have become commercially available [1]. Some well-known microwave computer-aided engineering (CAE) software packages are SUPERCOM-PACT [2], TOUCHSTONE [3], S-NODE [4], LINMIC [5], ACCAD [6], and others [7, 8, 9]. All of these programs have certain advantages and disadvantages. On the negative side is the fact that some changes required for many applications are not possible and often beyond the ability of most microwave engineers and designers. A knowledge of basic computational methods used in the CAD of microwave circuits is a necessity for a designer involved in the development of new software for new or modified applications.

The various chapters of this book are devoted to various computational tools used in computer-aided analysis and design of microwave circuits.

To analyze a network, its elements and topology are provided to the computer. The network equations should be set up in a suitable form for various kinds of circuit design and analysis. A wide variety of matrix representations for microwave elements and circuits are presented in Chapter 2. Chain matrix, scattering matrix, transfer scattering matrix, and admittance scattering matrix representations for microwave circuit elements are discussed. All of the discussed matrices for some basic circuit elements are presented in tables. Relationships among different multiport matrix representations are also given.

Computer-aided methods for linear microwave circuit analysis in the frequency domain are discussed in Chapter 3. Frequency-domain analysis is well established and the most widely used aspect of CAD. Microwave circuit analysis may be performed in terms of voltages and currents. In such case the nodal admittance matrix is the most suitable form of equation for circuits of any topology. The behavior of a microwave circuit may also be analyzed in terms of the normalized wave variables at the ports of component multiports described by the scattering matrices. Scattering parameters are used extensively in microwave design because they are easier to measure and use than other kinds of parameters. The analysis involves evaluation of the scattering parameters of individual circuit elements, and, together with information on circuit topology, setting them up in the form of the connection scattering matrix. Microwave circuits of any arbitrary topology may be analyzed by using this matrix formalism. Many microwave circuits, such as filters and amplifiers, can be expressed as a cascade connection of two-port elements. In such cases, the matrix representing the overall circuit can be obtained by the multiplication of chain matrices or transfer scattering matrices of individual two-port elements of the circuit. The generalized transfer scattering matrix approach and the multiport connection method allow analysis of circuits of any topology. All of these methods are discussed in detail in the book.

Admittance matrix, chain matrix, scattering matrix, and transfer scattering matrix representations for elements commonly encountered in microwave circuits are presented in tables. By using CAD methods, we can easily achieve goals that are very difficult or impossible to achieve in any other way; the evaluation of circuit function variations due to small changes in circuit element parameters or the presence of parasitics is an example of this capability. Such information could be generated through bench tests by introducing small changes in the parameters of circuit elements. This, of course, is not practical in MIC and MMIC design. Moreover, experimental study of parasitic influences on circuit characteristics is very difficult to perform because we are unable to control their values on the breadboard. Computer-aided sensitivity analysis can provide such information at negligible cost for any number of microwave circuit parameters. In Chapter 4, we present the direct and adjoint network methods for sensitivity analysis. Both approaches are useful for microwave circuits described by the nodal admittance matrix and the connection scattering matrix. The adjoint network and its modification, the adjoint matrix method, are discussed in detail. For this purpose, Tellegen's theorem for current and voltage variables and wave variables is introduced and used to derive the adjoint networks. Gradient vector computation of circuit functions is also discussed, as is second-order sensitivity computation. Other applications of sensitivity analysis appear in gradient optimization procedures and tolerance analysis and design of microwave circuits. The computational aspects of CAD sensitivity analysis using sparse matrix techniques are discussed in Chapter 8.

The method for evaluating group delay of microwave transmission functions based on the network sensitivity analysis principle is given. Derivation of sensitivity expressions for elements (components) commonly used in microwave circuits are given for both admittance and scattering matrix representations of circuit elements.

Computer-aided noise analysis methods of microwave circuits are considered in Chapter 5. First, we discuss noise analysis methods for circuits composed of interconnected two-ports. The next sections of the chapter deal with computer-aided methods for noise analysis of circuits of arbitrary topology. Noise properties of circuit elements are described by the noise correlation matrices. An algorithm for noise figure computation of circuits composed of two-ports connected in cascade is given. CAD-oriented algorithms for noise figure computation of circuits with any arbitrary topology also are discussed. Basic relationships for noisy circuits, noise representation, noise parameters, and correlation matrices for different circuit descriptions are discussed and presented in detail. Noise correlation matrices for common elements of microwave circuits are presented for both admittance and scattering matrix representations.

Many of the problems encountered in microwave circuit analysis and design can be reduced to solving a series of related systems of linear equations. Many numerical methods can be used to solve sets of linear equations generated for frequency-domain analysis, sensitivity analysis, or noise analysis of microwave circuits. Presentation and discussion of these methods in Chapter 6 concentrates on Gaussian elimination, LU decomposition, and bifactorization. Coverage contains conventional full-matrix techniques and their error mechanisms. Because microwave networks are often quite large, involving tens of interconnected elements or more, special methods are required to speed up the solution. Microwave circuits with, for instance, a total of 100 ports for all circuit elements, are common. Such a circuit would be described by a 100×100 matrix, but, fortunately, the matrices associated with microwave circuits tend to be very sparse (most of the matrix entries are zero). Conventional full-matrix programming techniques disregard this fact, which means that computation of the solution involves approximately $n^3/3$ long operations (multiplications and divisions). For a system of circuit equations with the coefficient matrix of dimension $n = 100$, this would represent about 1 million operations. Practical experience shows that the number of operations may be reduced to approximately $20n = 2000$ for our example. In Chapters 7 and 8, the techniques used for efficiently solving large sparse systems of equations are discussed. Coverage continues through general sparse matrix techniques into methods oriented toward sparse matrices (connection scattering matrices) associated with frequency-domain analysis, sensitivity analysis, and noise analysis of microwave circuits. Finally, the practical implementation of a sparse matrix equation solver is derived and presented. It uses the looping indexed code technique and is based on the bifactorization method. Appendix 2 provides a full FORTRAN source code listing of the sparse matrix solver.

Design of real microwave networks requires investigation of the effects of circuit parameter spreads on the performance of the circuit design. Tolerance analysis methods help address these problems. Chapter 9 presents the available tolerance analysis methods and provides a summary of their features. First, we present the deterministic tolerance analysis approach. Worst-case tolerance analysis methods using the differential sensitivity and large change sensitivity approach are discussed. The method of statistical moments and a Monte Carlo approach belong to the second class, called *statistical methods of tolerance analysis*. The outcome of the Monte Carlo tolerance analysis can also form the basis of effective methods of tolerance design. The discussion of the Monte Carlo analysis is necessarily detailed for better understanding. Because it is a statistical method, only estimates of the effect of circuit parameter tolerances can be obtained. We discuss how to calculate the degree of confidence for yield estimates. Algorithms for the generation of pseudorandom parameter values required by Monte Carlo analysis are also given.

Chapter 10 deals with computer-aided methods for tolerance design. Two principal approaches to tolerance design are presented: the *deterministic approach* and the *statistical exploration approach*. The *gravity method, parametric sampling method*, and *worst-case design by cut method* are discussed. All are based on Monte Carlo analysis. In the design procedure, we seek parameter values for which the circuit meets given design specifications. A sequence of analyses, interrupted by parameter modifications, is required. In CAD, changing the element parameters and observing the result by again analyzing the circuit can be done automatically by means of optimization methods. Microwave circuit designers should be aware of the theoretical aspects of optimization methods so as to use them effectively. Chapter 11 of the book covers basic concepts of optimization theory, including a description of some optimization algorithms especially suitable for computer-aided design of microwave circuits. Optimization algorithms find the minimum of the objective function. The CAD specialist and designer—in one person—should know how to define an appropriate objective function for a particular design problem. The most computationally efficient optimization algorithms use the objective function gradient information. Again, the CAD designer is responsible for supplying gradient vector data to the selected optimization algorithm by means of numerically efficient sensitivity computations. Microwave circuit designers should be able to find and choose appropriate minimization routines as well.

Some practical aspects of the implementation of optimization methods are discussed in Chapter 11. We believe that the subset of optimization theory considered here will be sufficient for applications in CAD of microwave circuits. The reader is referred to the literature for a broader view of the subject.

The book contains three appendices. Two of them contain background mathematical material needed for better understanding of the problems discussed in the book. Appendix 1 explains the basics of vector and matrix norms. Appendix 2 provides the FORTRAN source code and user information on the sparse matrix subroutine package. The package is designed for repeated solutions of systems of linear equations of microwave circuits described by the connection scattering matrix resulting from frequency-domain, sensitivity, and noise analyses of microwave circuits. Finally, Appendix 3 provides the basics of statistical analysis.

REFERENCES

[1]   J. Pustai, "High-Frequency Designers Sign up for Better CAE," *Microwaves & RF*, Vol. 27, No. 12, November 1988, pp. 136–151.
[2]   SUPERCOMPACT, Compact Software, Inc., Paterson, NJ.
[3]   TOUCHSTONE, EEsof, Inc., Westlake Village, CA.
[4]   S-NODE, Suncrest Software, Veradale, WA.
[5]   R.H. Jansen, "LINMIC, a CAD Package for the Layout-Oriented Design of Single- and Multilayer MICs/MMICs up to mm Wave Frequencies," *Microwave J.*, Vol. 29, No. 2, February 1986, pp. 151–161.
[6]   C. Beccari and C. Naldi, "ACCAD, a New Microwave Circuit CAD Package," Internal Symposium on Microwave Technology in Industrial Development, Compinos, Brazil, July 1985, pp. 269–272.
[7]   CIRCUIT MASTER, Hines Consulting Laboratory, Weston, MA.
[8]   MICROWAVE DESIGN SYSTEM, Hewlett-Packard Co., Customer Information Center, Cupertino, CA.
[9]   ANALOP, Analop Engineering, Milpitas, CA.

Chapter 2
Microwave Circuit Matrix Representations

A microwave circuit is a set of multiports connected to one another. The components (elements) of microwave circuits may be one-, two-, three-, or four-ports, and generally are multiports. The elements of microwave circuits and, of course, entire circuits can be described in terms of terminal voltages and currents at all ports of a circuit. This kind of circuit characterization is commonly used at lower frequencies and the typical matrix representations of circuits are the impedance matrix, admittance matrix, and hybrid matrices. As we will show, a much more convenient form for the description of microwave circuits uses normalized wave variables and the resulting scattering matrix representation for circuit elements as well as an entire circuit. The scattering matrix formulation is very convenient for the description and the computer-aided analysis and design of microwave circuits of any topology.

A large number of microwave circuits (e.g., transistor amplifiers, filters) may be expressed as a cascade connection of two-port elements. Computer-aided analysis of such circuits may be performed very efficiently if the individual two-ports are described in terms of the chain matrix parameters or transfer scattering matrix parameters.

The chain matrix, scattering matrix, and transfer scattering matrix descriptions for microwave circuits will be reviewed in this chapter. Relationships for transformations among these matrix representations also will be presented. Appropriate tables will contain $\mathbf{A}(ABCD)$, $\mathbf{S}$, $\mathbf{T}$, and $\mathbf{Y}$ matrices for some basic microwave circuit elements commonly encountered in practice. These matrix elements can be regarded as a library of necessary information and relations that will be used in computer programs developed by the reader of this book.

2.1 CHAIN MATRIX REPRESENTATION

The chain matrix $\mathbf{A}(ABCD)$ of a two-port circuit directly relates the output terminal voltage and current (V_2, I_2) to the input voltage and current (V_1, I_1) such that [1]

$$\begin{bmatrix} V_1 \\ I_1 \end{bmatrix} = \mathbf{A} \begin{bmatrix} V_2 \\ I_2 \end{bmatrix} = \begin{bmatrix} A & B \\ C & D \end{bmatrix} \begin{bmatrix} V_2 \\ I_2 \end{bmatrix} \qquad (2.1)$$

Figure 2.1 Chain matrix **A**(*ABCD*) representation.

Note that the currents are chosen to be in the same direction (see Figure 2.1). In (2.1), the quantities V_1, V_2, I_1, and I_2 are the actual voltages and currents (never normalized), and consequently they are continuous at the boundaries of the two-ports. In this manner, the output of one two-port becomes the input of the next one. This makes the **A**(*ABCD*) representation very useful when cascading two-port networks as shown in Figure 2.2. We can easily see that the resulting **A**(*ABCD*) matrix of the two two-ports in cascade is equal to the product of two individual *ABCD* matrices, as shown below:

$$\begin{bmatrix} A & B \\ C & D \end{bmatrix} = \mathbf{A}^{(1)}\mathbf{A}^{(2)} = \begin{bmatrix} A^{(1)} & B^{(1)} \\ C^{(1)} & D^{(1)} \end{bmatrix}\begin{bmatrix} A^{(2)} & B^{(2)} \\ C^{(2)} & D^{(2)} \end{bmatrix} \qquad (2.2)$$

For the case of N two-ports in cascade, the resultant matrix is given by

$$\begin{bmatrix} A & B \\ C & D \end{bmatrix} = \begin{bmatrix} A^{(1)} & B^{(1)} \\ C^{(1)} & D^{(1)} \end{bmatrix}\begin{bmatrix} A^{(2)} & B^{(2)} \\ C^{(2)} & D^{(2)} \end{bmatrix} \cdots \begin{bmatrix} A^{(N)} & B^{(N)} \\ C^{(N)} & D^{(N)} \end{bmatrix} \qquad (2.3)$$

Therefore, microwave circuits that can be represented as cascaded connections of two-ports can be analyzed very effectively by using the above relation.

Relations between the **A** and **Z** matrices can be derived very easily from (2.1).

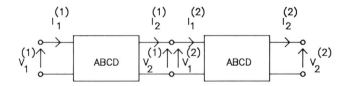

Figure 2.2 Cascaded connection of two-port networks.

The results are

$$Z_{11} = \frac{A}{C} \qquad Z_{12} = \frac{AD}{C} - B$$

$$Z_{21} = \frac{1}{C} \qquad Z_{22} = \frac{D}{C} \tag{2.4}$$

and, conversely,

$$A = \frac{Z_{11}}{Z_{21}} \qquad B = \frac{Z_{11}Z_{22} - Z_{12}Z_{21}}{Z_{21}}$$

$$C = \frac{1}{Z_{21}} \qquad D = \frac{Z_{22}}{Z_{21}} \tag{2.5}$$

For the reciprocal two-port networks ($Z_{12} = Z_{21}$), it follows that

$$AD - BC = 1 \tag{2.6}$$

For symmetrical networks ($Z_{11} = Z_{22}$ and $Z_{12} = Z_{21}$), we have

$$A = D \tag{2.7}$$

Similar relations between the **A** and **Y** matrices are as follows

$$Y_{11} = \frac{D}{B} = \frac{Z_{22}}{\Delta_z} \qquad\qquad Y_{21} = -\frac{1}{B} = \frac{-Z_{12}}{\Delta_z}$$

$$Y_{12} = C - \frac{AD}{B} = \frac{-Z_{12}}{\Delta_z} \qquad Y_{22} = \frac{A}{B} = \frac{Z_{11}}{\Delta_z} \tag{2.8}$$

where $\Delta_z = Z_{11}Z_{22} - Z_{12}Z_{21}$.

The usual network functions may be calculated from the overall $ABCD$ matrix of the circuit. Referring to Figure 2.1 for conventions, we derive the following relations.

The driving point impedance at the input port:

$$Z_{in} = \frac{AZ_L + B}{CZ_L + D} \tag{2.9}$$

The driving point impedance at the output port:

$$Z_{out} = \frac{DZ_s + B}{CZ_s + A} \tag{2.10}$$

The reflection coefficient at the input port:

$$\Gamma_{in} = \frac{Z_{in} - Z_s^*}{Z_{in} + Z_s} \tag{2.11}$$

The reflection coefficient at the output port:

$$\Gamma_{out} = \frac{Z_{out} - Z_L^*}{Z_{out} + Z_L} \tag{2.12}$$

The voltage gain:

$$A_v = \frac{V_2}{V_1} = \frac{Z_L}{AZ_L + B} \tag{2.13}$$

Transducer voltage gain:

$$A_{vs} = \frac{V_2}{E_s} = \frac{Z_L}{AZ_L + B + CZ_sZ_L + DZ_s} \tag{2.14}$$

Power gain:

$$G = \frac{P_L}{P_1} = \frac{\text{Re } Z_{in}}{\text{Re } Z_L} \left| \frac{Z_L}{AZ_L + B} \right|^2 \tag{2.15}$$

Transducer power gain:

$$G_T = \frac{P_L}{P_{AS}} = 4\frac{\text{Re } Z_s}{\text{Re } Z_L} \left| \frac{Z_L}{AZ_L + B + CZ_sZ_L + DZ_s} \right|^2 \tag{2.16}$$

In (2.15) and (2.16), P_L is the power delivered to the load, P_{AS} is the power

Table 2.1
Chain Matrices of Some Two-Port Elements of Microwave Circuits

Circuit Element	**A(** *ABCD* **)** *Matrix*
Z ⊶▭⊶ ⊶────⊶ Series impedance	$\begin{bmatrix} 1 & Z \\ 0 & 1 \end{bmatrix}$
Y Shunt admittance	$\begin{bmatrix} 1 & 0 \\ Y & 1 \end{bmatrix}$
l Z_0 $Z_0, l, \gamma = \alpha + j\beta$ Transmission line section	$\begin{bmatrix} \cosh(\gamma l) & Z_0 \sinh(\gamma l) \\ \dfrac{1}{Z_0} \sinh(\gamma l) & \cosh(\gamma l) \end{bmatrix}$
Z_0 l Open-ended shunt connected stub	$\begin{bmatrix} 1 & 0 \\ \dfrac{1}{Z_0} \tanh(\gamma l) & 1 \end{bmatrix}$
Z_0 l Short-ended shunt connected stub	$\begin{bmatrix} 1 & 0 \\ \dfrac{1}{Z_0} \coth(\gamma l) & 1 \end{bmatrix}$

Table 2.1 (*Continued*)

Circuit Element	A(*ABCD*) *Matrix*
  Open-ended series connected stub	$$\begin{bmatrix} 1 & Z_0 \coth(\gamma l) \\ 0 & 1 \end{bmatrix}$$
  Short-ended series connected stub	$$\begin{bmatrix} 1 & Z_0 \tanh(\gamma l) \\ 0 & 1 \end{bmatrix}$$
  Π = network	$$\begin{bmatrix} 1 + Z_2 Y_3 & Z_2 \\ Y_1(1 + Z_2 Y_3) + Y_3 & Y_1 Z_2 + 1 \end{bmatrix}$$
  T = network	$$\begin{bmatrix} 1 + Z_1 Y_2 & Z_3 + Z_1(Y_2 Z_3 + 1) \\ Y_2 & 1 + Y_2 Z_3 \end{bmatrix}$$
  Ideal transformer	$$\begin{bmatrix} p & 0 \\ 0 & \dfrac{1}{p} \end{bmatrix}$$

Table 2.1 (*Continued*)

Circuit Element	A(*ABCD*) *Matrix*
Attenuator $-\alpha$ dB	$$\frac{1}{2A}\begin{bmatrix} 1 & Z_0 \\ 1/Z_0 & 1 \end{bmatrix} \qquad A = 10^{-\alpha/20}$$
Gyrator	$$\begin{bmatrix} 0 & \alpha \\ 1/\alpha & 0 \end{bmatrix}$$
Distributed *RC* line section	$$\begin{bmatrix} \cosh\left(\sqrt{j\omega RC}\right) & \sqrt{\dfrac{R}{j\omega C}}\,\sinh\left(\sqrt{j\omega RC}\right) \\ \sqrt{\dfrac{j\omega C}{R}}\,\sinh\left(\sqrt{j\omega RC}\right) & \cosh\left(\sqrt{j\omega RC}\right) \end{bmatrix}$$ $R = r \cdot l -$ total resistance $C = c \cdot l -$ total capacitance
Voltage controlled voltage source	$$\begin{bmatrix} 1/k & 0 \\ 0 & 0 \end{bmatrix}$$
Voltage controlled current source	$$\begin{bmatrix} 0 & -1/g_m \\ 0 & 0 \end{bmatrix}$$
Current controlled voltage source	$$\begin{bmatrix} 0 & 0 \\ 1/r_m & 0 \end{bmatrix}$$

Table 2.1 (*Continued*)

Circuit Element	A(*ABCD*) *Matrix*
Current controlled current source	$\begin{bmatrix} 0 & 0 \\ 0 & -1/\beta \end{bmatrix}$
Transmission line junction	$\begin{bmatrix} 1 & 0 \\ 0 & 1 \end{bmatrix}$

available from the signal source, and P_1 is the power delivered to the input port of the circuit.

In Table 2.1, the *ABCD* matrices for some typical microwave circuit elements are presented.

2.2 SCATTERING MATRIX REPRESENTATION

In 1965, K. Kurokawa defined complex power wave variables and introduced generalized scattering parameters [2]. The scattering parameters for real and positive reference impedances of multiport ports have been defined earlier by P. Penfield [3, 4] and D Youla [5].

The scattering parameters describe the interrelationship of a new set of variables called *power waves* (a_i, b_i). The incoming and outgoing power waves at the ith port of the network are defined in terms of the terminal voltage V_i, the terminal current I_i and the arbitrary reference impedance Z_{N_i} (Re $Z_{N_i} > 0$), as follows [2]:

$$a_i = \frac{V_i + Z_{n_i} I_i}{2 \sqrt{\text{Re } Z_{N_i}}} \qquad (2.17)$$

$$b_i = \frac{V_i - Z_{N_i}^* I_i}{2 \sqrt{\text{Re } Z_{N_i}}} \qquad (2.18)$$

where the asterisk denotes the complex conjugate.

The inverse transformations to (2.17) and (2.18) have the form:

$$V_i = \frac{1}{\sqrt{\text{Re } Z_{N_i}}} \left(Z_{N_i}^* a_i + Z_{N_i} b_i \right) \tag{2.19}$$

$$I_i = \frac{1}{\sqrt{\text{Re } Z_{N_i}}} \left(a_i - b_i \right) \tag{2.20}$$

Therefore, for a fixed Z_{N_i}, if V_i and I_i are given, the power wave variables a_i and b_i can be calculated from (2.17) and (2.18), and *vice versa*, the power waves a_i and b_i can easily be converted into V_i and I_i using (2.19) and (2.20).

As a result of (2.17) to (2.18) we define the power wave reflection coefficient:

$$\Gamma = \frac{b_i}{a_i} = \frac{V_i - Z_{N_i}^* I_i}{V_i + Z_{N_i} I_i} = \frac{V_i/I_i - Z_{N_i}^*}{V_i/I_i + Z_{N_i}} \tag{2.21}$$

Because $V_i/I_i = Z_L$ (see Figure 2.3), then

$$\Gamma = \frac{Z_L - Z_{N_i}^*}{Z_L + Z_{N_i}} \tag{2.22}$$

where

Z_L = the load impedance (input impedance of the ith port),

Z_{N_i} = the reference impedance of the ith port.

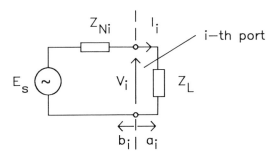

Figure 2.3 Signal source loaded with Z_L.

The physical interpretation of complex wave variables a_i and b_i comes from the following discussion. Referring to Figure 2.3, we have

$$V_i = E_s - Z_{N_i} I_i \qquad (2.23)$$

Inserting this into (2.17), we obtain

$$|a_i|^2 = \frac{|E_s|^2}{4 \operatorname{Re} Z_i} = P_{AS} \qquad (2.24)$$

Let us consider next a quantity $|a_i|^2 - |b_i|^2$. Direct substitution of (2.17) and (2.18) into this expression gives

$$|a_i|^2 - |b_i|^2 = \frac{1}{4 \operatorname{Re} Z_{N_i}} \left[\left(V_i + Z_{N_i} I_i \right) \left(V_i^* + Z_{N_i}^* I_i^* \right) \right.$$

$$\left. - \left(V_i - Z_{N_i}^* I_i \right) \left(V_i^* - Z_{N_i} I_i^* \right) \right]$$

$$= \frac{1}{4 \operatorname{Re} Z_i} \left(Z_{N_i} + Z_{N_i}^* \right) \left(V_i I_i^* + V_i^* I_i \right) = \operatorname{Re}\{V_i I_i^*\}$$

$$= P_L = P_{AS} - P_R \qquad (2.25)$$

The quantity $|a_i|^2 - |b_i|^2$ can be also written in the form

$$|a_i|^2 - |b_i|^2 = |a_i|^2 \left(1 - \left| \frac{b_i}{a_i} \right|^2 \right) = |a_i|^2 \left(1 - |\Gamma|^2 \right) \qquad (2.26)$$

Hence, from (2.24) to (2.26), the value of $|a_i|^2$ at the port of the load impedance Z_L is equal to the available power of the signal generator with the internal impedance equal to the port reference impedance Z_{N_i}.

The absolute value $|a_i|^2$ is the power that the generator sends toward a load, regardless of the load impedance. Equation (2.25) shows that the real power dissipated in the load impedance Z_L equals $|a_i|^2 - |b_i|^2$. Then, (2.24) to (2.26) can be interpreted as follows. The signal generator is sending the power $|a_i|^2$ toward the load, regardless of the load impedance. When the load is not matched (i.e., $\Gamma \neq 0$ (2.23), a part of the incident power is reflected back to the generator. In that case, when the load is matched (i.e., when $Z_L = Z_{N_i}^*$) the total power sent by the generator dissipates in the load.

Under the matching condition:

$$Z_L = Z_{N_i}^*$$ (2.27)

we have, $b_i = 0$, $\Gamma = 0$, and $P_L = |a_i|^2 = P_{AS}$.

Referring to Figure 2.4 the power wave scattering matrix of a multiport with n ports is defined by the equation:

$$\mathbf{b} = \mathbf{Sa}$$ (2.28)

where $\mathbf{S}$ is a square matrix of order n, and $\mathbf{a}$ and $\mathbf{b}$ are vectors, respectively, of input and output power wave variables at ports of the multiport.

When the reference impedances of the circuit ports Z_{N_i}, $i = 1, 2, \ldots, n$ are real and equal to the characteristic impedances Z_{0i} of the transmission lines connected to the multiport ports, the definitions of a_i and b_i simplify to those of complex input and output voltage (or current) waves in a transmission line, and the power wave scattering matrix $\mathbf{S}$ becomes the microwave scattering matrix.

By substituting

$$V_i = V_i^+ + V_i^-$$ (2.29)

$$I_i = I_i^+ + I_i^-$$ (2.30)

$$Z_{N_i} = Z_{0i} = \frac{V_i^+}{I_i^+} = \frac{V_i^-}{I_i^-}$$ (2.31)

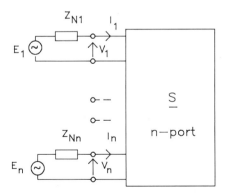

Figure 2.4 Multiport with n ports.

into (2.17) and (2.18), we obtain

$$a_i = \frac{V_i^+}{\sqrt{Z_{0i}}} = \sqrt{Z_{0i}}\, I_i^+ \qquad (2.32)$$

$$b_i = \frac{V_i^-}{\sqrt{Z_{0i}}} = \sqrt{Z_{0i}}\, I_i^- \qquad (2.33)$$

$$\Gamma = \frac{b_i}{a_i} = \frac{Z_L - Z_{0i}}{Z_L + Z_{0i}} \qquad (2.34)$$

In the above relations V_i^+ and I_i^+ denote the incoming voltage and current waves, respectively, whereas V_i^- and I_i^- are the outgoing voltage and current waves at the ith port of the circuit. Microwave scattering parameters can be measured by using any well-known methods and measurement systems [6].

2.2.1 Physical Interpretation of Scattering Parameters

Let P_i denote the real power delivered to the ith port of a multiport. According to (2.25), this power is given as

$$P_i = |a_i|^2 - |b_i|^2 \qquad (2.35)$$

The total power delivered to all n ports of the circuit describes the relation

$$P = \sum_{i=1}^{n} P_i = \sum_{i=1}^{n} |a_i|^2 - \sum_{i=1}^{n} |b_i|^2 = \mathbf{a}^+\mathbf{a} - \mathbf{a}^+\mathbf{a} \qquad (2.36)$$

where, the superscript $^+$ indicates the complex conjugate transposed matrix (vector).

By setting $\mathbf{b} = \mathbf{S}\mathbf{a}$ in the preceding equation, we have

$$P = \mathbf{a}^+[\mathbf{I} - \mathbf{S}^+\mathbf{S}]\mathbf{a} \qquad (2.37)$$

In a passive multiport, for which $P \geq 0$, the scattering matrix $\mathbf{S}$ must satisfy the condition

$$\mathbf{I} - \mathbf{S}^+\mathbf{S} \geq 0 \qquad (2.38)$$

for all real ω. In a case of a passive and lossless multiport, $P = 0$, and

$$\mathbf{I} - \mathbf{S}^+\mathbf{S} = 0 \qquad (2.39)$$

Therefore, the scattering matrix of a passive and lossless multiport is a unitary matrix (in (2.37) to (2.39) $\mathbf{I}$ indicates a unit matrix).

According to the matrix equation (2.28), the elements of the scattering matrix $\mathbf{S}$ can be derived from the relations

$$S_{ii} = \left. \frac{b_i}{a_i} \right|_{a_k = 0,\, k \neq i} \tag{2.40}$$

$$S_{ki} = \left. \frac{b_k}{a_i} \right|_{a_k = 0,\, k \neq i} \tag{2.41}$$

The condition $a_k = 0$, $k \neq i$ means that no circuit port, except port i, is excited by a signal generator and each simultaneously terminates in its reference impedance, Z_{N_i}, so that the waves b_k from those ports are totally absorbed in these terminations. Figure 2.5 presents such a circuit. Its kth port is excited by the voltage source E_k with the internal impedance equal to the reference impedance of this port. For all ports except the excited port, $V_k = -I_k Z_{N_k}$, and according to (2.17) to (2.18)

$$a_k = 0 \tag{2.42}$$

$$b_k = -\sqrt{\mathrm{Re}\, Z_{N_k}}\, I_k = \frac{\sqrt{\mathrm{Re}\, Z_{N_k}}}{Z_{N_k}}\, V_k \tag{2.43}$$

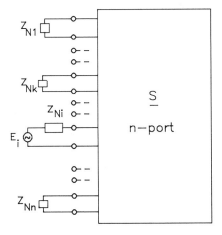

Figure 2.5 Multiport excited at the ith port and loaded with matched terminations at the other ports.

For the excited port, $E_i = V_i + Z_{N_i} I_i$ and, according to (2.17), the incoming wave is

$$a_i = \frac{E_i}{2 \sqrt{\text{Re } Z_{N_i}}} \tag{2.44}$$

To find the physical interpretation of the S_{ii} scattering parameter, let $Z_{\text{in } i}$ be the input impedance at the ith port of a circuit where all other ports are terminated with their reference impedances (Figure 2.5). Because $V_i = Z_{\text{in } i} I_i$, then, according to (2.17), we have

$$a_i = \frac{\left(Z_{\text{in } i} + Z_{N_i}\right) I_i}{2 \sqrt{\text{Re } Z_{N_i}}} \tag{2.45}$$

$$b_i = \frac{\left(Z_{\text{in } i} - Z_{N_i}^*\right) I_i}{2 \sqrt{\text{Re } Z_{N_i}}} \tag{2.46}$$

Thus, we have

$$S_{ii} = \frac{b_i}{a_i}\bigg|_{a_k = 0, \, k \neq i} = \frac{Z_{\text{in } i} - Z_{N_i}^*}{Z_{\text{in } i} + Z_{N_i}} \tag{2.47}$$

Therefore, S_{ii} is the power wave reflection coefficient at the ith port of a multiport that has its other ports terminated with their reference impedances.

The squared value of the magnitude of S_{ii} equals the power reflection coefficient:

$$|S_{ii}|^2 = \left|\frac{b_i}{a_i}\right|^2 \bigg|_{a_k = 0, \, k \neq i} = \frac{P_{R_i}}{P_{AS_i}}\bigg|_{a_k = 0, \, k \neq i} \tag{2.48}$$

where P_{AS_i} is the available power of the signal source with the internal impedance Z_{N_i} connected to the ith port, and P_{R_i} is the power reflected from ith port when all other ports are matched.

Let us consider now the physical interpretation of the S_{ki} scattering parameters. Using (2.43) and (2.44) in (2.41), we have

$$S_{ki} = \frac{b_k}{a_i}\bigg|_{a_k = 0, \, k \neq i} = -2 \sqrt{\text{Re } Z_{N_i} \, \text{Re } Z_{N_k}} \, \frac{I_k}{E_i} \tag{2.49}$$

When the reference impedances of the ith and kth ports are real and equal to each other ($Z_{N_i} = Z_{N_k} = R$), the transmission coefficient S_{ki} equals the voltage gain $2V_k/E_i$ between the ports i and k when all circuit ports are terminated at their reference impedances.

By setting the magnitude of S_{ki} to the second power, we obtain

$$|S_{ki}|^2 = \left|\frac{b_k}{a_i}\right|^2\bigg|_{a_k=0,\, k\neq i} = \frac{\text{Re } Z_k\, |I_k|^2}{|E_i|^2/(4\,\text{Re } Z_i)} = \frac{P_{Lk}}{P_{AS_i}} = G_T \quad (2.50)$$

where P_{AS_i} is the available power of the signal generator connected to the ith port, and P_{Lk} is the real power dissipated in the load $Z_{Lk} = Z_{Nk}$ of the kth port of the circuit.

In accordance with the above relations, $|S_{ki}|^2$ equals the transducer power gain G_T between the kth port and the ith port of a multiport having all its ports loaded with impedances equal to the reference impedances of these ports.

Group delay is an important circuit function for the filter and equalizer characterization.

For transmission from port i to port k, the group delay is defined as

$$\tau_{ki} = -\frac{\partial \Phi_{ki}}{\partial \omega} \quad (2.51)$$

where Φ_{ki} is the phase of S_{ki}.

The relation for τ_{ki} may be expressed as

$$\tau_{ki} = -\text{Im}\left[\frac{1}{S_{ki}}\frac{\partial S_{ki}}{\partial \omega}\right] \quad (2.52)$$

Some of the most important properties of the scattering matrix are

(1) The **S** matrix of a reciprocal network is symmetrical; that is,

$$\mathbf{S} = \mathbf{S}^T \quad (2.53)$$

where T denotes the operation of the transposition of a matrix. The condition (2.53) restricts the number of independent elements of the **S** matrix of the order n to $(n^2 + n)/2$.

(2) The **S** matrix of a lossless network is the unitary matrix

$$\mathbf{S}^+\mathbf{S} = \mathbf{I} \quad (2.54)$$

2.2.2 Change of Reference Impedance

Suppose that the port reference impedances of a multiport under consideration are changed from Z_{N_i} to Z'_{N_i}, $i = 1, 2, \ldots, n$. Then, according to (2.17) and (2.18), the incoming and outgoing waves have to be redefined accordingly. Of course, the scattering matrix $\mathbf{S'}$ connecting these new power wave vectors is different from the original one. It is expressed in terms of the original $\mathbf{S}$ matrix as [2]

$$\mathbf{S'} = \mathbf{A}^{-1}(\mathbf{S} - \boldsymbol{\Gamma}^+)(\mathbf{I} - \boldsymbol{\Gamma}\mathbf{S})^{-1}\mathbf{A}^+ \qquad (2.55)$$

where $\boldsymbol{\Gamma}$ is a diagonal matrix, the ith diagonal element of which equals

$$r_i = \frac{Z'_{N_i} - Z_{N_i}}{Z'_{N_i} - Z^*_{N_i}} \qquad (2.56)$$

A is also a diagonal matrix. It can be shown that the ith diagonal element A_i is

$$A_i = \frac{1 - r_i^*}{|1 - r_i|} \sqrt{|1 - r_i r_i^*|} \qquad (2.57)$$

The formula (2.57) finds practical applications. For example, in the design of two-port transistor amplifier whose source and load impedances are Z_{N1} and Z_{N2}, respectively, the matching conditions for the input and output ports are given by $S'_{11} = 0$ and $S'_{22} = 0$. Using (2.55), the condition $S'_{11} = 0$ yields

$$r_1^* = \frac{S_{11} - r_2(S_{11}S_{22} - S_{12}S_{21})}{1 - r_2 S_{22}} \qquad (2.58)$$

Similarly, $S'_{22} = 0$ yields

$$r_2^* = \frac{S_{22} - r_1(S_{11}S_{22} - S_{12}S_{21})}{1 - r_1 S_{11}} \qquad (2.59)$$

For the simultaneous match at the input and output of the two-port circuit, (2.58) and (2.59) have to be satisfied simultaneously.

The scattering matrices for some basic elements of microwave circuits are presented in Table 2.2. This table is to be treated as a "library" of $\mathbf{S}$ matrix representations that can be used in computer programs written by the reader.

2.3 TRANSFER SCATTERING MATRIX REPRESENTATION

2.3.1 Transfer Scattering Matrix of Two-Port Elements

By an analogy to the chain matrix $\mathbf{A}(ABCD)$, the transfer scattering matrix $\mathbf{T}$ of a two-port element directly relates the output port power wave variables (a_2, b_2) to the input port power wave variables (a_1, b_1). The transfer scattering matrix $\mathbf{T}$ defines the

Table 2.2
Scattering Matrices for Some Elements of Microwave Circuits

Circuit Element	Scattering Matrix **S**
Z_N $\quad$ Z	$\dfrac{Z - Z_N^*}{Z + Z_N}$
Z $Z_N \qquad Z_N$	$\dfrac{1}{Z + 2Z_N}\begin{bmatrix} Z & 2Z_N \\ 2Z_N & Z \end{bmatrix}$
Z_N $\quad$ Y	$\dfrac{1 - Z_N^* Y}{1 + Z_N Y}$
$Y_N \qquad Y \qquad Y_N$	$\dfrac{1}{Y + 2Y_N}\begin{bmatrix} -Y & 2Y_N \\ 2Y_N & -Y \end{bmatrix}$
$\overset{\longleftarrow \; l \; \longrightarrow}{Z_N \; Z_0 \qquad Z_N}$ Transmission line section $\gamma = \alpha + j\beta \qquad \beta = \dfrac{2\pi}{\lambda_g}$	$\dfrac{1}{Z_0^2 + Z_N^2 + 2Z_0 Z_N \coth(\gamma l)}$ $\cdot \begin{bmatrix} Z_0^2 - Z_N^2 & 2Z_0 Z_N \,\mathrm{csch}(\gamma l) \\ 2Z_0 Z_N \,\mathrm{csch}(\gamma l) & Z_0^2 - Z_N^2 \end{bmatrix}$
$Y_N \qquad Y_N$ Z_0 l Short-ended shunt connected stub	$\dfrac{1}{2Y_N + Y_0 \coth(\gamma l)}$ $\cdot \begin{bmatrix} -Y_0 \coth(\gamma l) & 2Y_N \\ 2Y_N & -Y_0 \coth(\gamma l) \end{bmatrix}$
$Y_N \qquad Y_N$ Z_0 l Open-ended shunt connected stub	$\dfrac{1}{2Y_N + Y_0 \tanh(\gamma l)}$ $\cdot \begin{bmatrix} -Y_0 \tanh(\gamma l) & 2Y_N \\ 2Y_N & -Y_0 \tanh(\gamma l) \end{bmatrix}$

Table 2.2 (*Continued*)

Circuit Element	Scattering Matrix **S**
Ideal transformer	$$\frac{1}{1+p^2}\begin{bmatrix} 1-p^2 & 2p \\ 2p & p^2-1 \end{bmatrix}$$
Gyrator	$$\frac{1}{\alpha^2+Z_N^2}\begin{bmatrix} \alpha^2-Z_N^2 & 2\alpha Z_N \\ -2\alpha Z_N & \alpha^2-Z_N^2 \end{bmatrix}$$
Z_{0e}, Z_{00}, l $\gamma_e = \alpha_e + j\beta_e, \ \beta_e = \dfrac{2\pi}{\lambda_{ge}}$ $\gamma_0 = \alpha_0 + j\beta_0, \ \beta_0 = \dfrac{2\pi}{\lambda_{g0}}$ A section of coupled transmission lines	$$\begin{bmatrix} S_{11} & S_{12} & S_{13} & S_{14} \\ S_{12} & S_{11} & S_{14} & S_{13} \\ S_{13} & S_{14} & S_{11} & S_{12} \\ S_{14} & S_{13} & S_{12} & S_{11} \end{bmatrix}$$ $S_{11} = \dfrac{S_{11e} + S_{110}}{2} \qquad S_{12} = \dfrac{S_{11e} - S_{110}}{2}$ $S_{13} = \dfrac{S_{12e} - S_{120}}{2} \qquad S_{14} = \dfrac{S_{12e} + S_{120}}{2}$ $S_{11e,0} = \dfrac{Z_{0e,0}^2 - Z_N^2}{Z_{0e,0}^2 + Z_N^2 + 2Z_{0e,0}Z_N \coth(\gamma_{e,0}l)}$ $S_{12e,0} = \dfrac{2Z_{0e,0}Z_N \operatorname{csch}(\gamma_{e,0}l)}{Z_{0e,0}^2 + Z_N^2 + 2Z_{0e,0}Z_N \coth(\gamma_{e,0}l)}$
Voltage controlled voltage source	$$\begin{bmatrix} 1 & 0 \\ 2k & -1 \end{bmatrix}$$
Voltage controlled current source	$$\begin{bmatrix} 1 & 0 \\ -2Z_N g_m & 1 \end{bmatrix}$$

Table 2.2 (*Continued*)

Circuit Element	Scattering Matrix S
Current controlled voltage source	$$\begin{bmatrix} -1 & 0 \\ \dfrac{2r_m}{Z_N} & -1 \end{bmatrix}$$
Current controlled current source	$$\begin{bmatrix} -1 & 0 \\ -2\beta & 1 \end{bmatrix}$$
Parallel connection of three ports	$$\frac{1}{3}\begin{bmatrix} -1 & 2 & 2 \\ 2 & -1 & 2 \\ 2 & 2 & -1 \end{bmatrix}$$
Series connection of three ports	$$\frac{1}{3}\begin{bmatrix} 1 & -2 & 2 \\ -2 & 1 & 2 \\ 2 & 2 & 1 \end{bmatrix}$$
Series connection of three ports	$$\frac{1}{3}\begin{bmatrix} 1 & 2 & 2 \\ 2 & 1 & -2 \\ 2 & -2 & 1 \end{bmatrix}$$
Ideal directional coupler	$$\begin{bmatrix} 0 & j\sqrt{1-c^2} & 0 & c \\ j\sqrt{1-c^2} & 0 & c & 0 \\ 0 & c & 0 & j\sqrt{1-c^2} \\ c & 0 & j\sqrt{1-c^2} & 0 \end{bmatrix}$$

Table 2.2 (*Continued*)

Circuit Element	Scattering Matrix **S**
Ideal directional coupler as a three-port	$$\begin{bmatrix} 0 & \sqrt{1-c^2} & 0 \\ \sqrt{1-c^2} & 0 & c \\ 0 & c & 0 \end{bmatrix}$$
An ideal directional coupler as a three-port	$$\begin{bmatrix} 0 & \sqrt{1-c^2} & c \\ \sqrt{1-c^2} & 0 & 0 \\ c & 0 & 0 \end{bmatrix}$$
Π-network	$\dfrac{1}{D}\begin{bmatrix} Y_N^2 - PY_N - T & 2Y_N Y_3 \\ 2Y_N Y_3 & Y_N^2 + PY_N - T \end{bmatrix}$ $D = Y_N^2 + QY_N + T \qquad P = Y_1 - Y_2$ $T = Y_1 Y_2 + Y_2 Y_3 + Y_1 Y_3 \qquad Q = Y_1 + Y_2 + 2Y_3$
T = network	$\dfrac{1}{D}\begin{bmatrix} -Z_N^2 + PZ_N + T & 2Z_N Z_3 \\ 2Z_N Z_3 & -Z_N^2 - PZ_N + T \end{bmatrix}$ $D = Z_N^2 + QZ_N + T \qquad P = Z_1 - Z_2$ $T = Z_1 Z_2 + Z_2 Z_3 + Z_1 Z_3 \qquad Q = Z_1 + Z_2 + 2Z_3$
Attenuator α dB	$$\begin{bmatrix} 0 & A \\ A & 0 \end{bmatrix}$$ $A = 10^{-\alpha/20}$

Table 2.2 (*Continued*)

Circuit Element	Scattering Matrix **S**
Distributed *RC* line section	As for a transmission line section with $Z_0 = \sqrt{\dfrac{R}{j\omega C}}$ and $\gamma l = \sqrt{j\omega RC}$ $R = r \cdot l$ — total resistance $C = c \cdot l$ — total capacitance
Short-circuited section of a waveguide $Z_N = Z_0$	$-\exp(-2\gamma l)$
Open-circuited section of a waveguide $Z_N = Z_0$	$\exp(-2\gamma l)$
Uniform waveguide section $Z_{N1} = Z_0 \quad Z_{N2} = Z_0$	$\begin{bmatrix} 0 & \exp(-\gamma l) \\ \exp(-\gamma l) & 0 \end{bmatrix}$
Short-ended series connected stub	$\dfrac{1}{2Z_N + Z_0 \tanh(\gamma l)}$ $\cdot \begin{bmatrix} Z_0 \tanh(\gamma l) & 2Z_N \\ 2Z_N & Z_0 \tanh(\gamma l) \end{bmatrix}$
Open-ended series connected stub	$\dfrac{1}{2Z_N + Z_0 \coth(\gamma l)}$ $\cdot \begin{bmatrix} Z_0 \coth(\gamma l) & 2Z_N \\ 2Z_N & Z_0 \coth(\gamma l) \end{bmatrix}$

equation [7, 8]:

$$\begin{bmatrix} a_1 \\ b_1 \end{bmatrix} = [\mathbf{T}] \begin{bmatrix} b_2 \\ a_2 \end{bmatrix} \tag{2.60}$$

The T parameters are defined in terms of wave variables normalized with respect to the reference impedances Z_{N1} and Z_{N2}, respectively, of the port 1 and port 2.

The transfer scattering parameters defined by (2.60) are related to the S parameters of a two-port in the following way:

$$T_{11} = \frac{-(S_{11}S_{22} - S_{12}S_{21})}{S_{21}} \tag{2.61}$$

$$T_{12} = \frac{S_{11}}{S_{21}} \tag{2.62}$$

$$T_{21} = \frac{-S_{22}}{S_{21}} \tag{2.63}$$

$$T_{22} = \frac{1}{S_{21}} \tag{2.64}$$

The reversed relations are

$$S_{11} = \frac{+T_{12}}{T_{22}} \tag{2.65}$$

$$S_{12} = T_{11} - \frac{T_{12}T_{21}}{T_{22}} \tag{2.66}$$

$$S_{21} = \frac{1}{T_{22}} \tag{2.67}$$

$$S_{22} = -\frac{T_{21}}{T_{22}} \tag{2.68}$$

The properties of the transfer scattering matrix of two-ports are as follows.
(1) For the reciprocal two-ports the T-parameters satisfy the relation:

$$T_{11}T_{22} - T_{12}T_{21} = 0 \tag{2.69}$$

which is analogous to the condition $AD - BC = 1$ for the chain $ABCD$ parameters.
(2) For the symmetrical two-ports, when both ports are mutually interchangeable, $S_{11} = S_{22}$, which in the domain of transfer scattering parameters is equivalent to

$$T_{21} = -T_{12} \tag{2.70}$$

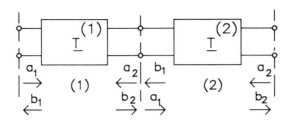

Figure 2.6 Cascaded connection of two-ports described by their transfer scattering matrices.

Referring to Figure 2.6 and expressions (2.17) and (2.18), we can easily show that when the reference impedances of the connected ports satisfy the relation:

$$Z_{N2}^{(1)} = Z_{N1}^{(2)*} \tag{2.71}$$

the incoming and outgoing waves at the connected ports are equal to each other; that is, $b_2^{(1)} = a_1^{(2)}$ and $a_2^{(1)} = b_1^{(2)}$.

When condition (2.71) is satisfied or, in particular, when the reference impedances of the connected ports are real and the same, the resultant **T** matrix of the two two-port elements in cascade is equal to the product of two individual **T** matrices as shown below:

$$\mathbf{T} = \mathbf{T}^{(1)} \cdot \mathbf{T}^{(2)} \tag{2.72}$$

The T-parameters of microwave circuit elements can be derived directly from their definition using basic circuit theory laws and transmission line equations. The T-parameters also can be calculated from the S-parameters using (2.61) to (2.64).

Table 2.3 presents transfer scattering matrices of some two-port elements often encountered in CAD microwave practice.

The **S** matrices are preferable for the characterization of microwave elements and circuits. Therefore, **T** matrices of the cascaded elements have to be computed from **S** matrices, and at the end of the analysis, the overall S matrix of the cascade must be obtained from the overall **T** matrix. The equations for both conversions are given in Section 2.5.

2.3.2 Generalized Transfer Scattering Matrix Representation

The transfer scattering matrix in its classical form describes two-port elements. It also can be defined for multiports with the same number of input and output ports [1]. Let us assume that such a multiport is described by the following scattering matrix (see Figure 2.7):

$$\begin{bmatrix} \mathbf{b}_{in} \\ \mathbf{b}_{out} \end{bmatrix} = \begin{bmatrix} \mathbf{S}_{11} & \mathbf{S}_{12} \\ \mathbf{S}_{21} & \mathbf{S}_{22} \end{bmatrix} \begin{bmatrix} \mathbf{a}_{in} \\ \mathbf{a}_{out} \end{bmatrix} \tag{2.73}$$

Table 2.3

Transfer Scattering Matrices for Some Elements of Microwave Circuits

Circuit Element	*Transfer Scattering Matrix* **T**
Z series element, Y_N	$\dfrac{1}{2}\begin{bmatrix} 2 - ZY_N & ZY_N \\ -ZY_N & 2 + ZY_N \end{bmatrix}$
Z_N shunt Y Z_N	$\dfrac{1}{2}\begin{bmatrix} 2 - YZ_N & -YZ_N \\ YZ_N & 2 + YZ_N \end{bmatrix}$
Z_N Z_0 Z_N $\gamma = \alpha + j\beta,\ \beta = \dfrac{2\pi}{\lambda_g}$ Transmission line section	$\cosh(\gamma l)$ $\cdot\begin{bmatrix} 1 + \dfrac{Z_N^2 + Z_0^2}{2Z_N Z_0}\tanh(\gamma l) & \dfrac{Z_N^2 - Z_0^2}{2Z_N Z_0}\tanh(\gamma l) \\ -\dfrac{Z_N^2 - Z_0^2}{2Z_N Z_0}\tanh(\gamma l) & 1 + \dfrac{Z_N^2 + Z_0^2}{2Z_n Z_0}\tanh(\gamma l) \end{bmatrix}$
Z_{N1} Z_{N2} Junction of two ports with different reference impedances	$\dfrac{1}{2\sqrt{\operatorname{Re} Z_{N1}\,\operatorname{Re} Z_{N2}}}\begin{bmatrix} Z_{N2}^* + Z_{N1}^* & Z_{N2} - Z_{N1}^* \\ Z_{N2}^* - Z_{N1} & Z_{N2} + Z_{N1} \end{bmatrix}$
Z_N Z_N Z_0 Open-ended shunt connected stub	$\dfrac{1}{2}\begin{bmatrix} 2 - YZ_N & -YZ_N \\ YZ_N & 2 + YZ_N \end{bmatrix}$ where $Y = Y_0\coth(\gamma l)$

Table 2.3 (*Continued*)

Circuit Element	Transfer Scattering Matrix **T**
  Short-ended shunt connected stub	$$\frac{1}{2}\begin{bmatrix} 2 - YZ_N & -YZ_N \\ YZ_N & 2 + YZ_N \end{bmatrix}$$ where $Y = Y_0 \tanh(\gamma l)$
  Open-ended series connected stub	$$\frac{1}{Z + 2Z_N}\begin{bmatrix} Z & 2Z_N \\ 2Z_N & Z \end{bmatrix}$$ where $Z = Z_0 \coth(\gamma l)$
  Short-ended series connected stub	$$\frac{1}{Z + 2Z_N}\begin{bmatrix} Z & 2Z_N \\ 2Z_N & Z \end{bmatrix}$$ where $Z = Z_0 \tanh(\gamma l)$
  Ideal transformer	$$\frac{1}{2p}\begin{bmatrix} p^2 + 1 & p^2 - 1 \\ p^2 - 1 & p^2 + 1 \end{bmatrix}$$
  Ideal transformer	$$\frac{1}{2p}\begin{bmatrix} p^2 + 1 & -(p^2 - 1) \\ -(p^2 - 1) & p^2 + 1 \end{bmatrix}$$

Table 2.3 (*Continued*)

Circuit Element	Transfer Scattering Matrix **T**
 Π = network	$\dfrac{1}{2Y_N Y_3}\begin{bmatrix} -Y_N^2 + QY_N - D & Y_N^2 - PY_N - D \\ -Y_N^2 - PY_N + D & Y_N^2 + QY_N + D \end{bmatrix}$ $D = Y_1 Y_2 + Y_2 Y_3 + Y_3 Y_1 \qquad P = Y_1 - Y_2$ $Q = Y_1 + Y_2 + 2Y_3$
 T = network	$\dfrac{1}{2Z_N Z_3}\begin{bmatrix} -Z_N^2 + QZ_N - D & -Z_N^2 + PZ_N + D \\ Z_N^2 + PZ_N - D & Z_N^2 + QZ_N + D \end{bmatrix}$ $D = Z_1 Z_2 + Z_2 Z_3 + Z_3 Z_1 \qquad p = Z_1 - Z_2$ $Q = Z_1 + Z_2 + 2Z_3$
 An attenuator α dB	$\begin{bmatrix} -A & 0 \\ 0 & A \end{bmatrix}$ where $A = 10^{\alpha/20}$
 Voltage controlled voltage source	$\dfrac{1}{2k}\begin{bmatrix} 1 & 1 \\ 1 & 1 \end{bmatrix}$
 Voltage controlled current source	$\dfrac{1}{2Z_N g_m}\begin{bmatrix} 1 & 1 \\ 1 & 1 \end{bmatrix}$
 Current controlled voltage source	$\dfrac{Z_N}{2r_m}\begin{bmatrix} -1 & -1 \\ 1 & 1 \end{bmatrix}$
 Current controlled current source	$\dfrac{1}{2\beta}\begin{bmatrix} -1 & 1 \\ 1 & -1 \end{bmatrix}$

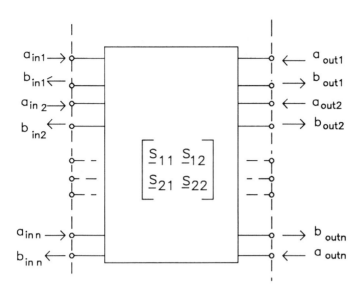

Figure 2.7 $2n$-multiport with n input ports and n output ports.

where

$$S_{11}, S_{12}, S_{21}, S_{22} = \text{scattering submatrices,}$$

$\mathbf{a}_{in}, \mathbf{b}_{in} = $ vectors, respectively, of incoming and outgoing waves at the n input ports of the $2n$-multiport,

$\mathbf{a}_{out}, \mathbf{b}_{out} = $ vectors, respectively, of incoming and outgoing waves at the n output ports of the $2n$-multiport.

Its transfer scattering matrix is given as

$$\begin{bmatrix} \mathbf{b}_{in} \\ \mathbf{a}_{in} \end{bmatrix} = \begin{bmatrix} S_{12} - S_{11}S_{21}^{-1}S_{22} & S_{11}S_{21}^{-1} \\ -S_{21}^{-1}S_{22} & S_{21}^{-1} \end{bmatrix} \begin{bmatrix} \mathbf{a}_{out} \\ \mathbf{b}_{out} \end{bmatrix} = \mathbf{T} \begin{bmatrix} \mathbf{a}_{out} \\ \mathbf{b}_{out} \end{bmatrix} \quad (2.74)$$

The reversed relation is

$$\begin{bmatrix} \mathbf{b}_{in} \\ \mathbf{b}_{out} \end{bmatrix} = \begin{bmatrix} T_{12}T_{22}^{-1} & T_{11} - T_{12}T_{22}^{-1}T_{21} \\ T_{22}^{-1} & -T_{22}^{-1}T_{21} \end{bmatrix} \begin{bmatrix} \mathbf{a}_{in} \\ \mathbf{a}_{out} \end{bmatrix} = \mathbf{S} \begin{bmatrix} \mathbf{a}_{in} \\ \mathbf{a}_{out} \end{bmatrix} \quad (2.75)$$

If the numbers of the input and output ports are not equal, the submatrix S_{21} in (2.73) is not a square matrix and its inverse S_{21}^{-1} does not exist. The transfer scattering matrix defined by (2.74) does not exist, either. This problem can be solved by proper extension of the scattering matrix (2.73) to get square submatrix S_{21}. The extended scattering matrix (2.73) will have now the following form [9]:

$$\begin{bmatrix} \mathbf{b}_{in} \\ \mathbf{b}_{out} \\ \mathbf{b}_e \end{bmatrix} = \begin{bmatrix} S_{11} & 0 & S_{12} \\ S_{21} & S_b & S_{22} \\ S_c & S_d & 0 \end{bmatrix} \begin{bmatrix} \mathbf{a}_{in} \\ \mathbf{a}_e \\ \mathbf{a}_{out} \end{bmatrix} \qquad (2.76)$$

where

$\mathbf{b}_e$ = a vector of additional virtual outgoing wave variables,

$\mathbf{a}_e$ = a vector of additional virtual incoming wave variables, all equal zero by definition, $\mathbf{a}_e = \mathbf{0}$. Virtual waves $\mathbf{a}_e$ and $\mathbf{b}_e$ correspond to virtual input ports that do not actually exist.

S_b, S_c, S_d = blocks of the extended square submatrix:

$$\mathbf{S}_{21}^e = \begin{bmatrix} S_{21} & S_b \\ S_c & S_d \end{bmatrix} \qquad (2.77)$$

The number of columns in the submatrix S_b equals the number of the new added virtual incoming wave variables located in the vector $\mathbf{a}_e$. Similarly the number of rows in the submatrix S_c defines the number of new added virtual outgoing wave variables $\mathbf{b}_e$ in (2.76). The extended scattering matrix S given by (2.76) is equivalent to the extended circuit in the form presented in Figure 2.8, which has virtual input and output ports that do not exist physically.

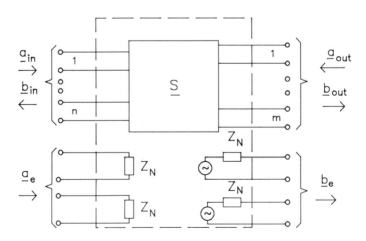

Figure 2.8 Extended circuit with virtual incoming wave variables $\mathbf{a}_e$ and virtual outgoing wave variables $\mathbf{b}_e$.

In general we have to consider three different configurations of a multiport.

(1) The number of the input ports is smaller than the number of output ports ($n < m$). The extended submatrix S_{21}^e will have the form:

$$S_{21}^e = \begin{bmatrix} S_1 & 0 \\ S_3 & I_{m-n} \end{bmatrix} \qquad (2.78)$$

where I_{m-n} is the $(m-n) \times (m-n)$ unit matrix and the generalized transfer scattering matrix is

$$\begin{bmatrix} b_{in} \\ a_{in} \\ 0 \end{bmatrix} = \begin{bmatrix} S_{12} - S_{11}\begin{bmatrix} S_1^{-1} & 0 \end{bmatrix}S_{22} & S_{11}S_1^{-1} & 0 \\ -\begin{bmatrix} S_1^{-1} & 0 \end{bmatrix} & S_1^{-1} & 0 \\ \begin{bmatrix} S_3 S_1^{-1} & -I_{m-n} \end{bmatrix} & -S_3 S_1^{-1} & I \end{bmatrix} \begin{bmatrix} a_{out} \\ b_{out} \end{bmatrix} \qquad (2.79)$$

(2) The number of input and output ports is the same. The transfer scattering matrix is given by (2.74).

(3) There are more input ports than output ports ($n > m$).

The extended submatrix S_{21} will have the following form:

$$S_{21}^e = \begin{bmatrix} S_1 & S_2 \\ 0 & I_{n-m} \end{bmatrix} \qquad (2.80)$$

and the generalized transfer scattering matrix is [9]:

$$\begin{bmatrix} b_{in} \\ a_{in} \end{bmatrix} = \begin{bmatrix} S_{12} - S_{11}\begin{bmatrix} S_1^{-1} \\ 0 \end{bmatrix}S_{22} & S_{11}\begin{bmatrix} S_1^{-1} \\ 0 \end{bmatrix} & S_{11}\begin{bmatrix} -S_1^{-1}S_2 \\ I \end{bmatrix} \\ -S_1^{-1}S_{22} & S_1^{-1} & -S_1^{-1}S_2 \\ 0 & 0 & I \end{bmatrix} \begin{bmatrix} a_{out} \\ b_{out} \\ b_e \end{bmatrix} \qquad (2.81)$$

From (2.76) and (2.80) we see that the elements of the b_e vector are equal to some selected incoming wave variables at the input of the multiport $b_{e1} = a_{in\,n+1}$, $b_{e2} = a_{in\,n+2}, \ldots, b_{en-m} = a_{in\,n}$. For computational convenience, by row and column permutations, we can modify the transfer scattering matrix to the form defined by the equation:

$$
\begin{bmatrix} b_{\text{in }1} \\ a_{\text{in }1} \\ b_{\text{in }2} \\ a_{\text{in }2} \\ \vdots \\ 0 \end{bmatrix} = \mathbf{T} \begin{bmatrix} a_{\text{out }1} \\ b_{\text{out }1} \\ a_{\text{out }2} \\ b_{\text{out }2} \\ \vdots \\ b_{e1} \\ \vdots \\ b_{en-m} \end{bmatrix}
\tag{2.82}
$$

The generalized transfer scattering matrix defined by equation (2.82) may be applied to the analysis of microwave circuits composed of N-ports connected in cascade. It is easy to show that if the reference impedances of all connected ports of two N-ports satisfy the relation (2.71), then the resultant generalized $\mathbf{T}$ matrix of the cascade is the product of two individual $\mathbf{T}$ matrices, as shown by (2.72).

Table 2.4 presents the generalized transfer scattering matrices of some microwave circuit elements often encountered in CAD microwave practice. To become more familiar with the presented generalized transfer scattering matrix formalism we will consider now the formulation and computation of generalized transfer scattering matrices for some simple microwave circuits.

Example 2.1

Let us consider a circuit presented in Figure 2.9. It consists of a transmission line section loaded by a one port with the reflection coefficient Γ. The generalized transfer scattering matrices of both elements are given in Table 2.4. We get the equation of the whole circuit by multiplying the generalized transfer scattering matrices of the cascaded elements:

$$
\begin{bmatrix} b_1 \\ a_1 \end{bmatrix} = \mathbf{T}^{(1)} \cdot \mathbf{T}^{(2)}[b_e]
$$

$$
= \frac{1}{S_{21}} \begin{bmatrix} -\Delta_s & S_{11} \\ -S_{22} & 1 \end{bmatrix} \begin{bmatrix} \Gamma \\ 1 \end{bmatrix} [b_e]
$$

The result is

$$
\begin{bmatrix} b_1 \\ a_1 \end{bmatrix} = \frac{1}{S_{21}} \begin{bmatrix} S_{11} - \Gamma \Delta_s \\ 1 - \Gamma S_{22} \end{bmatrix} [b_e]
$$

Table 2.4

Generalized Transfer Scattering Matrices of Some Microwave Circuit Elements

Circuit Element	Circuit Element Equation
 One-port as an output port load	$$\begin{bmatrix} b \\ a \end{bmatrix} = \begin{bmatrix} S \\ 1 \end{bmatrix} b_e$$
 One-port as an input port load	$$0 = \begin{bmatrix} -S & 1 \end{bmatrix} \begin{bmatrix} a \\ b \end{bmatrix}$$
 Two-port	$$\begin{bmatrix} b_1 \\ a_1 \end{bmatrix} = \frac{1}{S_{21}} \begin{bmatrix} -\Delta_s & S_{11} \\ -S_{22} & 1 \end{bmatrix} \begin{bmatrix} a_2 \\ b_2 \end{bmatrix}$$ $$\Delta_s = S_{11}S_{22} - S_{12}S_{21}$$
 Three-port	$$\begin{bmatrix} b_1 \\ a_1 \\ 0 \end{bmatrix} = \frac{1}{S_{21}} \begin{bmatrix} S_{11}S_{21} - S_{11}S_{22} & S_{11} & S_{13}S_{12} - S_{11}S_{23} & 0 \\ -S_{22} & 1 & -S_{23} & 0 \\ S_{22}S_{31} - S_{21}S_{32} & -S_{31} & S_{23}S_{31} - S_{21}S_{33} & S_{21} \end{bmatrix} \begin{bmatrix} a_2 \\ b_2 \\ a_3 \\ b_3 \end{bmatrix}$$

Table 2.4 (*Continued*)

Circuit Element	Circuit Element Equation
Three-port	$$\begin{bmatrix} b_1 \\ a_1 \\ b_2 \\ a_2 \end{bmatrix} = \frac{1}{S_{31}} \begin{bmatrix} S_{13}S_{21}-S_{11}S_{22} & S_{11} & S_{12}S_{31}-S_{11}S_{32} \\ -S_{33} & 1 & -S_{32} \\ S_{23}S_{31}-S_{21}S_{33} & S_{21} & S_{22}S_{31}-S_{21}S_{32} \\ 0 & 0 & S_{31} \end{bmatrix} \begin{bmatrix} a_3 \\ b_3 \\ b_e \end{bmatrix}$$
Multiport with all ports as input ports	$$\begin{bmatrix} b \\ a \end{bmatrix} = \begin{bmatrix} S \\ I \end{bmatrix} b_e$$
Multiport with all ports as output ports	$$0 = \begin{bmatrix} -S & I \end{bmatrix} \begin{bmatrix} a \\ b \end{bmatrix}$$

Independent signal generator	$$c = \begin{bmatrix} -S & 1 \end{bmatrix} \begin{bmatrix} a \\ b \end{bmatrix}$$
Independent signal generator	$$\begin{bmatrix} b \\ a \end{bmatrix} = \begin{bmatrix} 1 & S \\ 0 & 1 \end{bmatrix} \begin{bmatrix} c \\ b \end{bmatrix}$$
Independent signal generators	$$\begin{bmatrix} c_1 \\ c_2 \\ \cdot \\ c_n \end{bmatrix} = \begin{bmatrix} -S_1 & 0 & \cdot & 0 & 1 & 0 & \cdot & 0 \\ 0 & -S_2 & \cdot & & 0 & 1 & \cdot & \cdot \\ \cdot & \cdot & & & \cdot & 0 & & \cdot \\ 0 & 0 & & -S_n & 0 & \cdot & \cdot & 1 \end{bmatrix} \begin{bmatrix} a_1 \\ \cdot \\ a_n \\ b_1 \\ \cdot \\ b_n \end{bmatrix}$$

Table 2.4 (*Continued*)

Circuit Element	Circuit Element Equation
Independent signal generators	$$\begin{bmatrix} b_1 \\ \vdots \\ b_n \\ a_1 \\ \vdots \\ a_n \end{bmatrix} = \left[\begin{array}{cccc:cccc} 1 & 0 & 0 & S_1 & 0 & & & 0 \\ 0 & 1 & & & S_2 & & & \\ 0 & & 1 & & & S_n & & 0 \\ \hline & 0 & & & 1 & 0 & 0 & \\ & & & & 0 & 1 & & 0 \\ 0 & & & & 0 & & & 1 \end{array}\right]\begin{bmatrix} c_1 \\ \vdots \\ c_n \\ b_1 \\ \vdots \\ b_n \end{bmatrix}$$
Z	$$\begin{bmatrix} b_1 \\ a_1 \end{bmatrix} = \frac{1}{2}\begin{bmatrix} 2 - ZY_N & ZY_N \\ -ZY_N & 2 + ZY_N \end{bmatrix}\begin{bmatrix} a_2 \\ b_2 \end{bmatrix}$$
Y	$$\begin{bmatrix} b_1 \\ a_1 \end{bmatrix} = \frac{1}{2}\begin{bmatrix} 2 - YZ_N & -YZ_N \\ YZ_N & 2 + YZ_N \end{bmatrix}\begin{bmatrix} a_2 \\ b_2 \end{bmatrix}$$
Z_o, l $\gamma = \alpha + j\beta$, $\beta = \dfrac{2\pi}{\lambda}$ Transmission line	$$\cosh(\gamma l)\cdot\begin{bmatrix} 1 - \dfrac{Z_N^2 + Z_0^2}{2Z_N Z_0}\tanh(\gamma l) & \dfrac{Z_N^2 - Z_0^2}{2Z_N Z_0}\tanh(\gamma l) \\ -\dfrac{Z_N^2 - Z_0^2}{2Z_N Z_0}\tanh(\gamma l) & 1 + \dfrac{Z_N^2 + Z_0^2}{2Z_n Z_0}\tanh(\gamma l) \end{bmatrix}$$

$$\begin{bmatrix} b_1 \\ a_1 \\ 0 \end{bmatrix} = \begin{bmatrix} 1/2 & -1/2 & 1 & 0 \\ -1/2 & 1/2 & 0 & 1 \\ -1 & -1 & 1 & 1 \end{bmatrix} \begin{bmatrix} a_2 \\ b_2 \\ a_3 \\ b_3 \end{bmatrix}$$

Parallel T-junction

$$\begin{bmatrix} b_1 \\ a_1 \\ b_2 \\ a_2 \end{bmatrix} = \begin{bmatrix} 1/2 & 1/2 & 1 \\ 1/2 & 1/2 & -1 \\ 1 & 0 & -1 \\ 0 & 1 & 1 \end{bmatrix} \begin{bmatrix} a_3 \\ b_3 \\ b_e \end{bmatrix}$$

Parallel T-junction

$$\begin{bmatrix} b_1 \\ a_1 \\ 0 \end{bmatrix} = \begin{bmatrix} -1/2 & -1/2 & -1 & 0 \\ -1/2 & -1/2 & 0 & -1 \\ 1 & 1 & -1 & 1 \end{bmatrix} \begin{bmatrix} a_2 \\ b_2 \\ a_3 \\ b_3 \end{bmatrix}$$

Series T-junction

$$\begin{bmatrix} b_1 \\ a_1 \\ b_2 \\ a_2 \end{bmatrix} = \begin{bmatrix} -1/2 & 1/2 & -1 \\ 1/2 & -1/2 & -1 \\ -1 & 0 & 1 \\ 0 & -1 & 1 \end{bmatrix} \begin{bmatrix} a_3 \\ b_3 \\ b_e \end{bmatrix}$$

Series T-junction

Table 2.4 (*Continued*)

Circuit Element	Circuit Element Equation

$Z_{0e}, Z_{00},$

$\gamma_e = \alpha_e + j\beta_e, \quad \beta_e = \dfrac{2\pi}{\lambda_e}$

$\gamma_0 = \alpha_0 + j\beta_0, \quad \beta_0 = \dfrac{2\pi}{\lambda_0}$

A section of coupled transmission lines

$$
\begin{bmatrix} b_1 \\ a_1 \\ b_2 \\ a_2 \end{bmatrix} =
\begin{bmatrix}
a_{11} & a_{12} & a_{13} & a_{14} \\
-a_{12} & a_{22} & -a_{14} & a_{24} \\
a_{13} & a_{14} & a_{11} & a_{12} \\
-a_{14} & a_{24} & -a_{12} & a_{22}
\end{bmatrix}
\begin{bmatrix} a_3 \\ b_3 \\ a_4 \\ b_4 \end{bmatrix}
$$

$a_{11} = \dfrac{1}{2}(a_e - a_0), \quad a_{12} = \dfrac{1}{2}(b_e - b_0)$

$a_{13} = \dfrac{1}{2}(a_e + a_0), \quad a_{14} = \dfrac{1}{2}(b_e + b_0)$

$a_{22} = \dfrac{1}{2}(c_e - c_0), \quad a_{24} = \dfrac{1}{2}(c_e + c_0)$

$a_{e,0} = \cosh(\gamma_{e,0}l) - \dfrac{Z_{0e,0}^2 + Z_N^2}{2Z_{0e,0}Z_N} \cdot \sinh(\gamma_{e,0}l)$

$b_{e,0} = \dfrac{Z_{0e,0}^2 - Z_N^2}{2Z_{0e,0}Z_N} \sinh(\gamma_{e,0}l)$

$c_{e,0} = \cosh(\gamma_{e,0}l) + \dfrac{Z_{0e,0}^2 + Z_N^2}{2Z_{0e,0}Z_N} \cdot \sinh(\gamma_{e,0}l)$

Junction of two ports with different reference impedances

$$
\begin{bmatrix} b_1 \\ a_1 \end{bmatrix} =
\frac{1}{2\sqrt{\mathrm{Re}\,Z_{N1}\,\mathrm{Re}\,Z_{N2}}}
\begin{bmatrix}
Z_{N2}^* + Z_{N1} & Z_{N2} - Z_{N1}^* \\
Z_{N2}^* - Z_{N1} & Z_{N2} + Z_{N1}
\end{bmatrix}
\begin{bmatrix} a_2 \\ b_2 \end{bmatrix}
$$

The reflection coefficient at the input of the circuit is given by

$$\frac{b_1}{a_1} = \frac{S_{11} - \Gamma\Delta_s}{1 - \Gamma S_{22}}$$

Example 2.2

Figure 2.10 (a) presents a four-port that is a branch of a branch line directional coupler. The equivalent circuit of the branch in the form of a cascaded connection of multiports is shown in Figure 2.10 (b). The transfer scattering matrix of the circuit equals the product of the **T** matrices of the individual multiports in the cascade.

$$\begin{bmatrix} b_1 \\ a_1 \\ b_2 \\ a_2 \\ 0 \end{bmatrix} = \mathbf{T}^{(1)} \cdot \mathbf{T}^{(2)} \cdot \mathbf{T}^{(3)} \begin{bmatrix} a_3 \\ b_3 \\ a_4 \\ b_4 \\ b_e \end{bmatrix}$$

$$= \begin{bmatrix} 1/2 & -1/2 & 1 & 0 & \vdots & 0 & 0 \\ -1/2 & 1/2 & 0 & 1 & \vdots & 0 & 0 \\ \hdashline 0 & 0 & 0 & 0 & \vdots & 1 & 0 \\ 0 & 0 & 0 & 0 & \vdots & 0 & 0 \\ -1 & -1 & 1 & 1 & \vdots & 0 & 0 \end{bmatrix} \begin{bmatrix} 1 & 0 & 0 & 0 & 0 & 0 \\ 0 & 1 & 0 & 0 & 0 & 0 \\ 0 & 0 & T_{11} & T_{12} & 0 & 0 \\ 0 & 0 & T_{21} & T_{22} & 0 & 0 \\ 0 & 0 & 0 & 0 & 1 & 0 \\ 0 & 0 & 0 & 0 & 0 & 1 \end{bmatrix}$$

$$\cdot \begin{bmatrix} 1 & 0 & \vdots & 0 & 0 & 0 \\ 0 & 1 & \vdots & 0 & 0 & 0 \\ \hdashline 0 & 0 & \vdots & 1 & 0 & 1 \\ 0 & 0 & \vdots & 0 & 1 & -1 \\ 0 & 0 & \vdots & 1/2 & 1/2 & -1 \\ 0 & 0 & \vdots & 1/2 & 1/2 & 1 \end{bmatrix} \begin{bmatrix} a_3 \\ b_3 \\ a_4 \\ b_4 \\ b_e \end{bmatrix}$$

In $\mathbf{T}^{(2)}$, a block

$$\begin{bmatrix} T_{11} T_{12} \\ T_{21} T_{22} \end{bmatrix}$$

is the transfer scattering matrix **T** of the transmission line section.

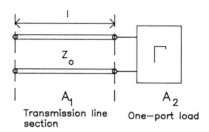

Figure 2.9 Circuit composed of a transmission line section loaded with a one-port.

The resulting overall matrix is

$$
\begin{bmatrix} b_1 \\ a_1 \\ b_2 \\ a_2 \\ 0 \end{bmatrix} =
\begin{bmatrix}
1/2 & -1/2 & T_{11} & T_{12} & T_{12} - T_{11} \\
-1/2 & 1/2 & T_{21} & T_{22} & T_{22} - T_{21} \\
0 & 0 & 1/2 & 1/2 & -1 \\
0 & 0 & 1/2 & 1/2 & 1 \\
-1 & -1 & T_{11} + T_{21} & T_{12} + T_{22} & T_{11} + T_{22}
\end{bmatrix}
\begin{bmatrix} a_3 \\ b_3 \\ a_4 \\ b_4 \\ b_e \end{bmatrix}
$$

and, after eliminating the additional row and the additional column, we have

$$
\begin{bmatrix} b_1 \\ a_1 \\ b_2 \\ a_2 \end{bmatrix} =
\begin{bmatrix}
1 - Q & -Q & P & P \\
Q & 1 + Q & -P & -P \\
P & P & 1 - Q & -Q \\
-P & -P & Q & 1 + Q
\end{bmatrix}
\begin{bmatrix} a_3 \\ b_3 \\ a_4 \\ b_4 \end{bmatrix}
$$

where $P = Z_0 \operatorname{csch}(\gamma 1)/(2Z_N)$ and $Q = Z_0 \coth(\gamma 1)/(2Z_N)$.

The elements T_{11}, T_{12}, T_{21}, and T_{22} of the **T** matrix for the transmission lin
section have been taken from Table 2.4.

Example 2.3

Figure 2.11 shows a microstrip band-stop filter and its equivalent circuit constructed i
the form of a cascaded connection of multiports. The overall transfer scattering matri

of the circuit is equal to the product of the **T** matrices of the individual multiports in the cascade.

$$
\begin{bmatrix} b_1 \\ a_1 \\ 0 \\ 0 \\ 0 \end{bmatrix} = \mathbf{T}^{(1)} \cdot \mathbf{T}^{(2)} \cdot \mathbf{T}^{(3)} \cdot \mathbf{T}^{(4)} \cdot \mathbf{T}^{(5)} \cdot \begin{bmatrix} a_2 \\ b_2 \\ b_{e1} \\ b_{e2} \\ b_{e3} \end{bmatrix}
$$

$$
= \begin{bmatrix} 1 & 0 & 0 & 0 & 0 & 0 \\ 0 & 1 & 0 & 0 & 0 & 0 \\ 0 & 0 & 1 & 0 & 0 & 0 \\ 0 & 0 & 0 & 1 & 0 & 0 \\ 0 & 0 & 0 & 0 & 0 & 1 \end{bmatrix} \begin{bmatrix} 1 & 0 & 0 & 0 & 0 & 0 & 0 & 0 \\ 0 & 1 & 0 & 0 & 0 & 0 & 0 & 0 \\ 0 & 0 & T_{11} & T_{12} & 0 & -1 & 0 & 0 \\ 0 & 0 & -T_{12} & T_{22} & -1 & 0 & 0 & 0 \\ 0 & 0 & 0 & 0 & 0 & 0 & 1 & 0 \\ 0 & 0 & 0 & 0 & 0 & 0 & 0 & 1 \end{bmatrix}
$$

$$
\cdot \begin{bmatrix} T_{11} & T_{12} & T_{13} & T_{14} & 0 & 0 & 0 & 0 \\ -T_{12} & T_{22} & T_{14} & T_{24} & 0 & 0 & 0 & 0 \\ T_{13} & T_{14} & T_{11} & T_{12} & 0 & 0 & 0 & 0 \\ -T_{14} & T_{24} & -T_{12} & T_{22} & 0 & 0 & 0 & 0 \\ 0 & 0 & 0 & 0 & T_{11} & T_{12} & T_{13} & T_{14} \\ 0 & 0 & 0 & 0 & -T_{12} & T_{22} & -T_{14} & T_{24} \\ 0 & 0 & 0 & 0 & T_{13} & T_{14} & T_{11} & T_{12} \\ 0 & 0 & 0 & 0 & -T_{14} & T_{24} & -T_{12} & T_{22} \end{bmatrix}
$$

$$
\cdot \begin{bmatrix} 1 & 0 & 0 & 0 & 0 & 0 \\ 0 & 1 & 0 & 0 & 0 & 0 \\ 0 & 0 & 0 & 0 & 0 & 1 \\ 0 & 0 & 0 & 0 & 1 & 0 \\ 0 & 0 & 0 & 0 & T_{11} & T_{12} \\ 0 & 0 & 0 & 0 & -T_{12} & T_{22} \\ 0 & 0 & 1 & 0 & 0 & 0 \\ 0 & 0 & 0 & 1 & 0 & 0 \end{bmatrix} \begin{bmatrix} 1 & 0 & 0 & 0 & 0 \\ 0 & 1 & 0 & 0 & 0 \\ 0 & 0 & 0 & 0 & 0 \\ 0 & 0 & 0 & 0 & 1 \\ 0 & 0 & 1 & 0 & 0 \\ 0 & 0 & 0 & 1 & 0 \end{bmatrix} \begin{bmatrix} b_1 \\ b_2 \\ b_{e1} \\ b_{e2} \\ b_{e3} \end{bmatrix}
$$

2.4 ADMITTANCE MATRIX REPRESENTATION

The last matrix representation we want to present is the admittance matrix. The admittance matrix **Y** and the impedance matrix **Z** of a multiport are defined by these

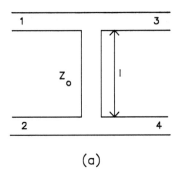

(a)

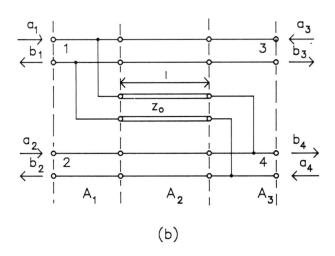

(b)

Figure 2.10 (a) A branch of the branch line directional coupler and (b) the equivalent cascade circuit.

equations:

$$\mathbf{I} = \mathbf{YV} \tag{2.83}$$

$$\mathbf{V} = \mathbf{ZI} \tag{2.84}$$

where $\mathbf{I}$ and $\mathbf{V}$ are vectors of terminal currents and terminal voltages of the multiport, respectively, and $\mathbf{Z} = \mathbf{Y}^{-1}$.

In Table 2.5 we introduce admittance matrices of some basic microwave circuit

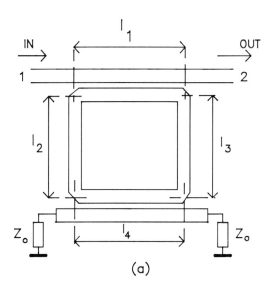

(a)

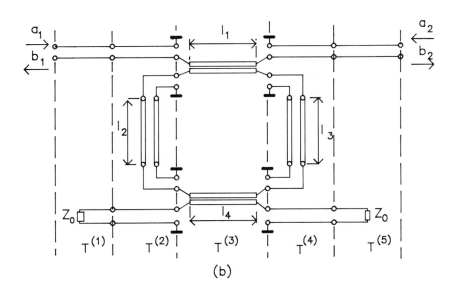

(b)

Figure 2.11 (a) Microwave band-stop filter and (b) the equivalent cascade circuit.

Table 2.5
Admittance Matrices of Some Basic Microwave Circuit Elements

Circuit Element	The Admittance Matrix $\mathbf{Y}$
 Y	$Y\begin{bmatrix} 1 & -1 \\ -1 & 1 \end{bmatrix}$
 Y	Does not exist
 Z_0 $\gamma = \alpha + j\beta,\ \beta = \dfrac{2\pi}{\lambda_g}$ Transmission line section	$Y_0\begin{bmatrix} \coth(\gamma l) & -\operatorname{csch}(\gamma l) \\ -\operatorname{csch}(\gamma l) & \coth(\gamma l) \end{bmatrix}$
 Y_0 Short-ended series connected stub	$Y\begin{bmatrix} 1 & -1 \\ -1 & 1 \end{bmatrix}$ where $Y = Y_0\coth(\gamma l)$
 Y_0 Open-ended series connected stub	$Y\begin{bmatrix} 1 & -1 \\ -1 & 1 \end{bmatrix}$ where $Y = Y_0\tanh(\gamma l)$
Short-ended parallel connected stub	Does not exist
Open-ended parallel connected stub	Does not exist
Ideal transformer	Does not exist

Table 2.5 (*Continued*)

Circuit Element	The Admittance Matrix **Y**
  Π = network	$\begin{bmatrix} Y_1 + Y_2 & -Y_2 \\ -Y_2 & Y_1 + Y_3 \end{bmatrix}$
  T = network	$\dfrac{1}{Z_1 Z_2 + Z_2 Z_3 + Z_1 Z_3} \begin{bmatrix} Z_2 + Z_3 & -Z_3 \\ -Z_3 & Z_1 + Z_3 \end{bmatrix}$
  Distributed RC-line section	$\sqrt{\dfrac{j\omega C}{R}} \begin{bmatrix} \coth\left(\sqrt{j\omega RC}\right) & -\cosh\left(\sqrt{j\omega RC}\right) \\ -\cosh\left(\sqrt{j\omega RC}\right) & \coth\left(\sqrt{j\omega RC}\right) \end{bmatrix}$ $C = c \cdot l$ – total capacitance $R = r \cdot l$ – total resistance
  Gyrator	$\begin{bmatrix} 0 & \alpha \\ -\alpha & 0 \end{bmatrix}$
  Y_{0e}, Y_{00}, l $\gamma_e = \alpha_e + j\beta_e, \qquad \beta_e = \dfrac{2\pi}{\lambda_{ge}}$ $\gamma_e = \alpha_e + j\beta_e, \qquad \beta_e = \dfrac{2\pi}{\lambda_{ge}}$ Section of coupled transmission lines	$\begin{bmatrix} Y_{11} & Y_{12} & Y_{13} & Y_{14} \\ Y_{12} & Y_{11} & Y_{14} & Y_{13} \\ Y_{13} & Y_{14} & Y_{11} & Y_{12} \\ Y_{14} & Y_{13} & Y_{12} & Y_{11} \end{bmatrix}$ $Y_{11} = \dfrac{1}{2}\left(Y_{0e}\coth(\gamma_e l) + Y_{00}\coth(\gamma_0 l)\right)$ $Y_{12} = \dfrac{1}{2}\left(Y_{0e}\coth(\gamma_e l) - Y_{00}\coth(\gamma_0 l)\right)$ $Y_{13} = \dfrac{-1}{2}\left(Y_{0e}\operatorname{csch}(\gamma_e l) - Y_{00}\operatorname{csch}(\gamma_0 l)\right)$ $Y_{14} = \dfrac{-1}{2}\left(Y_{0e}\operatorname{csch}(\gamma_e l) + Y_{00}\operatorname{csch}(\gamma_0 l)\right)$

elements. We will present more discussion on the admittance matrix representation of microwave circuits in Chapter 3.

2.5 RELATIONS BETWEEN DIFFERENT MATRIX REPRESENTATIONS OF MULTIPORTS

Relationships between the S Matrix and the Y and Z Matrices

Let us refer to the circuit shown in Figure 2.4 and assume that internal impedances of all signal generators connected to the network ports are equal to the reference impedances of individual ports. According to the definition of complex wave variables, (2.17) and (2.18), vectors of incoming power waves $\mathbf{a}$ and outgoing power waves $\mathbf{b}$ at the circuit ports can be written in the following form:

$$\mathbf{a} = \mathbf{F}(\mathbf{V} + \mathbf{GI}) \tag{2.85}$$

$$\mathbf{b} = \mathbf{F}(\mathbf{V} - \mathbf{G}^+\mathbf{I}) \tag{2.86}$$

In these equations $\mathbf{V}$ and $\mathbf{I}$ are vectors with ith elements that are equal to the terminal voltages V_i, $i = 1, 2, \ldots, n$, and to the terminal currents I_i, $i = 1, 2, \ldots, n$, respectively, at the ports of the n-port circuit. $\mathbf{F}$ and $\mathbf{G}$ are the diagonal matrices given by

$$\mathbf{F} = \begin{bmatrix} \dfrac{1}{2\sqrt{\operatorname{Re} Z_{N1}}} & 0 & \cdots & 0 \\ 0 & \dfrac{1}{2\sqrt{\operatorname{Re} Z_{N2}}} & & \vdots \\ \vdots & & \ddots & 0 \\ 0 & & \cdots & \dfrac{1}{2\sqrt{\operatorname{Re} Z_{Nn}}} \end{bmatrix} \tag{2.87}$$

$$\mathbf{Z} = \begin{bmatrix} Z_{N1} & 0 & \cdots & 0 \\ 0 & Z_{N2} & & \vdots \\ \vdots & & \ddots & 0 \\ 0 & \cdots & 0 & Z_{Nn} \end{bmatrix} \tag{2.88}$$

The admittance matrix $\mathbf{Y}$ and the impedance matrix $\mathbf{Z}$ of a multiport are defined by (2.83) and (2.84).

Elimination of **a**, **b**, and **V** from (2.83), (2.84), (2.85), and (2.86) yields

$$\mathbf{F}(\mathbf{Z} - \mathbf{G}^+)\mathbf{I} = \mathbf{SF}(\mathbf{Z} + \mathbf{G})\mathbf{I} \tag{2.89}$$

from which we have the relation between **S** and **Z**:

$$\mathbf{S} = \mathbf{F}(\mathbf{Z} - \mathbf{G}^+)(\mathbf{Z} + \mathbf{G})^{-1}\mathbf{F}^{-1} \tag{2.90}$$

Similarly, the impedance matrix **Z** can be expressed in terms of the scattering matrix **S**:

$$\mathbf{Z} = \mathbf{F}^{-1}(\mathbf{I} - \mathbf{S})^{-1}(\mathbf{SG} + \mathbf{G}^+)\mathbf{F} \tag{2.91}$$

The relations between **S** and **Y** matrices are found in the same way. We have

$$\mathbf{S} = \mathbf{F}(\mathbf{I} - \mathbf{G}^+\mathbf{Y})(\mathbf{I} + \mathbf{GY})^{-1}\mathbf{F}^{-1} \tag{2.92}$$

$$\mathbf{Y} = \mathbf{F}(\mathbf{I} - \mathbf{S})(\mathbf{SG} + \mathbf{G}^+)^{-1}\mathbf{F}^{-1} \tag{2.93}$$

In (2.91) to (2.93) **I** denotes the identity matrix.

Relationships between the **S** *Matrix and the* **T**(*ABCD*) *Matrix*

The relations describing the transformation between the **S** and **T**(*ABCD*) matrices are the following:

$$\mathbf{A} = \frac{Z_{N1}^* + Z_{N1}S_{11} - Z_{N1}^*S_{22} - Z_{N1}\Delta_s}{2S_{21}\sqrt{\mathrm{Re}\,\{Z_{N1}\}\,\mathrm{Re}\,\{Z_{N2}\}}} \tag{2.94}$$

$$\mathbf{B} = \frac{Z_{N1}^*Z_{N2}^* + Z_{N1}Z_{N2}^*S_{11} + Z_{N1}^*Z_{N2}S_{22} + Z_{N1}Z_{N2}\Delta_s}{2S_{21}\sqrt{\mathrm{Re}\,\{Z_{N1}\}\,\mathrm{Re}\,\{Z_{N2}\}}} \tag{2.95}$$

$$\mathbf{C} = \frac{1 - S_{11} - S_{22} + \Delta_s}{2S_{21}\sqrt{\mathrm{Re}\,\{Z_{N1}\}\,\mathrm{Re}\,\{Z_{N2}\}}} \tag{2.96}$$

$$\mathbf{D} = \frac{Z_{N2}^* - Z_{N2}^*S_{11} + Z_{N2}S_{22} - Z_{N2}\Delta s}{2S_{21}\sqrt{\mathrm{Re}\,\{Z_{N1}\}\,\mathrm{Re}\,\{Z_{N2}\}}} \tag{2.97}$$

where Z_{N1} and Z_{N2} are the reference impedances of two-port ports and $\Delta_s = S_{11}S_{22} - S_{12}S_{21}$.

The inverse relations are

$$S_{11} = \frac{AZ_{N2} + B - CZ_{N1}^* Z_{N2} - DZ_{N1}^*}{AZ_{N2} + B + CZ_{N1} Z_{N2} + DZ_{N1}} \tag{2.98}$$

$$S_{12} = \frac{2(AD - BC)\sqrt{\operatorname{Re}\{Z_{N1}\}\operatorname{Re}\{Z_{N2}\}}}{AZ_{N2} + B + CZ_{N1} Z_{N2} + DZ_{N1}} \tag{2.99}$$

$$S_{21} = \frac{2\sqrt{\operatorname{Re}\{Z_{N1}\}\operatorname{Re}\{Z_{N2}\}}}{AZ_{N2} + B + CZ_{N1} Z_{N2} + DZ_{N1}} \tag{2.100}$$

$$S_{22} = \frac{-AZ_{N2}^* + B - CZ_{N1} Z_{N2}^* + DZ_{N1}}{AZ_{N2} + B + CZ_{N1} Z_{N2} + DZ_{N1}} \tag{2.101}$$

REFERENCES

[1] R.W. Newcomb, *Linear Multiport Synthesis*, McGraw-Hill, New York, 1966.

[2] K. Kurokawa, "Power Waves and the Scattering Matrix," *IEEE Trans. Microwave Theory Tech.*, Vol. MTT-13, No. 2, March 1965, pp. 194–202.

[3] P. Penfield, "Noise in Negative Resistance Amplifiers," *IRE Trans. Circuit Theory*, Vol. CT-7, 1960, pp. 166–170.

[4] P. Penfield, "A Classification of Lossless Three Ports," *IRE Trans. Circuit Theory*, Vol. CT-9, 1963, pp. 215–223.

[5] D.C. Youla, "On Scattering Matrices Normalized to Complex Port Numbers," *Proc. IRE*, Vol. 49, 1961, s. 122.

[6] Hewlett-Packard, Application Note 117-1, 1969.

[7] E.S. Kuh and R.A. Rohrer, *Theory of Linear Active Networks*, Holden Day, San Francisco, 1967.

[8] H.J. Carlin and A.B. Giordano, *Network Theory*, Englwood Cliffs, NJ, Prentice-Hall, 1964.

[9] P. Miazga, "Computer Aided Analysis of Microwave Circuits by Means of Transfer Scattering Matrices," Ph.D. Dissertation, Warsaw University of Technology, Warsaw, Poland, 1989 [in Polish].

Chapter 3
Computer-Aided Analysis
of Microwave Circuits

We understand the analysis of a circuit to be the computation of a certain number of response functions in terms of the component parameters for a circuit of fixed topology. The response functions may be evaluated in terms of some node voltages and branch currents. This is usually done with circuits composed of lumped elements. Circuit equations also may be formulated by means of normalized wave variables. Microwave circuits may contain lumped as well as distributed components. The equivalent circuit models of active devices are mostly the origins of lumped components in microwave circuits. Obviously, the lumped resistances, capacitances, and inductances can be present in these circuits [1–3]. Lumped-element circuits are analyzed by using the nodal equations scheme [4, 5] and the overall circuit is characterized in terms of the admittance matrix. The distributed elements are present as sections of transmission lines, directional couplers, circulators, *et cetera*. In the second case, the components are multiports connected through pairs of ports, and the overall circuit is usually characterized in terms of scattering matrices, transfer scattering matrices, or the *ABCD* parameters [6, 7]. Because a microwave circuit may contain lumped as well as distributed elements, a convenient way for computer-aided analysis to characterize both types of elements is as either lumped equivalents or multiport equivalents. As is well known, distributed components can be modeled only approximately as lumped equivalents, whereas multiterminal lumped elements can be transformed exactly into multiport circuits [8]. As a result, microwave circuits are normally treated as multiports.

3.1 MICROWAVE CIRCUIT ANALYSIS IN TERMS OF VOLTAGES AND CURRENTS

The solution of an electrical circuit in terms of voltages and currents may be achieved in several ways depending on the variables assumed to be unknown.

3.1.1 Nodal Admittance Matrix Method

The most frequently used method for computer-aided analysis of linear circuits in the frequency domain is the nodal admittance matrix method in which the voltages V_N of all nodes are assumed as unknowns. A set of equations is constructed based on the fact that the sum of the currents flowing into any node in the circuit is zero. In microwave circuits the nodal equations are usually set up in the definite form; that is, the voltage of each node is taken with respect to a distinguished node of the circuit, which is called the *reference node*.

The equations are easily set up in the form of a matrix expression [4, 5]:

$$\mathbf{Y}\mathbf{V}_N = \mathbf{I}_{N_0} \tag{3.1}$$

where

$\mathbf{Y}$ = a square nodal admittance matrix (a degree of this matrix equals the number of the nodes in the analyzed circuit);
$\mathbf{V}_N$ = a vector of node voltages taken with respect to a point outside the circuit;
$\mathbf{I}_{N_0}$ = a vector of terminal currents of the independent current sources connected between the nodes and the reference node of the circuit.

Example 3.1

Consider a circuit presented in Figure 3.1. The circuit equations derive from the application of Kirchhoff's current law to each node of the circuit.

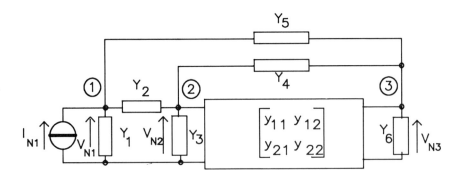

Figure 3.1 A circuit to illustrate the principles of the formulation of the nodal voltage equations.

Node 1:

$$Y_1 V_{N1} + Y_2(V_{N1} - V_{N2}) + Y_5(V_{N1} - V_{N3}) = I_{N1}$$

Node 2:

$$Y_3 V_{N2} + Y_2(V_{N2} - V_{N1}) + Y_4(V_{N2} - V_{N3}) + y_{11}V_{N2} + y_{12}V_{N3} = 0$$

Node 3:

$$Y_6 V_{N3} + Y_4(V_{N3} - V_{N2}) + Y_5(V_{N3} - V_{N1}) + y_{21}V_{N2} + y_{22}V_{N3} = 0$$

These equations may be written in the matrix form in the following way:

$$
\begin{array}{c}
\text{Node numbers} \\
\begin{array}{ccc} 1 & 2 & 3 \end{array} \\
\begin{array}{c} 1 \\ 2 \\ 3 \end{array}
\begin{bmatrix}
Y_1 + Y_2 + Y_5 & -Y_2 & -Y_5 \\
-Y_2 & Y_2 + Y_3 + Y_4 + y_{11} & -Y_4 + y_{12} \\
-Y_5 & -Y_4 + y_{21} & Y_4 + Y_5 + Y_6 + y_{22}
\end{bmatrix}
\begin{bmatrix}
V_{N1} \\ V_{N2} \\ V_{N3}
\end{bmatrix}
=
\begin{bmatrix}
I_{N1} \\ 0 \\ 0
\end{bmatrix}
\end{array}
$$

The nodal admittance matrix of a circuit may be derived using the following rules:

(1) *An Admittance Y.* A passive admittance connected between nodes i and j contributes a term into the nodal admittance matrix **Y** given by

$$
\begin{array}{c}
\text{column} \\
\begin{array}{cc} (i) & (j) \end{array} \\
\text{row}
\begin{array}{c} (i) \\ (j) \end{array}
\begin{bmatrix}
\cdots \ \dot{Y} \cdots -\dot{Y} \cdots \\
\cdots -\dot{Y} \cdots \ \dot{Y} \cdots
\end{bmatrix}
\end{array}
\tag{3.2}
$$

If the node i is the reference node, then the admittance **Y** appears only on the main diagonal of the **Y** matrix in the element Y_{jj}:

$$
(j)
\begin{array}{c}
(j) \\
\begin{bmatrix}
\cdots \ \dot{Y} \cdots
\end{bmatrix}
\end{array}
\tag{3.3}
$$

(2) *An Independent Current Source I.* An independent current source connected between nodes i and j, in the direction $i \to j$, contributes a term to the right-hand vector $\mathbf{I}_N$ as

$$
\begin{array}{c} (i) \\ \\ (j) \end{array}
\begin{bmatrix} \vdots \\ -I \\ \vdots \\ I \\ \vdots \end{bmatrix}
\tag{3.4}
$$

If node i is the reference node in the circuit, an independent current source I appears in $\mathbf{I}_N$ as

$$
(j)
\begin{bmatrix} \vdots \\ \vdots \\ I \\ \vdots \\ \vdots \end{bmatrix}
\tag{3.5}
$$

(3) *An Independent Voltage Source, E.* An independent voltage source connected between nodes i and j, in the direction $i \to j$, contributes a term in $\mathbf{I}_N$ as

$$
\begin{array}{c} (i) \\ \\ (j) \end{array}
\begin{bmatrix} \vdots \\ yE \\ \vdots \\ -yE \\ \vdots \end{bmatrix}
\tag{3.6}
$$

where y is the series admittance of the source. Note that the nodal admittance matrix does not lend itself to circuit descriptions that include an independent voltage source unless it has finite series admittance.

(4) *A Voltage-Controlled Current Source.* Let the current source be connected between nodes m and l, and let its current be gV_{kl}, where V_{kl} is the voltage between nodes k and l. According to Figure 3.2, the components of $\mathbf{Y}$ corresponding to this

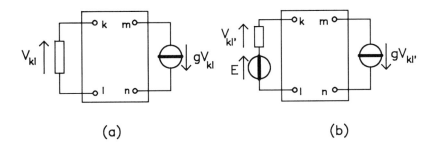

Figure 3.2 Voltage-controlled current source: (a) voltage source controlled by a branch voltage; (b) voltage source controlled by a voltage across the series admittance of an independent voltage source.

source are

$$
\begin{array}{cc}
(k) & (l)
\end{array}
$$

$$
\begin{array}{c}
(m) \\
\\
(n)
\end{array}
\left[
\begin{array}{ccccc}
\vdots & & \vdots & \\
\cdots & g & \cdots & -g & \cdots \\
\vdots & & \vdots & \\
\cdots & -g & \cdots & g & \cdots \\
\vdots & & \vdots &
\end{array}
\right]
\qquad (3.7)
$$

If the controlling voltage is a voltage across the series admittance of the independent voltage source, such a voltage-controlled current source also contributes to the vector $\mathbf{I}_N$ of the right-hand side of the (3.1):

$$
\begin{array}{c}
(m) \\
\\
(n)
\end{array}
\left[
\begin{array}{c}
\vdots \\
gE \\
\vdots \\
-gE \\
\vdots
\end{array}
\right]
\qquad (3.8)
$$

(5) *A Current-Controlled Current Source.* Let the source be connected between nodes m and n and let its current be βI, where I is the current in the element connected between nodes k and l. According to Figure 3.3, the components of the

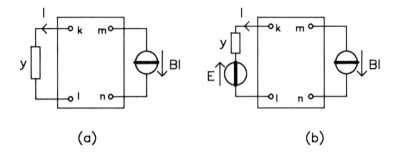

(a)   (b)

Figure 3.3 Current-controlled current source: (a) current source controlled by the branch current; (b) current source controlled by the current flowing through the series admittance of an independent voltage source.

admittance matrix corresponding to this current source are

$$
\begin{array}{cc}
\quad (k) \qquad (l)
\end{array}
$$

$$
\begin{array}{c}
(m) \\
\\
(n)
\end{array}
\left[
\begin{array}{ccc}
\vdots & & \vdots \\
\cdots \quad \beta y \cdots & -\beta y \cdots \\
\vdots & & \vdots \\
\cdots \quad -\beta y \cdots & \beta y \cdots \\
\vdots & & \vdots
\end{array}
\right]
\tag{3.9}
$$

If the controlling current flows through the admittance y associated with an independent voltage source (see Figure 3.3 (b)), the contribution to the vector $\mathbf{I}_N$ is given by

$$
\begin{array}{c}
(m) \\
\\
(n)
\end{array}
\left[
\begin{array}{c}
\vdots \\
+\beta y E \\
\vdots \\
-\beta y E \\
\vdots
\end{array}
\right]
\tag{3.10}
$$

(6) *A Current-Controlled Voltage Source.* In accordance with Figure 3.4, the

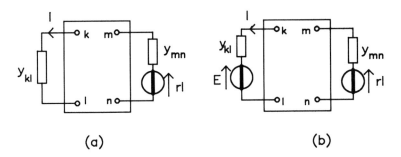

Figure 3.4 Current-controlled voltage source: (a) voltage source controlled by a branch current; (b) voltage source controlled by a current flowing through the series admittance of the independent voltage source.

contribution of this element to the admittance matrix is

$$
\begin{array}{cc}
(k) & (l)
\end{array}
$$

$$
\begin{array}{c}
(m) \\
\\
(n)
\end{array}
\left[
\begin{array}{cccc}
\cdots & -y_{mn}{}^r y_{kl} & \cdots & y_{mn}{}^r y_{kl} & \cdots \\
& \vdots & & \vdots & \\
\cdots & y_{mn}{}^r y_{kl} & \cdots & -y_{mn}{}^r y_{kl} & \cdots \\
\end{array}
\right]
\tag{3.11}
$$

If controlling current I flows through the series admittance of an independent voltage source (Figure 3.4), the contribution to the vector $\mathbf{I}_N$ is

$$
\begin{array}{c}
(m) \\
\\
(n)
\end{array}
\left[
\begin{array}{c}
\vdots \\
-y_{mn}{}^r y_{kl} E \\
\vdots \\
+y_{mn}{}^r y_{kl} E \\
\vdots
\end{array}
\right]
\tag{3.12}
$$

As in the case of an independent voltage source, a controlled voltage source cannot be included in a circuit described by the nodal admittance matrix unless it has finite series impedance.

(7) *Voltage-Controlled Voltage Source.* With reference to Figure 3.5, a voltage source controlled by a branch voltage contributes to $\mathbf{Y}$ as

$$
\begin{array}{cc}
(k) & (l)
\end{array}
$$

$$
\begin{array}{c}
(t) \\
(u)
\end{array}
\left[
\begin{array}{ccc}
\vdots & & \vdots \\
\cdots \; -ky_{tu} \; \cdots & & ky_{tu} \; \cdots \\
\vdots & & \vdots \\
\cdots \; ky_{tu} \; \cdots & & -ky_{tu} \; \cdots \\
\vdots & & \vdots
\end{array}
\right]
\tag{3.13}
$$

If the controlling voltage is a voltage across the admittance y associated with an independent voltage source (see Figure 3.5 (b)), the contribution to the vector $\mathbf{I}_N$ is given by

$$
\begin{array}{c}
(t) \\
(u)
\end{array}
\left[
\begin{array}{c}
\vdots \\
-ky_{kl}E \\
\vdots \\
ky_{kl}E \\
\vdots
\end{array}
\right]
\tag{3.14}
$$

(8) *Multiterminal Elements.* Such elements as sections of transmission lines, sections of coupled transmission lines, MESFETs, and bipolar transistors are usually described by their terminal parameters. Devices represented by $\mathbf{y}$ matrices are entered into the matrix $\mathbf{Y}$ easily if the matrix $\mathbf{y}$ is available in its indefinite form.

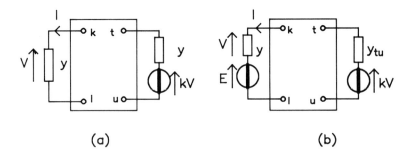

(a)   (b)

Figure 3.5 Voltage-controlled voltage source: (a) voltage source controlled by a branch voltage; (b) voltage source controlled by a voltage across the series admittance of an independent voltage source.

Consider, as an example, a three-terminal device, whose indefinite matrix **y** defines this equation (see Figure 3.6):

$$
\begin{bmatrix} y_{11} & y_{12} & y_{13} \\ y_{21} & y_{22} & y_{23} \\ y_{31} & y_{32} & y_{33} \end{bmatrix} \begin{bmatrix} V_1 \\ V_2 \\ V_3 \end{bmatrix} = \begin{bmatrix} I_1 \\ I_2 \\ I_3 \end{bmatrix} \tag{3.15}
$$

If node 1 of the terminal device is connected to node i of the circuit, node 2 to node j, and node 3 to node k, the terms of the three terminal device admittance matrix **y** are added to the nodal admittance matrix **Y** of the entire circuit as

$$
\begin{array}{ccc} (i) & (j) & (k) \end{array}
$$

$$
\begin{array}{c} (i) \\ \\ (j) \\ \\ (k) \end{array}
\begin{bmatrix}
\cdots\ y_{11}\ \cdots & y_{12}\ \cdots & y_{13}\ \cdots \\
\\
\cdots\ y_{21}\ \cdots & y_{22}\ \cdots & y_{23}\ \cdots \\
\\
\cdots\ y_{31}\ \cdots & y_{32}\ \cdots & y_{33}\ \cdots
\end{bmatrix} \tag{3.16}
$$

As we see, a multiterminal device (element) does not have to be replaced by its equivalent circuit composed of two-terminal elements.

When a multiterminal device is described by its definite admittance matrix $\mathbf{y}_d$, the elements of $\mathbf{y}_d$ may be entered into the matrix **Y** of the whole circuit only if the reference node of the device agrees with the reference node of the whole circuit.

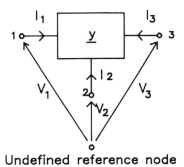

Undefined reference node

Figure 3.6 Three-terminal device and its terminal voltages and currents.

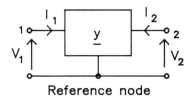

Figure 3.7 Three-terminal device with one terminal taken as its reference node.

Figure 3.7 presents a three-terminal device as a four-pole with the terminal 3 taken as the reference node. The definite admittance matrix of the device is given by the following equation:

$$\begin{bmatrix} I_1 \\ I_2 \end{bmatrix} = \begin{bmatrix} y_{11} & y_{12} \\ y_{21} & y_{22} \end{bmatrix} \begin{bmatrix} V_1 \\ V_2 \end{bmatrix} \tag{3.17}$$

If the reference node of the four-pole agrees with the reference node of the whole circuit, and if node 1 of the terminal device is connected to the node i of the circuit and node 2 to the node j, the terms of the definite admittance matrix y_d of the device are added to the nodal admittance matrix $\mathbf{Y}$ of the whole circuit:

$$
\begin{array}{cc}
& (i) \qquad (j) \\
\begin{array}{c} (i) \\[20pt] (j) \end{array}
&
\begin{bmatrix}
\cdots & \vdots & \cdots & \vdots & \\
\cdots & y_{11} & \cdots & y_{12} & \cdots \\
& \vdots & & \vdots & \\
\cdots & y_{21} & \cdots & y_{22} & \cdots \\
& \vdots & & \vdots &
\end{bmatrix}
\end{array}
\tag{3.18}
$$

The conversion of a multiterminal element admittance matrix from indefinite form into definite form, and *vice versa*, can be performed very simply because the indefinite admittance matrix of an n terminal circuit satisfies the following relations:

$$\sum_{i=1}^{n} y_{ij} = 0, \qquad \text{for } j = 1, 2, \ldots, n \tag{3.19}$$

$$\sum_{j=1}^{n} y_{ij} = 0, \qquad \text{for } i = 1, 2, \ldots, n. \tag{3.20}$$

which indicate that the sums of admittances in each column (3.19) and each row (3.20) of the indefinite admittance matrix $\mathbf{Y}$ are equal to zero.

Relation (3.19) can be proved by considering the case where a voltage source V_j is applied between the reference node and the jth node, while the other nodes are short-circuited to the reference node. Because the Kirchhoff current law requires the sum of all currents flowing from the reference node to be zero, the sum $\sum_{i=1}^{n} y_{ij}$ must be equal to zero.

Consider another case where the voltages applied between the reference node and all circuit nodes are changed by a constant ΔV. This cannot change the currents flowing into all nodes from the reference node, which means that the equation $\mathbf{Y}\,\Delta\mathbf{V} = \mathbf{0}$ must hold ($\Delta\mathbf{V}$ is a constant vector, all the elements of which are equal to ΔV). For the row i of the indefinite admittance matrix, we have $\Delta V \sum_{j=1}^{n} y_{ij} = 0$. As ΔV may be any nonzero constant, the sum must be equal to zero. This proves relation (3.20).

3.1.2 Numerical Considerations

The nodal admittance matrix $\mathbf{Y}$ and the right-hand vector $\mathbf{I}_N$ of (3.1) are constructed in accordance with the rules given earlier. The resulting matrix $\mathbf{Y}$ and the vector $\mathbf{I}_N$ are computed at the frequency of interest. The entries Y_{ij} of the nodal admittance matrix may become very small or very large as the circuit function (response) is computed for different values of frequency. The most important are values of the main diagonal entries Y_{ii} of the nodal admittance matrix. For example, if an inductor and a capacitor are connected to the node i, at a series resonant frequency of these elements, the entry Y_{ii} becomes undefined.

Second, the nodal admittance matrix of a circuit with dependent sources may become singular and the set of circuit equations cannot be solved. For example, the circuit presented in Figure 3.8 does not have a solution when the transconductance $g_m = 7\ 1/\Omega$. In microwave circuits, we assume very often a lossy circuit. The determinant of the nodal admittance matrix of such a circuit is a polynomial in the frequency variable ω. The matrix $\mathbf{Y}$ becomes singular at the zeros of the polynomial, and the system of circuit equations has no solution.

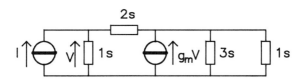

Figure 3.8 This circuit has no solution if $g_m = 7$ s.

Computer-aided analysis of microwave circuits in the frequency domain based on the nodal admittance matrix requires multiple computation of the solution of a system of linear equations with complex coefficients. An important problem is the numerical instability of a solution process performed by computer, which may occur if the so-called condition number of the coefficient matrix of the linear equation system is too large. This numerical problem will be discussed in Chapter 6.

The nodal admittance matrix in general is a sparse matrix, which means that many entries of the matrix are equal to zero, particularly for large circuits. In conventional numerical procedures used to solve a system of linear equations, the arithmetic operations are performed on all nonzero and zero valued entries of the coefficient matrix. The whole solution procedure involves $n^3/3 + n^2 - n/3$ complex number multiplications and divisions [5] (n = the order of the coefficient matrix). To save computation time and minimize storage requirements, the sparsity of the coefficient matrix must be taken into account in the solution procedure. These problems will be discussed in Chapter 7.

3.1.3 Computation of Circuit Functions

The solution of the nodal matrix expression (3.1) provides us with the numerical values of all node voltages of a circuit. In most practical cases, as for example filter or amplifier design, there is one source of excitation in the circuit and we are interested in one or two voltages in the circuit.

For the sake of clarity in the considerations to follow, we assume that an independent source of one ampere has been connected between node 1 and the reference node and that the output node is node n.

The node voltage equations are now

$$\mathbf{YV} = \mathbf{I}_N = \begin{bmatrix} 1 \\ 0 \\ \vdots \\ 0 \end{bmatrix} \tag{3.21}$$

Let $\mathbf{e}_i$ be a vector of zero entries except the ith, which is one. The postmultiplication of a matrix by $\mathbf{e}_i$ results in extracting the ith column from the matrix. If matrix $\mathbf{Z}$ is the inverse of $\mathbf{Y}$, then $\mathbf{YZ} = \mathbf{I}$, where $\mathbf{I}$ is an identity matrix of order n. By postmultiplication of $\mathbf{YZ}$ by $\mathbf{e}_1$, we have

$$\mathbf{YZe}_1 = \begin{bmatrix} 1 \\ 0 \\ \vdots \\ 0 \end{bmatrix} \tag{3.22}$$

By comparing (3.21) and (3.22), we see that the solution vector $\mathbf{V}$ of (3.21) is simply the first column of $\mathbf{Z}$:

$$\mathbf{V} = \mathbf{Z}\mathbf{e}_1 \qquad (3.23)$$

We define a transfer function $H(j\omega)$ as the ratio of the output voltage V_k to the input current I_1 of the independent current source connected between node 1 and the reference node:

$$H(j\omega) = \frac{V_k}{I_1} = |H(j\omega)|e^{j\phi(\omega)} \qquad (3.24)$$

The magnitude of $H(j\omega)$ is called the *gain of the circuit* and $\phi(\omega)$ is the *phase*. The term $|H(j\omega)|$ is directly related to the available power gain:

$$G_T = \frac{P_L}{P_{SA}} = 4\,\text{Re}\{Y_s\}\,\text{Re}\{Y_L\}|H(j\omega)|^2 \qquad (3.25)$$

In (3.25), P_L is the active power dissipated in the load admittance Y_L connected between the output node and the reference node, and P_{SA} is the available power of the current source with the internal admittance Y_s.

In filter design problems, the quantity "insertion loss" is more commonly used:

$$\text{IL} = -20\log\left(|Y_s + Y_L| \cdot |H(j\omega)|\right) \qquad (3.26)$$

Other circuit functions are also important. We have the input impedance

$$Y_{in} = \frac{I_1}{V_1} - Y_s \qquad (3.27)$$

or, equivalently, the reflection coefficient

$$\Gamma_{in} = \frac{Y_{in} - Y_s}{Y_{in} + Y_s} = 1 - 2\frac{V_1}{I_1}Y_s \qquad (3.28)$$

The discussed network functions can be computed, once the node voltages V_1 and V_k are determined.

Group Delay

The group delay is a very important circuit function in the filter or equalizer design. It is defined as

$$\tau(\omega) = -\frac{d\phi}{d\omega} \qquad (3.29)$$

where ϕ is the phase of the transfer function.

The group delay, τ, is a measure of the deviation of the phase function from being a linear function of frequency. In practice, we often want the group delay to be constant over a range of frequency.

The group delay may be computed by using numerical differentiation:

$$\tau(\omega) = -\frac{\phi(\omega + \Delta\omega) - \phi(\omega - \Delta\omega)}{2\,\Delta\omega} \tag{3.30}$$

where $\Delta\omega$ is a small increment of frequency. Thus, for each frequency point of interest, computation of $\tau(\omega)$ by using (3.30) requires two additional solutions of the node equations, which significantly increases the computational effort. Furthermore, the relative error of $\tau(\omega)$ as computed from (3.30) can be unacceptable if the phase difference is too small. We recommend that ϕ be computed at a frequency increased by 0.01% (a 1.0001 factor) and decreased by 0.01%.

A very effective and accurate method for computing the group delay will be discussed in Chapter 4.

3.1.4 Multiport Connection Method Based on an Indefinite Admittance Matrix

The multiport connection method applicable to the indefinite admittance matrix was proposed by M. A. Murray-Lasso [9, 10]. The indefinite admittance matrix is simply the nodal admittance matrix defined in Section 3.1.1, but with the reference node outside the circuit. The method reduces computation time and memory space requirements in the analysis of large microwave circuits by dividing a circuit into subcircuits and separately computing the indefinite admittance matrices of the successive circuits connected to one another. In each step of the connection of a subcircuit, the indefinite admittance matrix relative to the external nodes of the resultant circuit is computed. In the last step of the procedure (connection of the last subcircuit), the indefinite admittance matrix is computed relative to the external nodes of the whole circuit. Thus, the method reduces the number of circuit equations before their solution.

Let us consider two subcircuits with the indefinite admittance matrices $\mathbf{Y}_1$ and $\mathbf{Y}_2$, respectively. As shown in Figure 3.9, the two subcircuits are connected by their internal nodes. The other nodes are external nodes of the resultant circuit.

The indefinite admittance matrix of the resultant circuit, with the form as in Figure 3.10, can be ordered for a set of matrix equations in the form

$$\begin{bmatrix} \mathbf{I}_e \\ \mathbf{I}_i \end{bmatrix} = \begin{bmatrix} \mathbf{Y}_{ee} & \mathbf{Y}_{ei} \\ \mathbf{Y}_{ie} & \mathbf{Y}_{ii} \end{bmatrix} \begin{bmatrix} \mathbf{V}_e \\ \mathbf{V}_i \end{bmatrix} \quad \begin{array}{l} \text{1st set of equations} \\ \text{2nd set of equations} \end{array} \tag{3.31}$$

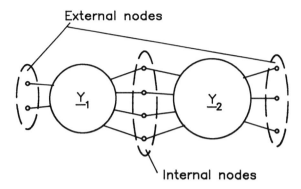

Figure 3.9 Two subcircuits connected by internal nodes.

where

$\mathbf{I}_e$ and $\mathbf{V}_e$ = the vectors of the currents and voltages relative to the external nodes of the resultant circuit,

$\mathbf{I}_i$ and $\mathbf{V}_i$ = the vectors of the currents and voltages relative to the internal nodes of the resultant circuit.

Now we can eliminate vector $\mathbf{V}_i$ from both sets of equations of (3.31). Because independent current sources are not connected to the internal nodes of the circuit, $\mathbf{I}_i = \mathbf{0}$, and from the second set of equations (3.31):

$$\mathbf{V}_i = -\mathbf{Y}_{ii}^{-1}\mathbf{Y}_{ie}\mathbf{V}_e \qquad (3.32)$$

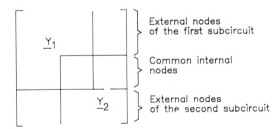

Figure 3.10 The indefinite admittance matrix of a circuit composed of two subcircuits connected by common internal nodes.

Substituting $\mathbf{V}_i$ given by (3.32) into the first set of equations in (3.31):

$$\mathbf{I}_e = \mathbf{Y}_e \mathbf{V}_e \qquad (3.33)$$

where

$$\mathbf{Y}_e = \mathbf{Y}_{ee} - \mathbf{Y}_{ei} \mathbf{Y}_{ii}^{-1} \mathbf{Y}_{ie} \qquad (3.34)$$

is the indefinite admittance matrix of the resultant circuit with respect to the external nodes of both connected subcircuits. By subsequent connection of the succeeding subcircuits up to the last one, and by repeating the procedure described by the equation (2.34), we can compute the indefinite admittance matrix for the external nodes of the complete circuit.

This procedure requires computation of the inverse matrix $\mathbf{Y}_{ii}^{-1}$. The number of multiplications and divisions required for the computation of the inverse matrix is proportional to n^3 (where n is the order of the matrix). The total number of long numerical operations to be performed for the computation of the final indefinite admittance matrix for the given circuit topology depends on the order according to which the subcircuits are connected to each other.

As has been proved, we obtain the optimal sequence of connected subcircuits—by which we mean the minimum number of arithmetic operations—when we connect two subcircuits to form a circuit that has the smallest number of external nodes [6].

3.1.5 Chain Matrix Method and Its Modifications

The chain matrix of a two-port presented in Figure 3.11 defines the relation (2.1). The chain matrix of two cascaded two-ports equals the product of the chain matrices of the connected two-ports. Because many microwave circuits can be modeled as cascaded connections of the component two-ports, the chain matrix can be used very effectively in the computer-aided analysis of microwave circuits. The chain matrix of the complete circuit can be computed by successive multiplication of the chain matrices of the cascaded two-port elements of the circuit. Equation (2.3) presents this rule mathematically. The knowledge of the resultant *ABCD* chain matrix of the whole circuit allows us to compute particular circuit functions.

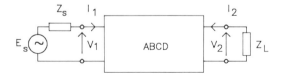

Figure 3.11 A two-port described by the chain matrix **A**(*ABCD*).

We define the transfer function $H(j\omega)$ as the ratio of the output voltage V_{out} to the voltage E_s of the independent voltage signal source:

$$H(j\omega) = \frac{V_{out}}{E_s} = |H(j\omega)|e^{j\phi(\omega)}$$

$$= \frac{Z_L}{AZ_L + B + CZ_sZ_L + DZ_s} \tag{3.35}$$

where $H(j\omega)$ equals the transducer voltage gain.

The magnitude of $H(j\omega)$ is directly related to the transducer power gain:

$$G_T = \frac{P_L}{P_{AS}} = 4\,\text{Re}\{Z_s\}\,\text{Re}\{Z_L\}|H(j\omega)|^2 \tag{3.36}$$

where P_L is the active power dissipated in Z_L, and P_{AS} is the available power of the signal source with the internal impedance Z_s.

The insertion loss of the circuit is

$$IL = \frac{(\text{Re } Z_L)^2}{|Z_s + Z_L|^2}|H(j\omega)|^2 \tag{3.37}$$

The voltage gain is equal to

$$A_v = \frac{V_2}{V_1} = \frac{Z_L}{AZ_L + B} \tag{3.38}$$

The power gain is

$$G = \frac{P_L}{P_1} = \frac{\text{Re}\{Z_{in}\}}{\text{Re}\{Z_L\}}\left|\frac{Z_L}{AZ_L + B}\right|^2 \tag{3.39}$$

In (3.39), P_1 is the power delivered to the input port of the circuit. The input impedance is

$$Z_{in} = \frac{AZ_L + B}{CZ_L + D} \tag{3.40}$$

The output impedance is

$$Z_{out} = \frac{DZ_s + B}{CZ_s + A} \qquad (3.41)$$

The input port reflection coefficient (with respect to the Z_s impedance) is

$$\Gamma_{in} = \frac{Z_{in} - Z_s^*}{Z_{in} + Z_s} \qquad (3.42)$$

The output port reflection coefficient (with respect to the Z_L impedance) is

$$\Gamma_{out} = \frac{Z_{out} - Z_L^*}{Z_{out} + Z_L} \qquad (3.43)$$

The basic advantages of the discussed analysis method are its algorithmic simplicity (multiple repetitions of the 2×2 matrix multiplication) and small computational effort, in both computing time and memory space requirements. The basic shortcoming of the method is its limited capability in the analysis of circuits composed of cascaded two-ports.

Any other, noncascaded connection of two-ports requires appropriate matrix representation for connected two-ports. The simplicity of these connection rules has brought about their widespread use in many general and special-purpose programs [11–17]. In the method discussed, the matrix for the external ports of the whole circuit is computed each time by successive connection of the two-ports in pairs and determining the matrix of the resulting two-port according to appropriate connection rules. Table 3.1 presents some of the most frequently used rules for connections of two-ports.

Particular matrix representations of two-ports are defined by the four equations to follow.

Admittance matrix **Y**:

$$\begin{bmatrix} I_1 \\ I_2 \end{bmatrix} = \begin{bmatrix} y_{11} & y_{12} \\ y_{21} & y_{22} \end{bmatrix} \begin{bmatrix} V_1 \\ V_2 \end{bmatrix} = \mathbf{Y} \begin{bmatrix} V_1 \\ V_2 \end{bmatrix} \qquad (3.44)$$

Impedance matrix **Z**:

$$\begin{bmatrix} V_1 \\ V_2 \end{bmatrix} = \begin{bmatrix} z_{11} & z_{12} \\ z_{21} & z_{22} \end{bmatrix} \begin{bmatrix} I_1 \\ I_2 \end{bmatrix} = \mathbf{Z} \begin{bmatrix} V_1 \\ V_2 \end{bmatrix} \qquad (3.45)$$

Mixed matrix **H**:

$$\begin{bmatrix} V_1 \\ I_2 \end{bmatrix} = \begin{bmatrix} h_{11} & h_{12} \\ h_{21} & h_{22} \end{bmatrix} \begin{bmatrix} I_1 \\ V_2 \end{bmatrix} = \mathbf{H} \begin{bmatrix} I_1 \\ V_2 \end{bmatrix} \qquad (3.46)$$

Table 3.1
Connection Rules of Two-Ports

Connection Type		Matrix Operation
Cascaded		$\mathbf{A} = \mathbf{A}^{(1)} \cdot \mathbf{A}^{(2)}$
Parallel-parallel		$\mathbf{Y} = \mathbf{Y}^{(1)} + \mathbf{Y}^{(2)}$
Series-series		$\mathbf{Z} = \mathbf{Z}^{(1)} + \mathbf{Z}^{(2)}$
Series-parallel		$\mathbf{H} = \mathbf{H}^{(1)} + \mathbf{H}^{(2)}$
Parallel-series		$\mathbf{G} = \mathbf{G}^{(1)} + \mathbf{G}^{(2)}$

Mixed matrix **G**:

$$\begin{bmatrix} I_1 \\ V_2 \end{bmatrix} = \begin{bmatrix} g_{11} & g_{12} \\ g_{21} & g_{22} \end{bmatrix} \begin{bmatrix} V_1 \\ I_2 \end{bmatrix} = \mathbf{G} \begin{bmatrix} V_1 \\ I_2 \end{bmatrix} \tag{3.47}$$

The matrix representation required by the given type of connection may not exist for some elements in a circuit. Moreover, numerous matrix transformations may cause a loss of accuracy in element values of the resulting final circuit.

Example 3.2

Consider the procedure for computation of the resulting final chain matrix of the circuit shown in Figure 3.12. We shall proceed in the following way:

(1) Computation of the $\mathbf{A}_{E1}$ and $\mathbf{A}_{E2}$ chain matrices of elements $E1$ and $E2$.

(2) Computation of the matrix $\mathbf{A}_{E12}$ of the two-port $E12$ resulting from cascaded connection of elements $E1$ and $E2$: $\mathbf{A}_{E12} = \mathbf{A}_{E1} \cdot \mathbf{A}_{E2}$.

(3) Transformation of chain matrix $\mathbf{A}_{E12}$ into admittance matrix $\mathbf{Y}_{E12}$.

(4) Computation of admittance matrix $\mathbf{Y}_{E3}$ of element $E3$.

(5) Computation of matrix $\mathbf{Y}_{E123}$ of the two-port $E123$ resulting from the parallel-parallel connection of elements $E12$ and $E3$: $\mathbf{Y}_{E123} = \mathbf{Y}_{E12} + \mathbf{Y}_{E3}$.

(6) Transformation of admittance matrix $\mathbf{Y}_{E123}$ into the chain matrix $\mathbf{A}_{123}$.

(7) Computation of mixed matrix $\mathbf{H}_{E4}$ of the element $E4$.

(8) Computation of mixed matrix $\mathbf{H}_{E5}$ of the element $E5$.

(9) Computation of matrix $\mathbf{H}_{E45}$ of the two-port $E45$ resulting from the series-parallel connection of two-ports $E4$ and $E5$: $\mathbf{H}_{E45} = \mathbf{H}_{E4} + \mathbf{H}_{E5}$.

(10) Transformation of mixed matrix $\mathbf{H}_{E45}$ into chain matrix $\mathbf{A}_{E45}$.

(11) Computation of matrix $\mathbf{A}_{E12345}$ of the two-port $E12345$ resulting from the cascaded connection of elements $E123$ and $E45$: $\mathbf{A}_{E12345} = \mathbf{A}_{E123} \cdot \mathbf{A}_{E45}$.

Despite its obvious restrictions and limitations, the simplicity of element connection rules and the uncomplicated logic needed in the method's implementation as a computer program have led to its widespread application in CAD of microwave circuits. Table 3.2 presents transformation relations for different matrix representations of two-ports [18].

The transformations from $\mathbf{S}$ to $\mathbf{A}$, to $\mathbf{Y}$ and to $\mathbf{Z}$ matrices, and *vice versa* are discussed in Chapter 2.

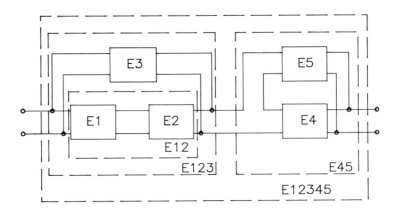

Figure 3.12 A circuit showing different types of connections for component two-ports.

Table 3.2
Transformation Relations for Chain, Admittance,
Impedance, and Mixed Matrix Representations of Two-Ports

	A(ABCD)	Y	Z	H	G
A ABCD	$\begin{matrix} A & B \\ C & D \end{matrix}$	$\begin{matrix} \dfrac{-y_{22}}{y_{21}} & \dfrac{-1}{y_{21}} \\[4pt] \dfrac{-\Delta_y}{y_{21}} & \dfrac{-y_{11}}{y_{21}} \end{matrix}$	$\begin{matrix} \dfrac{z_{11}}{z_{21}} & \dfrac{\Delta_z}{z_{21}} \\[4pt] \dfrac{1}{z_{21}} & \dfrac{z_{22}}{z_{21}} \end{matrix}$	$\begin{matrix} \dfrac{-\Delta_h}{h_{21}} & \dfrac{-h_{11}}{h_{21}} \\[4pt] \dfrac{-h_{22}}{h_{21}} & \dfrac{-1}{h_{21}} \end{matrix}$	$\begin{matrix} \dfrac{1}{g_{21}} & \dfrac{g_{22}}{g_{21}} \\[4pt] \dfrac{g_{11}}{g_{21}} & \dfrac{\Delta_g}{g_{21}} \end{matrix}$
Y	$\begin{matrix} \dfrac{D}{B} & \dfrac{-\Delta}{B} \\[4pt] \dfrac{-1}{B} & \dfrac{A}{B} \end{matrix}$	$\begin{matrix} y_{11} & y_{12} \\ y_{21} & y_{22} \end{matrix}$	$\begin{matrix} \dfrac{z_{22}}{\Delta_z} & \dfrac{-z_{12}}{\Delta_z} \\[4pt] \dfrac{-z_{21}}{\Delta_z} & \dfrac{z_{11}}{\Delta_z} \end{matrix}$	$\begin{matrix} \dfrac{1}{h_{11}} & \dfrac{-h_{12}}{h_{11}} \\[4pt] \dfrac{h_{21}}{h_{11}} & \dfrac{\Delta_h}{h_{11}} \end{matrix}$	$\begin{matrix} \dfrac{\Delta_g}{g_{22}} & \dfrac{g_{12}}{g_{22}} \\[4pt] \dfrac{-g_{12}}{g_{22}} & \dfrac{1}{g_{22}} \end{matrix}$
Z	$\begin{matrix} \dfrac{A}{C} & \dfrac{\Delta}{C} \\[4pt] \dfrac{1}{C} & \dfrac{D}{C} \end{matrix}$	$\begin{matrix} \dfrac{y_{22}}{\Delta_y} & \dfrac{-y_{12}}{\Delta_y} \\[4pt] \dfrac{-y_{21}}{\Delta_y} & \dfrac{y_{11}}{\Delta_y} \end{matrix}$	$\begin{matrix} z_{11} & z_{12} \\ z_{21} & z_{22} \end{matrix}$	$\begin{matrix} \dfrac{\Delta_h}{h_{22}} & \dfrac{h_{12}}{h_{22}} \\[4pt] \dfrac{-h_{21}}{h_{22}} & \dfrac{1}{h_{22}} \end{matrix}$	$\begin{matrix} \dfrac{1}{g_{11}} & \dfrac{g_{12}}{g_{11}} \\[4pt] \dfrac{g_{21}}{g_{11}} & \dfrac{\Delta_g}{g_{11}} \end{matrix}$
H	$\begin{matrix} \dfrac{B}{D} & \dfrac{\Delta}{D} \\[4pt] \dfrac{-1}{D} & \dfrac{A}{D} \end{matrix}$	$\begin{matrix} \dfrac{1}{y_{11}} & \dfrac{-y_{12}}{y_{11}} \\[4pt] \dfrac{y_{21}}{y_{11}} & \dfrac{\Delta_y}{y_{11}} \end{matrix}$	$\begin{matrix} \dfrac{\Delta_z}{z_{22}} & \dfrac{z_{12}}{z_{22}} \\[4pt] \dfrac{-z_{21}}{z_{22}} & \dfrac{1}{z_{22}} \end{matrix}$	$\begin{matrix} h_{11} & h_{12} \\ h_{21} & h_{22} \end{matrix}$	$\begin{matrix} \dfrac{g_{22}}{\Delta_g} & \dfrac{-g_{12}}{\Delta_g} \\[4pt] \dfrac{-g_{21}}{\Delta_g} & \dfrac{g_{11}}{\Delta_g} \end{matrix}$
G	$\begin{matrix} \dfrac{C}{A} & \dfrac{-\Delta}{A} \\[4pt] \dfrac{1}{A} & \dfrac{B}{A} \end{matrix}$	$\begin{matrix} \dfrac{\Delta_y}{y_{22}} & \dfrac{y_{12}}{y_{22}} \\[4pt] \dfrac{-y_{21}}{y_{22}} & \dfrac{1}{y_{22}} \end{matrix}$	$\begin{matrix} \dfrac{1}{z_{11}} & \dfrac{-z_{12}}{z_{11}} \\[4pt] \dfrac{z_{21}}{z_{11}} & \dfrac{\Delta_z}{z_{11}} \end{matrix}$	$\begin{matrix} \dfrac{h_{22}}{\Delta_h} & \dfrac{-h_{12}}{\Delta_h} \\[4pt] \dfrac{-h_{21}}{\Delta_h} & \dfrac{h_{11}}{\Delta_h} \end{matrix}$	$\begin{matrix} g_{11} & g_{12} \\ g_{21} & g_{22} \end{matrix}$

Where $\Delta = AD - BC$, $\Delta_z = z_{11}z_{22} - z_{12}z_{21}$, $\Delta_y = y_{11}y_{22} - y_{12}y_{21}$, $\Delta_h = h_{11}h_{22} - h_{12}h_{21}$, $\Delta_g = g_{11}g_{22} - g_{12}g_{21}$.

3.2 MICROWAVE CIRCUIT ANALYSIS IN TERMS OF WAVE VARIABLES

3.2.1 Connection Scattering Matrix Method

The properties of microwave circuits may also be analyzed in terms of wave variables at the ports of component multiports. A microwave circuit is an interconnection of m elements (multiports) as shown in Figure 3.13. For the kth circuit element we have a set of linear equations:

$$\mathbf{b}^{(k)} = \mathbf{S}^{(k)}\mathbf{a}^{(k)} \tag{3.48}$$

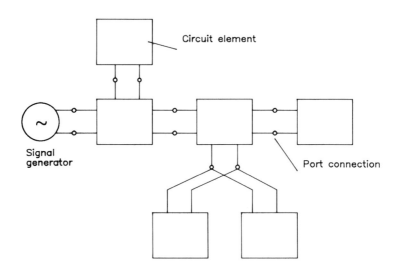

Figure 3.13 Microwave circuit containing interconnected multiports and independent signal generators.

where $\mathbf{S}^{(k)}$ is its scattering matrix, and $\mathbf{a}^{(k)}$ and $\mathbf{b}^{(k)}$ are vectors of incoming and outgoing wave variables, respectively, at their ports.

An independent signal generator is determined by the equation

$$b_g = S_g a_g + c_g \tag{3.49}$$

illustrated in Figure 3.14.

If we substitute into the equation:

$$V_g = E_g + I_g Z_g \tag{3.50}$$

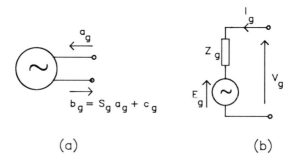

(a)   (b)

Figure 3.14 Independent signal generator: (a) incoming and outgoing waves at the generator port; (b) equivalent signal voltage source.

the relations:

$$V_g = \frac{1}{\sqrt{\mathrm{Re}\, Z_N}} (Z_N^* a_g + Z_N b_g) \tag{3.51}$$

$$I_g = \frac{1}{\sqrt{\mathrm{Re}\, Z_N}} (a_g - b_g) \tag{3.52}$$

which relate the voltage V_g across and the current I_g flowing into the independent generator port with the incoming wave a_g and outgoing wave b_g at that port, we have

$$b_g = \frac{Z_g - Z_N^*}{Z_g + Z_N} a_g + \frac{\sqrt{\mathrm{Re}\, Z_N}\, E_g}{Z_g + Z_N} \tag{3.53}$$

where Z_N is the reference impedance of the generator port.

If we compare (3.49) and (3.53), we see that

$$S_g = \frac{Z_g - Z}{Z_g + Z_N} \tag{3.54}$$

is the reflection coefficient of the independent generator port and

$$c_g = \frac{\sqrt{\mathrm{Re}\, Z_N}}{Z_g + Z_N} \tag{3.55}$$

is the independent complex wave impressed by the generator.

The magnitude of c_g set to the second power equals the power transferred from the generator to the load whose impedance equals the reference impedance Z_N of the generator port.

Considering a whole microwave circuit with m elements, we have a set of linear equations in the matrix form [19, 20]

$$\mathbf{Sa + c = b} \tag{3.56}$$

where

$$
\mathbf{a} = \begin{bmatrix} \mathbf{a}^{(1)} \\ \mathbf{a}^{(2)} \\ \vdots \\ \mathbf{a}^{(k)} \\ \vdots \\ \mathbf{a}^{(m)} \end{bmatrix}, \quad
\mathbf{b} = \begin{bmatrix} \mathbf{b}^{(1)} \\ \mathbf{b}^{(2)} \\ \vdots \\ \mathbf{b}^{(k)} \\ \vdots \\ \mathbf{b}^{(m)} \end{bmatrix}, \quad
\mathbf{c} = \begin{bmatrix} \mathbf{c}^{(1)} \\ \mathbf{c}^{(2)} \\ \vdots \\ \mathbf{c}^{(k)} \\ \vdots \\ \mathbf{c}^{(m)} \end{bmatrix}
\tag{3.57}
$$

$$
\mathbf{S} = \begin{bmatrix}
\mathbf{S}^{(1)} & \mathbf{0} & \cdots & \mathbf{0} & \cdots & \mathbf{0} \\
\mathbf{0} & \mathbf{S}^{(2)} & & & & \vdots \\
\vdots & & \ddots & & & \\
\mathbf{0} & & \cdots & \mathbf{S}^{(k)} & \cdots & \mathbf{0} \\
\vdots & & & & & \vdots \\
\mathbf{0} & & \cdots & \mathbf{0} & \cdots & \mathbf{S}^{(m)}
\end{bmatrix}
\tag{3.58}
$$

In (3.58) $\mathbf{S}^{(1)}, \mathbf{S}^{(2)}, \ldots, \mathbf{S}^{(m)}$ are the scattering matrices of multiports or the reflection coefficients of independent generators and loads. They describe all m elements of a microwave circuit. The vectors $\mathbf{a}^{(1)}, \mathbf{a}^{(2)}, \ldots, \mathbf{a}^{(m)}$, and $\mathbf{b}^{(1)}, \mathbf{b}^{(2)}, \ldots, \mathbf{b}^{(m)}$, and $\mathbf{c}^{(1)}, \mathbf{c}^{(2)}, \ldots, \mathbf{c}^{(m)}$ represent wave variables related to these elements.

Taking into account the connections among the m individual elements of the circuit, we have to introduce the constraints on vectors $\mathbf{a}$ and $\mathbf{b}$ that can be represented by a matrix equation:

$$
\mathbf{b} = \boldsymbol{\Gamma}\mathbf{a}
\tag{3.59}
$$

where $\boldsymbol{\Gamma}$ is called the *connection matrix*.

Substituting (3.59) into (3.56), we obtain

$$
\mathbf{W}\mathbf{a} = \mathbf{c}
\tag{3.60}
$$

where

$$
\mathbf{W} = \boldsymbol{\Gamma} - \mathbf{S}
\tag{3.61}
$$

The coefficient matrix $\mathbf{W}$ in the preceding equation is called the *connection scattering matrix* of the circuit. The right-hand vector $\mathbf{c}$ is the vector of the impressed waves of the independent signal generators in the circuit. The solution of the matrix equation (3.60) is the vector of incoming waves at all ports of the analyzed circuit:

$$\mathbf{a} = \mathbf{W}^{-1}\mathbf{c} \tag{3.62}$$

Substituting (3.62) into (3.59) we can compute a vector of outgoing waves in all ports of the circuit:

$$\mathbf{b} = \boldsymbol{\Gamma}\,\mathbf{W}^{-1}\mathbf{c} \tag{3.63}$$

The set of linear equations (3.60) with connection scattering matrix $\mathbf{W}$ as the coefficient matrix makes possible computation of the incoming and outgoing waves at all circuit ports when the excitations $\mathbf{c}$ of the circuit are known. The connection scattering matrix represents the circuit in full because values of the scattering parameters of all circuit elements are given in the matrix $\mathbf{S}$ and information on the circuit topology in the matrix $\boldsymbol{\Gamma}$.

To find the principles of the formation of the connection matrix $\boldsymbol{\Gamma}$ we consider the simple case of a connection of two ports, shown in Figure 3.15.

The connection between the ith and jth ports of the circuit imposes the following relations in the plane of the port connection

$$V_i = V_j \quad \text{and} \quad -I_i = I_j \tag{3.64}$$

Substituting into (3.64) the relations for currents (2.19) and voltages (2.20) as functions of incoming and outgoing waves on both sides of the port connection, we

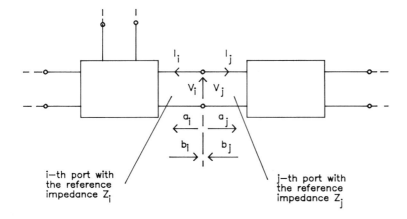

Figure 3.15 Connection of two ports in a microwave circuit.

obtain

$$
\begin{bmatrix} b_i \\ b_j \end{bmatrix} = \frac{1}{Z_i + Z_j} \begin{bmatrix} Z_j - Z_i^* & 2\sqrt{\mathrm{Re}\,Z_i\,\mathrm{Re}\,Z_j} \\ 2\sqrt{\mathrm{Re}\,Z_i\,\mathrm{Re}\,Z_j} & Z_i - Z_j^* \end{bmatrix} \begin{bmatrix} a_i \\ a_j \end{bmatrix} \tag{3.65}
$$

The coefficient matrix in this equation is the connection matrix $\boldsymbol{\Gamma}$ for one pair of the connected ports. In the same way we express the connection matrices of each of the connected pairs of ports from which, after appropriate ordering, we can formulate the connection matrix $\boldsymbol{\Gamma}$ of the whole circuit.

From (3.65), we observe that if

$$
Z_i = Z_j^* \tag{3.66}
$$

the connection of two ports is a nonreflecting connection, which means that the outgoing and incoming waves at ports i and j connected together satisfy the following relations:

$$
a_i = b_j \quad \text{and} \quad a_j = b_i \tag{3.67}
$$

It follows from these considerations that when the reference impedances of the connected ports satisfy the condition of (3.66)—and of course when the reference impedances of the connected ports are real and equal to each other—the connection matrix of such a pair of ports simplifies to the form

$$
\boldsymbol{\Gamma} = \begin{bmatrix} 0 & 1 \\ 1 & 0 \end{bmatrix} \tag{3.68}
$$

Also, the elements of the connection matrix of the whole circuit are equal to ones and zeros. As we see, connection matrix $\boldsymbol{\Gamma}$ is a sparse matrix in which most of the elements are equal to zero. Only the elements γ_{ij} and γ_{ji} corresponding to pairs of the connected ports equal ones.

The algorithm of the formation of connection matrix $\boldsymbol{\Gamma}$ of a microwave circuit is not complicated when the connections between element ports are restricted to the connections of pairs of ports. When three or more ports are connected in one plane the form of connection matrix $\boldsymbol{\Gamma}$ and the algorithm of its formation are much more complicated. Therefore, it is more convenient to limit the number of the connected ports in one plane to two. Whenever three or more ports in the analyzed circuit are connected in one plane, it is appropriate to treat such connection as a separate circuit element with a given scattering matrix **S**. For example, the parallel connection of three ports in one plane (presented in Figure 3.16) treated as a circuit element has the scattering matrix (when the reference impedances of all ports are real and equal to each

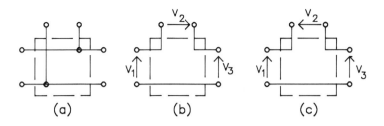

Figure 3.16 Connections of three ports in one plane: (a) parallel connection; (b) series connection.

other) given by

$$\mathbf{S} = \frac{1}{3} \begin{bmatrix} -1 & 2 & 2 \\ 2 & -1 & 2 \\ 2 & 2 & -1 \end{bmatrix} \tag{3.69}$$

We should treat the series connection of three ports in one plane in similar way. It, too, is a circuit element and its scattering matrix (see Figure 3.16(a) and (b), respectively) is equal to

$$\mathbf{S} = \frac{1}{3} \begin{bmatrix} 1 & -2 & 2 \\ -2 & 1 & 2 \\ 2 & 2 & 1 \end{bmatrix}$$

$$\mathbf{S} = \frac{1}{3} \begin{bmatrix} 1 & 2 & 2 \\ 2 & 1 & -2 \\ 2 & -2 & 1 \end{bmatrix} \tag{3.70}$$

Computer-aided analysis methods based on the connection scattering matrix require multiple computation of the solution of a system of linear equations in which the order of the coefficient matrix equals the number of ports of all circuit elements. Conventional numerical procedures for the solution of sets of linear equations require us to perform $n^3/3 + n^2 - n/3$ multiplications and divisions (n is the order of the coefficient matrix). The execution time of multiple repeated numerical solutions of network equations (3.60) for different circuit parameters, various frequencies, or for different excitations of the circuit (different $\mathbf{c}$ vectors) can be reduced considerably by using the sparse matrix technique, which also reduces the memory space required for storing the elements of the coefficient matrix of a system of linear equations to be solved.

Example 3.3

Consider the circuit presented in Figure 3.17, a microwave reflection type of amplifier with a negative resistance device. The scattering matrix $\mathbf{S}$ and connection matrix $\boldsymbol{\Gamma}$ of the circuit are as follows:

$$\mathbf{S} = \begin{array}{c} \\ 1 \\ 2 \\ 3 \\ 4 \\ 5 \\ 6 \\ 7 \\ 8 \end{array} \begin{bmatrix} S_{11} & S_{12} & 0 & 0 & 0 & 0 & 0 & 0 \\ S_{21} & S_{22} & 0 & 0 & 0 & 0 & 0 & 0 \\ 0 & 0 & S_{33} & 0 & 0 & 0 & 0 & 0 \\ 0 & 0 & 0 & S_{44} & S_{45} & S_{46} & 0 & 0 \\ 0 & 0 & 0 & S_{54} & S_{55} & S_{56} & 0 & 0 \\ 0 & 0 & 0 & S_{64} & S_{65} & S_{66} & 0 & 0 \\ 0 & 0 & 0 & 0 & 0 & 0 & S_{77} & 0 \\ 0 & 0 & 0 & 0 & 0 & 0 & 0 & S_{88} \end{bmatrix}$$

with columns labeled $1\ 2\ 3\ 4\ 5\ 6\ 7\ 8$.

$$\mathbf{\Gamma} = \begin{array}{c} \\ 1 \\ 2 \\ 3 \\ 4 \\ 5 \\ 6 \\ 7 \\ 8 \end{array} \begin{bmatrix} 0 & 0 & 1 & 0 & 0 & 0 & 0 & 0 \\ 0 & 0 & 0 & 1 & 0 & 0 & 0 & 0 \\ 1 & 0 & 0 & 0 & 0 & 0 & 0 & 0 \\ 0 & 1 & 0 & 0 & 0 & 0 & 0 & 0 \\ 0 & 0 & 0 & 0 & 0 & 0 & 1 & 0 \\ 0 & 0 & 0 & 0 & 0 & 0 & 0 & 1 \\ 0 & 0 & 0 & 0 & 1 & 0 & 0 & 0 \\ 0 & 0 & 0 & 0 & 0 & 1 & 0 & 0 \end{bmatrix}$$

with columns labeled $1\ 2\ 3\ 4\ 5\ 6\ 7\ 8$.

The connection scattering matrix $\mathbf{W}$ and the vector of excitations $\mathbf{c}$ have the following forms:

$$\mathbf{W} = \begin{array}{c} \\ 1 \\ 2 \\ 3 \\ 4 \\ 5 \\ 6 \\ 7 \\ 8 \end{array} \begin{bmatrix} -S_{11} & -S_{12} & 1 & 0 & 0 & 0 & 0 & 0 \\ -S_{21} & -S_{22} & 0 & 1 & 0 & 0 & 0 & 0 \\ 1 & 0 & -S_{33} & 0 & 0 & 0 & 0 & 0 \\ 0 & 1 & 0 & -S_{44} & -S_{45} & -S_{46} & 0 & 0 \\ 0 & 0 & 0 & -S_{54} & -S_{55} & -S_{56} & 1 & 0 \\ 0 & 0 & 0 & -S_{64} & -S_{65} & -S_{66} & 0 & 1 \\ 0 & 0 & 0 & 0 & 1 & 0 & -S_{77} & 0 \\ 0 & 0 & 0 & 0 & 0 & 1 & 0 & -S_{88} \end{bmatrix} \qquad \mathbf{c} = \begin{array}{c} \\ 1 \\ 2 \\ 3 \\ 4 \\ 5 \\ 6 \\ 7 \\ 8 \end{array} \begin{bmatrix} 0 \\ 0 \\ 0 \\ 0 \\ 0 \\ 0 \\ 0 \\ 1 \end{bmatrix}$$

with columns labeled $1\ 2\ 3\ 4\ 5\ 6\ 7\ 8$.

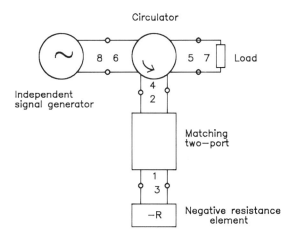

Figure 3.17 Microwave reflection type amplifier with a negative resistance element.

Excitation vector **c** has only one nonzero element, c_8. It corresponds to the generator port numbered 8.

3.2.2 Multiport Connection Method

The scattering matrix of a microwave circuit composed of elements (multiports) connected to each other by pairs of ports is computed by partitioning the circuit ports into two categories: external ports and internal ports. Ports connected to independent signal generators or loads are external ports; the other ports in the circuit are internal ports. Figure 3.18 illustrates the concept of the external and internal ports of a circuit. The independent signal generators and loads can be treated as outside the circuit. Each element in the circuit is determined by its scattering matrix **S**. For the entire circuit we can write the equation:

$$\mathbf{b} = \mathbf{S}\mathbf{a} \qquad (3.71)$$

which is equivalent to (3.56) with $\mathbf{c} = 0$.

The rows and columns of (3.71) can be reordered so that the incoming and outgoing waves in circuit ports are separated into two sets: the first corresponding to the external ports, and the second to the internal ports of the circuit. According to this the system of equations, (3.71) can be written as

$$\begin{bmatrix} \mathbf{b}_e \\ \mathbf{b}_i \end{bmatrix} = \begin{bmatrix} \mathbf{S}_{ee} & \mathbf{S}_{ei} \\ \mathbf{S}_{ie} & \mathbf{S}_{ii} \end{bmatrix} \begin{bmatrix} \mathbf{a}_e \\ \mathbf{a}_i \end{bmatrix} \qquad (3.72)$$

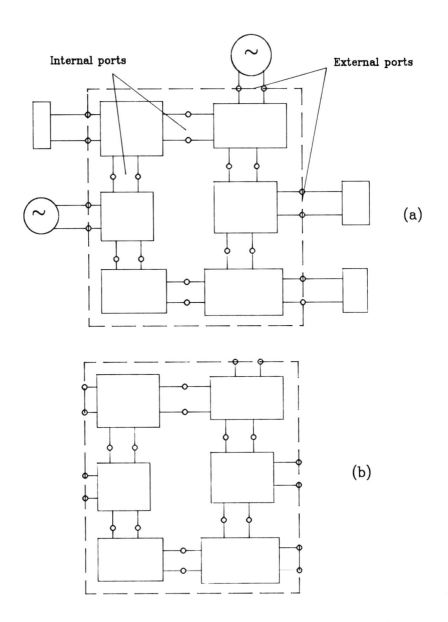

Figure 3.18 A network of multiports connected by internal ports: (a) with independent signal generators and loads connected to the external ports; (b) without signal generators and loads.

where $\mathbf{a}_e, \mathbf{b}_e$ and $\mathbf{a}_i, \mathbf{b}_i$ are vectors of incoming and outgoing waves, respectively, in the external and internal ports of the circuit.

The connections of pairs of internal ports impose restrictions on the vectors of incoming and outgoing waves in the form:

$$\mathbf{b}_i = \boldsymbol{\Gamma}_i \mathbf{a}_i \qquad (3.73)$$

where $\boldsymbol{\Gamma}_i$ is the connection matrix defined in Section 3.2.1 related to connections of internal ports of the circuit.

From (3.72) and (3.73), by first eliminating $\mathbf{b}_i$, we obtain

$$\mathbf{a}_i = (\boldsymbol{\Gamma}_i - \mathbf{S}_{ii})^{-1} \mathbf{S}_{ie} \mathbf{a}_e \qquad (3.74)$$

and, after next eliminating $\mathbf{a}_i$,

$$\mathbf{b}_e = \left[\mathbf{S}_{ee} + \mathbf{S}_{ei}(\boldsymbol{\Gamma}_i - \mathbf{S}_{ii})^{-1} \mathbf{S}_{ie} \right] \mathbf{a}_e \qquad (3.75)$$

The coefficient matrix in this equation:

$$\mathbf{S}_e = \mathbf{S}_{ee} + \mathbf{S}_{ei}(\boldsymbol{\Gamma}_i - \mathbf{S}_{ii})^{-1} \mathbf{S}_{ie} \qquad (3.76)$$

is the scattering matrix of the whole circuit referred to the external ports.

The computation of the incoming wave variables of the external ports of the circuit requires the solution of the matrix equation:

$$(\boldsymbol{\Gamma}_e - \mathbf{S}_e) \mathbf{a}_e = \mathbf{c} \qquad (3.77)$$

where

$\boldsymbol{\Gamma}_e$ = the connection matrix related to connections of external ports,
$\mathbf{c}$ = the vector of excitations.

The vector of outgoing wave variables at the external ports of a circuit can be computed by multiplying the right-hand side of the $\boldsymbol{\Gamma}_e$ matrix by the vector of incoming wave variables $\mathbf{a}_e$:

$$\mathbf{b}_e = \boldsymbol{\Gamma}_e \mathbf{a}_e \qquad (3.78)$$

Matrix $(\boldsymbol{\Gamma}_i - \mathbf{S}_{ii})$ in (3.76) and matrix $(\boldsymbol{\Gamma}_e - \mathbf{S}_{ee})$ in (3.77) have the same characteristics as matrix $\mathbf{W} = \boldsymbol{\Gamma} - \mathbf{S}$ in the equation (3.60). All of them are very sparse and their sparse structures depend only on the topology of the circuit.

After finding the solution $\mathbf{a}_e$ of (3.77), we can use equations (3.74) and (3.73) to obtain the wave variables at the internal ports of the circuit.

As we will show in Chapter 4 this information together with the adjoint network approach lead to a very efficient computation of the network scattering matrix sensitivities.

Example 3.4

In this example we consider the example circuit shown in Figure 3.19. The circuit ports are numbered as shown in the figure.

The ordered scattering matrix equation of the form (3.72) is

$$
\mathbf{b}_e \left\{ \begin{bmatrix} b_6 \\ b_5 \end{bmatrix} \right. \\
\mathbf{b}_i \left\{ \begin{bmatrix} b_1 \\ b_2 \\ b_3 \\ b_4 \end{bmatrix} \right.
=
\begin{bmatrix}
S_{66} & S_{65} & 0 & 0 & 0 & S_{64} \\
S_{56} & S_{55} & 0 & 0 & 0 & S_{54} \\
0 & 0 & S_{11} & S_{12} & 0 & 0 \\
0 & 0 & S_{21} & S_{22} & 0 & 0 \\
0 & 0 & 0 & 0 & S_{33} & 0 \\
S_{46} & S_{45} & 0 & 0 & 0 & S_{44}
\end{bmatrix}
\begin{bmatrix} a_6 \\ a_5 \\ a_1 \\ a_2 \\ a_3 \\ a_4 \end{bmatrix}
\left. \begin{matrix} \\ \end{matrix} \right\} \mathbf{a}_e \\
\left. \begin{matrix} \\ \\ \\ \end{matrix} \right\} \mathbf{a}_i
$$

The connection matrix $\boldsymbol{\Gamma}_i$ for the internal ports can be written as

$$
\mathbf{b}_i =
\begin{bmatrix} b_1 \\ b_2 \\ b_3 \\ b_4 \end{bmatrix}
=
\begin{bmatrix}
0 & 0 & 1 & 0 \\
0 & 0 & 0 & 1 \\
1 & 0 & 0 & 0 \\
0 & 1 & 0 & 0
\end{bmatrix}
\begin{bmatrix} a_1 \\ a_2 \\ a_3 \\ a_4 \end{bmatrix}
= \boldsymbol{\Gamma}_i \mathbf{a}_i
$$

The scattering matrix of the whole circuit related to the external ports 6 and 5 is given by

$$
\mathbf{S}_e = \mathbf{S}_{ee} + \mathbf{S}_{ei}(\boldsymbol{\Gamma}_i - \mathbf{S}_{ii})^{-1}\mathbf{S}_{ie}
$$

$$
\mathbf{S}_e =
\begin{bmatrix} S_{66} & S_{65} \\ S_{56} & S_{55} \end{bmatrix}
+
\begin{bmatrix} 0 & 0 & 0 & S_{64} \\ 0 & 0 & 0 & S_{54} \end{bmatrix}
$$

$$
\cdot
\begin{bmatrix}
-S_{11} & -S_{12} & 1 & 0 \\
-S_{21} & -S_{22} & 0 & 1 \\
1 & 0 & -S_{33} & 0 \\
0 & 1 & 0 & -S_{44}
\end{bmatrix}^{-1}
\cdot
\begin{bmatrix}
0 & 0 \\
0 & 0 \\
0 & 0 \\
S_{46} & S_{45}
\end{bmatrix}
$$

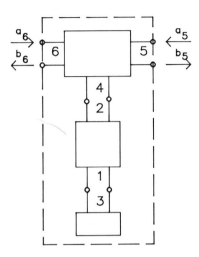

Figure 3.19 The multiport circuit for Example 3.4.

The incoming wave variables at the internal ports are found by using (3.74), which has the form

$$\mathbf{a}_i = (\boldsymbol{\Gamma}_i - \mathbf{S}_{ii})^{-1} \mathbf{S}_{ie} \mathbf{a}_e$$

$$
\begin{bmatrix} a_1 \\ a_2 \\ a_3 \\ a_4 \end{bmatrix} =
\begin{bmatrix}
-S_{11} & -S_{12} & 1 & 0 \\
-S_{21} & -S_{22} & 0 & 1 \\
1 & 0 & -S_{33} & 0 \\
0 & 1 & 0 & -S_{44}
\end{bmatrix}
\begin{bmatrix}
0 & 0 \\
0 & 0 \\
0 & 0 \\
S_{46} & S_{45}
\end{bmatrix}
\begin{bmatrix} a_6 \\ a_5 \end{bmatrix}
$$

To compute the $\mathbf{S}_e$ matrix of a circuit, we must invert the matrix $(\boldsymbol{\Gamma} - \mathbf{S}_{ii})$ of the order equal to the number of interconnected internal ports. In large networks the order of the matrix to be inverted is large, and the computational effort is large, too. Such effort is reduced greatly by connecting two multiports at a time and computing each $\mathbf{S}_e$ matrix of the resulting subcircuit. For a circuit composed of m component multiports, $(m - 1)$ applications of (3.76) computes the overall $\mathbf{S}_e$ matrix of the complete circuit.

The order according to which the circuit components are connected to each other determines the total number of long arithmetic operations (multiplications and divisions) required to compute the $\mathbf{S}_e$ matrix for a circuit with a given topology.

The algorithm for near optimal ordering of the connection sequence proposed by V. Monaco and P. Tiberio [19] calls for, at each step, connecting the two components whose resulting multiport has the fewest ports. A circuit composed of five multiport elements and the successive alteration in the topology of the circuit after each connection are shown in Figure 3.20. Components D and E are connected first forming DE, a multiport with three external ports. The components C and DE are combined next. Then we connect A and CDE. In the last step the B component is connected to the $ACDE$ multiport finally yielding the $\mathbf{S}_e$ matrix of the overall three-port circuit.

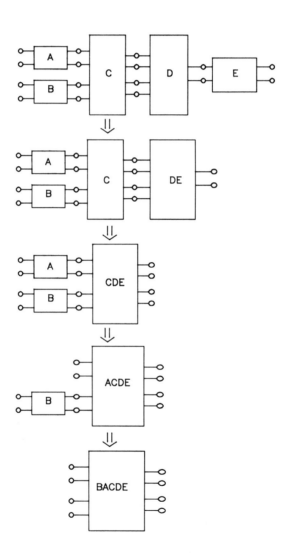

Figure 3.20 Near optimal order of multiport connections.

The efficient computation of (3.76) is very important because it is used repeatedly to obtain the S_e matrix for combining two components connected together at each step of the analysis. The computational efficiency of the method can be improved by proper ordering and numbering of the internal ports of two multiports being interconnected. The internal ports of these multiports have to be divided into two groups. One group, called q, will contain internal ports belonging to the first component; the second group, called r, will contain internal ports belonging to the second connected component. Further, both groups of ports have to be ordered so that a port q_1 is connected with port r_1, q_2 with r_2, and so on. Equation (3.72) for both connected multiports can be written in the following form:

$$
\begin{bmatrix} \mathbf{b}_e \\ \hline \mathbf{b}_q \\ \mathbf{b}_r \end{bmatrix} = \begin{bmatrix} \mathbf{S}_{ee} & \mathbf{S}_{eq} & \mathbf{S}_{er} \\ \hline \mathbf{S}_{qe} & \mathbf{S}_{qq} & \mathbf{0} \\ \mathbf{S}_{re} & \mathbf{0} & \mathbf{S}_{rr} \end{bmatrix} \begin{bmatrix} \mathbf{a}_e \\ \hline \mathbf{a}_q \\ \mathbf{a}_r \end{bmatrix} \tag{3.79}
$$

In (3.79), $\mathbf{a}_e$ and $\mathbf{b}_e$ are vectors of wave variables at the external ports, and $\mathbf{a}_q, \mathbf{b}_q$ and $\mathbf{a}_r, \mathbf{b}_r$ wave variables at the internal ports, respectively, at the internal ports of the first and of the second element. The submatrices $\mathbf{S}_{qr}$ and $\mathbf{S}_{rq}$ in (3.79) are null matrices because q ports and r ports belong to different components.

For such port ordering and numbering, the connection matrix related to the internal ports is in the form:

$$
\boldsymbol{\Gamma}_i = \begin{bmatrix} \mathbf{0} & \mathbf{I} \\ \mathbf{I} & \mathbf{0} \end{bmatrix} \tag{3.80}
$$

and the $(\boldsymbol{\Gamma}_i - \mathbf{S}_{ii})$ matrix to be inverted in (3.76) is

$$
\boldsymbol{\Gamma}_i - \mathbf{S}_{ii} = \begin{bmatrix} -\mathbf{S}_{qq} & \mathbf{I} \\ \mathbf{I} & -\mathbf{S}_{rr} \end{bmatrix} \tag{3.81}
$$

The inverse matrix $(\boldsymbol{\Gamma}_i - \mathbf{S}_{ii})^{-1}$ is now given by

$$
(\boldsymbol{\Gamma}_i - \mathbf{S}_{ii})^{-1} = \begin{bmatrix} \mathbf{S}_{rr}(\mathbf{I} - \mathbf{S}_{qq}\mathbf{S}_{rr})^{-1} & (\mathbf{I} - \mathbf{S}_{rr}\mathbf{S}_{qq})^{-1} \\ \hline (\mathbf{I} - \mathbf{S}_{qq}\mathbf{S}_{rr})^{-1} & \mathbf{S}_{qq}(\mathbf{I} - \mathbf{S}_{rr}\mathbf{S}_{qq})^{-1} \end{bmatrix}
$$

$$
= \begin{bmatrix} \mathbf{M}_{11} & \mathbf{M}_{12} \\ \mathbf{M}_{21} & \mathbf{M}_{22} \end{bmatrix} \tag{3.82}
$$

The computation requires the inverse of two matrices $(\mathbf{I} - \mathbf{S}_{qq}\mathbf{S}_{rr})$ and $(\mathbf{I} - \mathbf{S}_{rr}\mathbf{S}_{qq})$, each two times smaller than the order of the original matrix $(\boldsymbol{\Gamma}_i - \mathbf{S}_{ii})$.

The computational efficiency of (3.76) can be improved further. The external ports can be divided into two groups: p_1 ports belonging to the first connected component and p_2 ports belonging to the second connected component. In such a case the submatrices $\mathbf{S}_{ee}$, $\mathbf{S}_{ei}$, and $\mathbf{S}_{ie}$ have the forms:

$$\mathbf{S}_{ee} = \begin{bmatrix} \mathbf{S}_{1e} & \mathbf{0} \\ \mathbf{0} & \mathbf{S}_{2e} \end{bmatrix} \tag{3.83}$$

$$\mathbf{S}_{ei} = \begin{bmatrix} \mathbf{S}_{eq} & \mathbf{S}_{er} \end{bmatrix} = \begin{bmatrix} \mathbf{S}_{1eq} & \mathbf{0} \\ \mathbf{0} & \mathbf{S}_{2er} \end{bmatrix} \tag{3.84}$$

$$\mathbf{S}_{ie} = \begin{bmatrix} \mathbf{S}_{qe} \\ \mathbf{S}_{re} \end{bmatrix} = \begin{bmatrix} \mathbf{S}_{1qe} & \mathbf{0} \\ \mathbf{0} & \mathbf{S}_{2re} \end{bmatrix} \tag{3.85}$$

In these relations, subscripts 1 and 2 refer to port groups p_1 and p_2, respectively. The resulting scattering matrix is now given by the relation

$$\mathbf{S}_e = \begin{bmatrix} \mathbf{S}_{1e} + \mathbf{S}_{1eq}\mathbf{M}_{11}\mathbf{S}_{1qe} & \mathbf{S}_{1eq}\mathbf{M}_{12}\mathbf{S}_{2re} \\ \mathbf{S}_{2er}\mathbf{M}_{21}\mathbf{S}_{1qe} & \mathbf{S}_{2e} + \mathbf{S}_{2er}\mathbf{M}_{22}\mathbf{S}_{2re} \end{bmatrix} \tag{3.86}$$

Less computation is needed to compute the $\mathbf{S}_e$ matrix by using (3.86) than by directly using (3.76), because two inverted matrices in (3.82) are of the order two times smaller than the matrix to be inverted in (3.76). Also of smaller orders are matrices to be multiplied by themselves.

The wave variables at the internal connected ports can be computed by using (3.74), which now is expressed as

$$\begin{bmatrix} \mathbf{a}_q \\ \mathbf{a}_r \end{bmatrix} = \begin{bmatrix} \mathbf{M}_{11}\mathbf{S}_{1qe} & \mathbf{M}_{12}\mathbf{S}_{2re} \\ \mathbf{M}_{21}\mathbf{S}_{1qe} & \mathbf{M}_{22}\mathbf{S}_{2re} \end{bmatrix} \begin{bmatrix} \mathbf{a}_{1e} \\ \mathbf{a}_{2e} \end{bmatrix} \tag{3.87}$$

In this relation $\mathbf{a}_{1e}$ and $\mathbf{a}_{2e}$ are the vectors of incoming waves at the external ports, respectively, of the first and second multiports.

Example 3.5

Let us consider the circuit presented in Figure 3.21. The multiport connection method for this circuit goes as follows. The unordered scattering matrix equation $\mathbf{Sa} = \mathbf{b}$ has the form

$$
\begin{bmatrix}
S_{11} & S_{12} & S_{13} & 0 & 0 & 0 & 0 \\
S_{21} & S_{22} & S_{23} & 0 & 0 & 0 & 0 \\
S_{31} & S_{32} & S_{33} & 0 & 0 & 0 & 0 \\
0 & 0 & 0 & S_{44} & S_{45} & S_{46} & S_{47} \\
0 & 0 & 0 & S_{54} & S_{55} & S_{56} & S_{57} \\
0 & 0 & 0 & S_{64} & S_{65} & S_{66} & S_{67} \\
0 & 0 & 0 & S_{74} & S_{75} & S_{76} & S_{77}
\end{bmatrix}
\begin{bmatrix}
a_1 \\ a_2 \\ a_3 \\ a_4 \\ a_5 \\ a_6 \\ a_7
\end{bmatrix}
=
\begin{bmatrix}
b_1 \\ b_2 \\ b_3 \\ b_4 \\ b_5 \\ b_6 \\ b_7
\end{bmatrix}
$$

The same system of equations ordered according to the discussed principles is

$$
\begin{bmatrix}
S_{11} & 0 & 0 & S_{12} & S_{13} & 0 & 0 \\
0 & S_{66} & S_{67} & 0 & 0 & S_{64} & S_{65} \\
0 & S_{76} & S_{77} & 0 & 0 & S_{74} & S_{75} \\
S_{21} & 0 & 0 & S_{22} & S_{23} & 0 & 0 \\
S_{31} & 0 & 0 & S_{32} & S_{33} & 0 & 0 \\
0 & S_{46} & S_{47} & 0 & 0 & S_{44} & S_{45} \\
0 & S_{56} & S_{57} & 0 & 0 & S_{54} & S_{55}
\end{bmatrix}
\begin{bmatrix}
a_1 \\ a_6 \\ a_7 \\ a_2 \\ a_3 \\ a_4 \\ a_5
\end{bmatrix}
=
\begin{bmatrix}
b_1 \\ b_6 \\ b_7 \\ b_2 \\ b_3 \\ b_4 \\ b_5
\end{bmatrix}
\begin{matrix}
\left.\vphantom{\begin{matrix}b_1\\b_6\\b_7\end{matrix}}\right\}\mathbf{a}_e, \mathbf{b}_e \\
\left.\vphantom{\begin{matrix}b_2\\b_3\end{matrix}}\right\}\mathbf{a}_{ei}, \mathbf{b}_{ei} \\
\left.\vphantom{\begin{matrix}b_4\\b_5\end{matrix}}\right\}\mathbf{a}_{ie}, \mathbf{b}_{ie}
\end{matrix}
$$

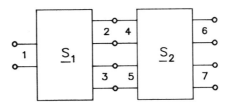

Figure 3.21 Connection of two multiports.

The submatrices S_{ee}, S_{ei}, and S_{ie} are

$$S_{ee} = \begin{bmatrix} S_{11} & 0 & 0 \\ 0 & S_{65} & S_{67} \\ 0 & S_{76} & S_{77} \end{bmatrix}$$

$$S_{ei} = \begin{bmatrix} S_{12} & S_{13} & 0 & 0 \\ 0 & 0 & S_{64} & S_{65} \\ 0 & 0 & S_{74} & S_{75} \end{bmatrix}$$

$$S_{ie} = \begin{bmatrix} S_{21} & 0 & 0 \\ S_{31} & 0 & 0 \\ 0 & S_{46} & S_{47} \\ 0 & S_{74} & S_{75} \end{bmatrix}$$

The matrix $\boldsymbol{\Gamma}_i - \mathbf{S}_{ii}$ which has to be inverted is of the fourth order and given as

$$\boldsymbol{\Gamma}_i - \mathbf{S}_{ii} = \begin{bmatrix} S_{22} & S_{23} & 1 & 0 \\ S_{32} & S_{33} & 0 & 1 \\ 1 & 0 & S_{44} & S_{45} \\ 0 & 1 & S_{54} & S_{55} \end{bmatrix}$$

3.2.3 Transfer Scattering Matrix Method

A circuit composed of cascaded two-ports or cascaded $2n$-ports is analyzed very conveniently by using transfer scattering matrices, since the matrices need be multiplied only to obtain the overall matrix of the whole circuit. Figure 3.22 presents the cascaded connection of two two-ports and two $2n$-ports (multiports with an equal number of input and output ports). In the first case, if port 2 of the first element (1) and port 1 of the second two-port (2) have the same real reference impedances, we have

$$b_2^{(1)} = a_1^{(2)} \quad \text{and} \quad b_1^{(2)} = a_2^{(1)} \tag{3.88}$$

and transfer scattering matrix $\mathbf{T}$ of the cascade equals the product of the $\mathbf{T}$ matrices of the individual components:

$$\mathbf{T} = \mathbf{T}^{(1)} \cdot \mathbf{T}^{(2)} \tag{3.89}$$

We can easily show that for the cascade of m two-ports the overall $\mathbf{T}$ matrix of

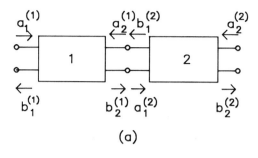

(a)

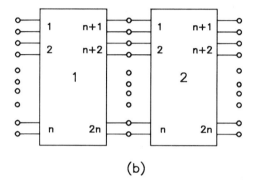

(b)

Figure 3.22 (a) Cascaded connection of two-ports; (b) cascaded connection of two $2n$-ports.

the circuit is

$$\mathbf{T} = \mathbf{T}^{(1)} \cdot \mathbf{T}^{(2)} \cdot \mathbf{T}^{(3)} \cdot \cdots \cdot \mathbf{T}^{(m)} \tag{3.90}$$

The superscripted terms on the right-hand side of (3.89) and (3.90) indicate the $\mathbf{T}$ matrices for the corresponding components.

The preceding relations are applicable to the cascaded $2n$-ports. If the reference impedances of the connected ports of both elements are real and equal to each other ($Z_{Nn+1}^{(1)} = Z_{N1}^{(2)}$, $Z_{Nn+2}^{(1)} = Z_{N2}^{(2)}, \ldots, Z_{N2n}^{(1)} = Z_{Nn}^{(2)}$, in Figure 3.22), the overall $\mathbf{T}$ matrix of the cascaded connection equals the product of the $\mathbf{T}$ matrices of the individual components. This rule can be extended to the long cascade of the $2n$-ports.

When the reference impedances are not real and not equal to each other, the overall $\mathbf{T}$ matrix of the circuit is

$$\mathbf{T} = \mathbf{T}^{(1)} \cdot \mathbf{T}_f^{(1,2)} \cdot \mathbf{T}^{(2)} \cdot \mathbf{T}_f^{(2,3)} \cdot \cdots \cdot \mathbf{T}^{(m-1)} \cdot \mathbf{T}_f^{(m-1,m)} \cdot \mathbf{T}^{(m)} \tag{3.91}$$

where, as in (3.90), $\mathbf{T}^{(1)}, \mathbf{T}^{(2)}, \ldots, \mathbf{T}^{(m)}$ are transfer scattering matrices of the individual components, and $\mathbf{T}_J^{(1, 2)}, \mathbf{T}_J^{(2, 3)}, \ldots, \mathbf{T}_J^{(m-1, m)}$ are transfer scattering matrices of port junctions, respectively, of the first and second element, the second and third element, and so on.

For the junction of two ports, its transfer scattering matrix (see Figure 3.23(a)) is

$$\mathbf{T}_J = \frac{1}{2\sqrt{\operatorname{Re} Z_{N1} \operatorname{Re} Z_{N2}}} \begin{bmatrix} Z_{N2}^* + Z_{N1}^* & Z_{N2} - Z_{N1}^* \\ Z_{N2}^* - Z_{N1} & Z_{N2} + Z_{N1} \end{bmatrix} \quad (3.92)$$

where Z_{N1} and Z_{N2} are the reference impedances of the connected ports. The transfer scattering matrix for the multiport junction can be derived easily from (3.92) and Figure 3.23.

3.2.4 Generalized Transfer Scattering Matrix Method

The overall transfer scattering matrix of the cascaded $2n$ multiports equals the product of the $\mathbf{T}$ matrices of the individual multiports. In multiport circuits described by the generalized transfer scattering matrices $\mathbf{T}$, the cascaded connection of multiport elements corresponds to the product of their appropriately ordered $\mathbf{T}$ matrices [20].

To find the rules of CAD analysis of cascaded circuits described by the generalized transfer scattering matrices $\mathbf{T}$ let us consider a cascade connection of two multiports, shown in Figure 3.24. We introduce the following vectors of wave variables

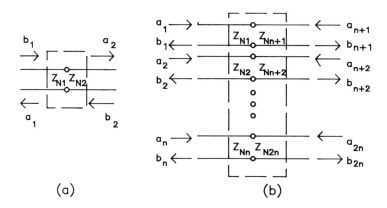

(a)   (b)

Figure 3.23 (a) Port junction; (b) junction of many ports.

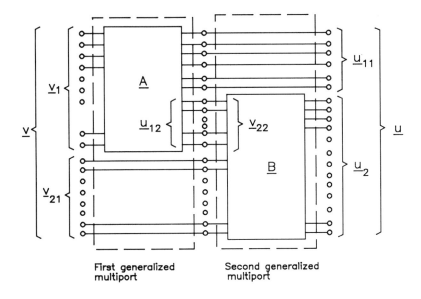

Figure 3.24 General case of a cascaded connection of two multiports: A and B are the generalized transfer scattering matrices of the first and second multiport, respectively.

related to the input and output ports of two connected multiports:

$\mathbf{v}_1 = [b_{IN1}, a_{IN1}, b_{IN2}, a_{IN2}, \dots]^T$ = a vector of wave variables at input ports of the first element;

$\mathbf{u}_{11} = [a_{OUT1}, b_{OUT1}, a_{OUT2}, b_{OUT2}, \dots]^T$ = a vector of wave variables at the first element output ports not connected with the input ports of the second element;

$\mathbf{u}_{12} = [a_{OUTn}, b_{OUTn}, a_{OUTn+1}, b_{OUTn+1}, \dots]^T$ = a vector of wave variables at the first element output ports connected with the appropriate input ports of the second element;

$\mathbf{v}_{22} = [a_{IN1}, b_{IN1}, a_{IN2}, b_{IN2}, \dots]^T$ = a vector of wave variables at the second element input ports connected with the appropriate output ports of the first element;

$\mathbf{v}_{21} = [a_{INl}, b_{INl}, a_{INl+1}, b_{INl+1}, \dots]^T$ = a vector of wave variables at the input ports of the second element;

$\mathbf{u}_2 = [a_{OUT1}, b_{OUT1}, a_{OUT2}, b_{OUT2}, \dots]^T$ = a vector of wave variables at the output ports of the second element.

The cascaded connection of two multiports introduces the following constraint on the elements of the vectors $\mathbf{u}_{12}$ and $\mathbf{v}_{22}$ (we assume here that reference impedances of connected ports are real and equal to each other):

$$\mathbf{u}_{12} = \mathbf{v}_{22} \tag{3.93}$$

The generalized transfer scattering matrices defined in Chapter 2 by (2.82) can be written in the block form. For the first element,

$$\begin{bmatrix} \mathbf{v}_1 \\ \mathbf{0}_1 \end{bmatrix} = \begin{bmatrix} \mathbf{A}_{11} & \mathbf{A}_{12} & \mathbf{A}_{1e} \\ \mathbf{A}_{01} & \mathbf{A}_{02} & \mathbf{A}_{0e} \end{bmatrix} \begin{bmatrix} u_{11} \\ u_{12} \\ b_{e1} \end{bmatrix} \tag{3.94}$$

and for the second element,

$$\begin{bmatrix} \mathbf{v}_{21} \\ \mathbf{v}_{22} \\ \mathbf{0}_2 \end{bmatrix} = \begin{bmatrix} \mathbf{B}_{11} & \mathbf{B}_{1e} \\ \mathbf{B}_{21} & \mathbf{B}_{2e} \\ \mathbf{B}_{01} & \mathbf{B}_{02} \end{bmatrix} \begin{bmatrix} \mathbf{u}_2 \\ \mathbf{b}_{e2} \end{bmatrix} \tag{3.95}$$

where

$\mathbf{0}_1, \mathbf{0}_2$ = null vectors corresponding to the incoming waves at the additional virtual input ports, respectively, of the first and second elements,

$\mathbf{b}_{e1}, \mathbf{b}_{e2}$ = vectors of the outgoing waves at the additional virtual output ports, respectively, of the first and second elements.

In accordance with Figure 3.24, a vector of incoming and outgoing waves at input ports of the cascade is

$$\mathbf{v} = \begin{bmatrix} \mathbf{v}_1^T & \mathbf{v}_{21}^T & \mathbf{0}_1^T & \mathbf{0}_2^T \end{bmatrix}^T \tag{3.96}$$

Similarly, a vector of incoming and outgoing waves at the output ports of the cascade is

$$\mathbf{u} = \begin{bmatrix} \mathbf{u}_{11}^T & \mathbf{u}_2^T & \mathbf{b}_{e1}^T & \mathbf{b}_{e2}^T \end{bmatrix}^T \tag{3.97}$$

We now are able to write generalized transfer scattering matrices of two general multiports connected in cascade. First, the $\mathbf{T}$ matrices of both elements have to be ordered in accordance with the $\mathbf{v}$ and $\mathbf{u}$ vectors given by (3.96) and (3.97), respectively.

For the first element, we shall have

$$
\begin{bmatrix} \mathbf{v}_1 \\ \mathbf{v}_{21} \\ \mathbf{0}_1 \\ \mathbf{0}_2 \end{bmatrix} = \begin{bmatrix} \mathbf{A}_{11} & \mathbf{A}_{12} & \mathbf{A}_{1e} & 0 & 0 \\ 0 & 0 & 0 & \mathbf{I} & 0 \\ \mathbf{A}_{01} & \mathbf{A}_{02} & \mathbf{A}_{02} & 0 & 0 \\ 0 & 0 & 0 & 0 & \mathbf{I} \end{bmatrix} \begin{bmatrix} \mathbf{u}_{11} \\ \mathbf{u}_{12} \\ \mathbf{b}_{e1} \\ \mathbf{v}_{21} \\ \mathbf{0}_2 \end{bmatrix}
\tag{3.98}
$$

and for the second element,

$$
\begin{bmatrix} \mathbf{u}_{11} \\ \mathbf{v}_{22} \\ \mathbf{b}_{e1} \\ \mathbf{v}_{21} \\ \mathbf{0}_2 \end{bmatrix} = \begin{bmatrix} \mathbf{I} & 0 & 0 & 0 \\ 0 & \mathbf{B}_{21} & 0 & \mathbf{B}_{2e} \\ 0 & 0 & \mathbf{I} & 0 \\ 0 & \mathbf{B}_{11} & 0 & \mathbf{B}_{1e} \\ 0 & \mathbf{B}_{01} & 0 & \mathbf{B}_{0e} \end{bmatrix} \begin{bmatrix} \mathbf{u}_{11} \\ \mathbf{u}_2 \\ \mathbf{b}_{e1} \\ \mathbf{b}_{e2} \end{bmatrix}
\tag{3.99}
$$

By direct multiplication of the matrices given by (3.98) and (3.99) we get the transfer scattering matrix of the cascade shown in Figure 3.24. It has the form [20]:

$$
\begin{bmatrix} \mathbf{v}_1 \\ \mathbf{v}_{21} \\ \mathbf{0}_1 \\ \mathbf{0}_2 \end{bmatrix} = \begin{bmatrix} \mathbf{A}_{11} & \mathbf{A}_{12}\mathbf{B}_{21} & \mathbf{A}_{1e} & \mathbf{A}_{12}\mathbf{B}_{2e} \\ 0 & \mathbf{B}_{11} & 0 & \mathbf{B}_{1e} \\ \mathbf{A}_{01} & \mathbf{A}_{02}\mathbf{B}_{21} & \mathbf{A}_{0e} & \mathbf{A}_{02}\mathbf{B}_{2e} \\ 0 & \mathbf{B}_{01} & 0 & \mathbf{B}_{0e} \end{bmatrix} \begin{bmatrix} \mathbf{u}_{11} \\ \mathbf{u}_2 \\ \mathbf{b}_{e1} \\ \mathbf{b}_{e2} \end{bmatrix}
$$

$$
= \mathbf{T} \begin{bmatrix} \mathbf{u}_{11} \\ \mathbf{u}_2 \\ \mathbf{b}_{e1} \\ \mathbf{b}_{e2} \end{bmatrix}
\tag{3.100}
$$

In the particular case of the cascaded connection of two $2n$-ports (Figure 3.22(b)) this relation reduces to the form:

$$
\mathbf{v}_1 = \mathbf{A}_{12}\mathbf{B}_{21}\mathbf{u}_2
\tag{3.101}
$$

which is equivalent to (3.89).

This method can be used for the analysis of microwave circuits with any topology. The analyzed network has to be ordered to the form of subnetworks connected in cascade, as shown in Figure 3.25. In the network we have to separate a block of signal generators connected to the external input ports from a block of loads connected to the external output ports of the circuit. The other elements of the circuit are located in the subnetworks, called *transmission layers*, W.

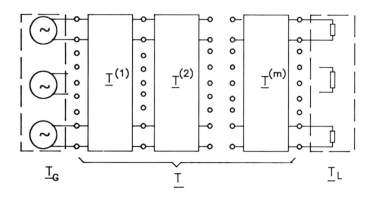

Figure 3.25 Cascade model of a microwave circuit with an arbitrary internal topology.

The elements belonging to a transmission layer are

- multiport elements not connected to each other,
- connections between elements belonging to the different transmission layers,
- connections of pairs of ports with different reference impedances,
- series and parallel T-junctions of ports.

Figure 3.26 presents an example of a circuit ordered in a form of the cascaded connection of transmission layers. The transfer scattering matrix of a transmission layer will consist of the transfer scattering matrices of multiports belonging to the particular

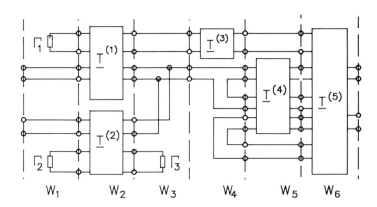

Figure 3.26 Cascade form of a circuit with transmission layers.

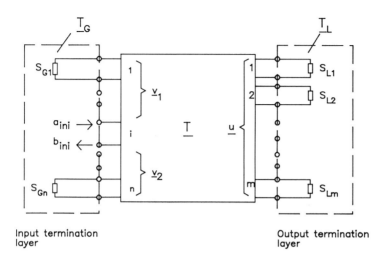

Figure 3.27 A circuit described by the A matrix with terminations at input and output ports.

layer, transfer scattering matrices of port junctions (unit matrices when reference impedances of the interconnected ports are real and equal to each other), and zero blocks (matrix blocks with all elements equal to zero).

The resulting transfer scattering matrix of the cascaded connection of layers equals

$$\mathbf{T} = \mathbf{T}^{(1)}\mathbf{T}^{(2)} \ldots \mathbf{T}^{(m)} \tag{3.102}$$

with the $\mathbf{T}$ matrices multiplied in accordance with (3.100).

Computation of Circuit Functions

To compute the reflection coefficient at the ith input port and the transmission coefficients between the ith input port and the output ports, we assume that all input and all output ports are terminated with one-port loads. At the input ports these loads are characterized by the reflection coefficients S_{Gj} ($j = 1, 2, \ldots, n$; $n \neq i$). At the output ports the reflection coefficients of loads are S_{Lj} ($j = 1, 2, \ldots, m$). Figure 3.27 presents the multiport with loads at its ports. Because the ith port is the distinguished input port, we partition the transfer scattering matrix $\mathbf{T}$ of the circuit in the following way [20]:

$$
\begin{bmatrix} \mathbf{v}_1 \\ b_{\mathrm{IN}i} \\ a_{\mathrm{IN}i} \\ \mathbf{v}_2 \\ \mathbf{0} \end{bmatrix} = \begin{bmatrix} \mathbf{T}_1 & \mathbf{T}_{1e} \\ \mathbf{T}_b & \mathbf{T}_{be} \\ \mathbf{T}_a & \mathbf{T}_{ae} \\ \mathbf{T}_2 & \mathbf{T}_{2e} \\ \mathbf{T}_0 & \mathbf{T}_{0e} \end{bmatrix} \begin{bmatrix} \mathbf{u} \\ \mathbf{b}_e \end{bmatrix} = \mathbf{T} \begin{bmatrix} \mathbf{u} \\ \mathbf{b}_e \end{bmatrix} \tag{3.103}
$$

where

$\mathbf{v}_1 = [\,b_{\mathrm{IN}1}, a_{\mathrm{IN}1}, \dots, b_{\mathrm{IN}i-1}, a_{\mathrm{IN}i-1}]^T = $ a vector of wave variables at the input ports $1, 2, \dots, i-1$,

$b_{\mathrm{IN}i}, a_{\mathrm{IN}i} = $ outgoing and incoming wave variables at the ith input port,

$\mathbf{v}_2 = [\,b_{\mathrm{IN}i+1}, a_{\mathrm{IN}i+1}, \dots, b_{\mathrm{IN}n}, a_{\mathrm{IN}n}]^T = $ a vector of wave variables at the output ports $i+1, i+2, \dots, n$,

$\mathbf{u} = [\,a_{\mathrm{OUT}1}, b_{\mathrm{OUT}1}, \dots, a_{\mathrm{OUT}m}, b_{\mathrm{OUT}m}]^T = $ a vector of wave variables at the output ports.

The transfer scattering matrix of the output port loads is

$$
\begin{bmatrix} \\ \mathbf{u} \\ \\ \end{bmatrix} = \begin{bmatrix} S_{L1} & 0 & 0 & \cdots & 0 \\ 1 & 0 & 0 & & 0 \\ 0 & S_{L2} & 0 & & \vdots \\ 0 & 1 & 0 & & \\ 0 & 0 & S_{L3} & & \\ & & 1 & & \vdots \\ & & 0 & S_{LM} & \\ 0 & \cdots & & 0 & 1 \end{bmatrix} \begin{bmatrix} \\ \mathbf{b}_{eL} \\ \\ \end{bmatrix} = \mathbf{T}_L \mathbf{b}_{eL} \tag{3.104}
$$

The vector of wave variables $\mathbf{b}_{eL}$ at the additional virtual output ports of the output termination layer is equal to the vector of outgoing wave variables $b_{\mathrm{OUT}j}$, $j = 1, 2, \dots, m$ at the output ports of the circuit. The input port loads have the following transfer scattering matrix (see Table 2.4):

$$
\begin{bmatrix} \mathbf{0} \\ b_i \\ a_i \\ \mathbf{0} \end{bmatrix} = \begin{bmatrix} \mathbf{T}_{G1} & 0 & 0 & 0 \\ 0 & 1 & 0 & 0 \\ 0 & 0 & 1 & 0 \\ 0 & 0 & 0 & \mathbf{T}_{G2} \end{bmatrix} \begin{bmatrix} \mathbf{v}_1 \\ b_{\mathrm{IN}i} \\ a_{\mathrm{IN}i} \\ \mathbf{v}_2 \end{bmatrix} = \mathbf{T}_G \begin{bmatrix} \mathbf{v}_1 \\ b_{\mathrm{IN}i} \\ b_{\mathrm{IN}i} \\ \mathbf{v}_2 \end{bmatrix} \tag{3.105}
$$

where submatrices $\mathbf{T}_{G1}$ and $\mathbf{T}_{G2}$ are

$$
\mathbf{T}_{G1} = \begin{bmatrix}
-S_{G1} & 1 & 0 & 0 & 0 & 0 & \cdots & & 0 & 0 \\
0 & 0 & -S_{G2} & 1 & 0 & 0 & & & \vdots & \vdots \\
\vdots & \vdots & 0 & 0 & -S_{G3} & 1 & & & & \\
& & 0 & & 0 & 0 & & & & \\
\vdots & & & & & & & & 0 & 0 \\
0 & 0 & 0 & 0 & 0 & 0 & \cdots & 0 & -S_{Gi-1} & 1
\end{bmatrix}
$$

$$(3.106)$$

$$
\mathbf{T}_{G2} = \begin{bmatrix}
-S_{Gi+1} & 1 & 0 & 0 & 0 & 0 & \cdots & & 0 & 0 \\
0 & 0 & -S_{Gi+2} & 1 & 0 & 0 & & & \vdots & \vdots \\
0 & 0 & 0 & 0 & -S_{Gi+3} & 1 & & & & \\
\vdots & \vdots & 0 & 0 & 0 & 0 & & & & \\
& & & & & & & & 0 & 0 \\
0 & 0 & 0 & 0 & 0 & 0 & \cdots & 0 & -S_{Gn} & 1
\end{bmatrix}
$$

$$(3.107)$$

The matrix equation of the circuit with the input and output port loads received by combining the equations (3.103), (3.104), and (3.105) (by multiplication $\mathbf{T}_G \cdot \mathbf{T} \cdot \mathbf{T}_L$) has the form [20]:

$$
\begin{bmatrix}
\mathbf{0} \\
b_{\mathrm{IN}i} \\
a_{\mathrm{IN}i} \\
\mathbf{0} \\
\mathbf{0}
\end{bmatrix}
=
\begin{bmatrix}
\mathbf{T}_{G1}\mathbf{T}_{1e} & \mathbf{T}_{G1}\mathbf{T}_1\mathbf{T}_L \\
\mathbf{T}_{be} & \mathbf{T}_b\mathbf{T}_L \\
\mathbf{T}_{ae} & \mathbf{T}_a\mathbf{T}_L \\
\mathbf{T}_{G2}\mathbf{T}_{2e} & \mathbf{T}_{G2}\mathbf{T}_2\mathbf{T}_L \\
\mathbf{T}_{\mathrm{OUT}e} & \mathbf{T}_0\mathbf{T}_L
\end{bmatrix}
\begin{bmatrix}
\mathbf{b}_e \\
\\
\mathbf{b}_{\mathrm{OUT}}
\end{bmatrix}
$$

$$(3.108)$$

Dividing the equation (3.108) by $a_{\mathrm{IN},i}$ and then reordering its rows we get the relation

$$
\begin{bmatrix}
\Gamma_i \\
\\
\delta_i
\end{bmatrix}
=
\begin{bmatrix}
\mathbf{T}_{be} & \mathbf{T}_b\mathbf{T}_L \\
\overline{\mathbf{T}_{G1}\mathbf{T}_{1e}} & \overline{\mathbf{T}_{G1}\mathbf{T}_1\mathbf{T}_L} \\
\mathbf{T}_{ae} & \mathbf{T}_a\mathbf{T}_L \\
\mathbf{T}_{G2}\mathbf{T}_{2e} & \mathbf{T}_{G2}\mathbf{T}_2\mathbf{T}_L \\
\mathbf{T}_{0e} & \mathbf{T}_0\mathbf{T}_L
\end{bmatrix}
\begin{bmatrix}
\\
\beta_i \\
\\
\end{bmatrix}
=
\begin{bmatrix}
\mathbf{Q}_i \\
\overline{\quad} \\
\mathbf{P}_i
\end{bmatrix}
\begin{bmatrix}
\\
\beta_i \\
\\
\end{bmatrix}
$$

$$(3.109)$$

where

$\Gamma_i = b_i / a_i$ = reflection coefficient at the ith port,

$\beta_i = [\mathbf{b}_e^T / a_i, \mathbf{b}_{OUT}^T / a_i]^T$ = a vector of transmission coefficients between the ith input port and the output ports and some of the internal ports of the circuit,

δ_i = a vector, the elements of which are all null except for a 1 in ith position.

A vector of transmission coefficients β_i is a solution vector of a system of linear equations

$$\mathbf{P}_i\beta_i = \delta_i \qquad (3.110)$$

where $\mathbf{P}_i$ is the lower submatrix of the coefficient matrix in the (3.109). The matrix $\mathbf{P}_i$ is square and nonsingular.

According to (3.109) the reflection coefficient at the ith input port equals

$$\Gamma_i = \mathbf{Q}_i\beta_i \qquad (3.111)$$

where $\mathbf{Q}_i$ is the upper submatrix of the coefficient matrix in (3.109).

The computation of the reflection coefficient and transmission coefficients for other input ports requires repeating the procedure described by (3.109) to (3.111).

Example 3.6

Let us consider the Wilkinson power divider circuit shown in Figure 3.28(a). Its transmission layer model will look like the circuit presented in Figure 3.28(b).

The transfer scattering matrix of the whole circuit is a product of transfer scattering matrices of successive transfer layers numbered 1 through 5. The largest number of input ports in a layer is three (layers 4 and 5) and the largest number of output ports in a layer also is three (layers 3 and 4).

The transfer scattering matrix equations for the transmission layers, starting from the last layer, are as follows.
Fifth layer:

$$
\begin{bmatrix} b_1 \\ a_1 \\ b_4 \\ a_4 \\ b_5 \\ a_5 \end{bmatrix}
=
\begin{bmatrix}
1 & 0 & \vdots & 0 & 0 & 0 \\
0 & 1 & \vdots & 0 & 0 & 0 \\
\hdashline
0 & 0 & \vdots & 0.5 & 0.5 & 1 \\
0 & 0 & \vdots & 0.5 & 0.5 & -1 \\
0 & 0 & \vdots & 1 & 0 & -1 \\
0 & 0 & \vdots & 0 & 1 & 1
\end{bmatrix}
\begin{bmatrix} a_2 \\ b_2 \\ a_3 \\ b_3 \\ b_e \end{bmatrix}
= \mathbf{T}^{(5)}
\begin{bmatrix} a_2 \\ b_2 \\ a_3 \\ b_3 \\ b_e \end{bmatrix}
$$

Connection (top, with ↓ arrow)
Parallel T (bottom, with ↑ arrow)

Fourth layer:

$$
\begin{bmatrix} b_1 \\ a_1 \\ b_3 \\ a_3 \\ b_5 \\ a_5 \end{bmatrix} =
\begin{array}{cc}
\text{Connection} & \text{Series } R \\
\downarrow &
\end{array}
\begin{bmatrix}
1 & 0 & 0 & 0 & 0 & 0 \\
0 & 1 & 0 & 0 & 0 & 0 \\
0 & 0 & 1-0.5R & 0.5R & 0 & 0 \\
0 & 0 & -0.5R & 1+0.5R & 0 & 0 \\
0 & 0 & 0 & 0 & 1 & 0 \\
0 & 0 & 0 & 0 & 0 & 1
\end{bmatrix}
\begin{bmatrix} a_2 \\ b_2 \\ a_4 \\ b_4 \\ a_6 \\ b_6 \end{bmatrix}
$$

$$\uparrow$$
$$\text{Connection}$$

$$
= \mathbf{T}^{(4)} \begin{bmatrix} a_2 \\ b_2 \\ a_4 \\ b_4 \\ a_6 \\ b_6 \end{bmatrix}
$$

Third layer:

$$
\begin{bmatrix} b_1 \\ a_1 \\ b_3 \\ a_3 \\ 0 \end{bmatrix} =
\begin{bmatrix}
0.5 & -0.5 & 1 & 0 & 0 & 0 \\
-0.5 & 0.5 & 0 & 1 & 0 & 0 \\
0 & 0 & 0 & 0 & 1 & 0 \\
0 & 0 & 0 & 0 & 0 & 1 \\
-1 & -1 & 1 & 1 & 0 & 0
\end{bmatrix}
\begin{bmatrix} a_2 \\ b_2 \\ a_5 \\ b_5 \\ a_4 \\ b_4 \end{bmatrix}
$$

$$\uparrow$$
$$
\begin{array}{cc}
\text{Parallel } T & \text{Connection}
\end{array}
$$

$$
= \mathbf{T}^{(3)} \begin{bmatrix} a_2 \\ b_2 \\ a_5 \\ b_5 \\ a_4 \\ b_4 \end{bmatrix}
$$

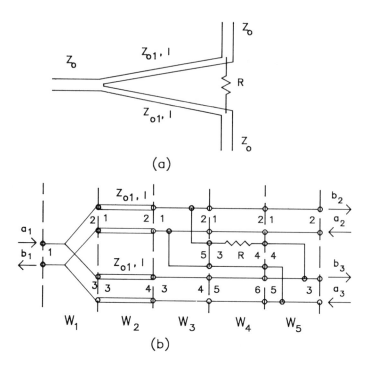

(a)

(b)

Figure 3.28 (a) A microstrip circuit of the Wilkinson power divider and (b) its transmission layer model.

Second layer:

$$
\begin{bmatrix} b_1 \\ a_1 \\ b_3 \\ a_3 \\ 0 \end{bmatrix} = \begin{bmatrix} a_{11}^{(1)} & a_{12}^{(1)} & 0 & 0 & 0 \\ -a_{12}^{(1)} & a_{22} & 0 & 0 & 0 \\ 0 & 0 & a_{11}^{(2)} & a_{12}^{(2)} & 0 \\ 0 & 0 & -a_{12}^{(2)} & a_{22}^{(2)} & 0 \\ 0 & 0 & 0 & 0 & 1 \end{bmatrix} \begin{bmatrix} a_2 \\ b_2 \\ a_4 \\ b_4 \\ b_e \end{bmatrix}
$$

$$
= \mathbf{T}^{(2)} \begin{bmatrix} a_2 \\ b_2 \\ a_4 \\ b_4 \\ b_e \end{bmatrix}
$$

Transmission line ↓ (above matrix)

Transmission line ↑ (below matrix)

First layer:

Parallel T
↓

$$
\begin{bmatrix} b_1 \\ a_1 \\ 0 \\ 0 \end{bmatrix} = \left[\begin{array}{cccc|c} 0.5 & -0.5 & 1 & 0 & 0 \\ -0.5 & 0.5 & 0 & 1 & 0 \\ -1 & -1 & 1 & 1 & 0 \\ \hline 0 & 0 & 0 & 0 & 1 \end{array} \right] \begin{bmatrix} a_2 \\ b_2 \\ a_3 \\ b_3 \\ b_e \end{bmatrix} = \mathbf{T}^{(1)} \begin{bmatrix} a_2 \\ b_2 \\ a_3 \\ b_3 \\ b_e \end{bmatrix}
$$

By using the computer-aided analysis method with the generalized transfer scattering matrix we also can compute the power wave variables at the output ports of the circuit (see Figure 3.27).

The transfer scattering matrix of the output port loads $\mathbf{T}_L$ is given by the equation (3.104) whereas the matrix $\mathbf{T}_G$ of signal generators has the form

$$
\begin{bmatrix} c_1 \\ c_2 \\ \vdots \\ c_n \end{bmatrix} = \begin{bmatrix} -S_{G1} & 1 & 0 & 0 & \cdots & 0 & 0 & 0 \\ 0 & 0 & -S_{G2} & 1 & \cdots & 0 & 0 & 0 \\ \vdots & \vdots & & & & \vdots & \vdots & \vdots \\ & & & & & 0 & 0 & 0 \\ 0 & 0 & 0 & 0 & \cdots & 0 & -S_{Gn} & 1 \end{bmatrix} \mathbf{v} \qquad (3.112)
$$

where
    $\mathbf{c} = [c_1, c_2, \ldots, c_n]^T$ = a vector of impressed waves of signal generators,
    S_{Gi} = a reflection coefficient of the ith signal generator,
    $\mathbf{v}$ = a vector of incoming and outgoing waves at the signal generator ports.

The transfer scattering matrix $\mathbf{T}$ of the circuit (see Figure 3.27) is

$$
\begin{bmatrix} \mathbf{v} \\ \mathbf{0} \end{bmatrix} = \begin{bmatrix} \mathbf{T}_1 & \mathbf{T}_{1e} \\ \mathbf{T}_0 & \mathbf{T}_{0e} \end{bmatrix} \begin{bmatrix} \mathbf{u} \\ \mathbf{b}_e \end{bmatrix} \qquad (3.113)
$$

By multiplying the matrices $\mathbf{T}_G \cdot \mathbf{T} \cdot \mathbf{T}_L$ (equations (3.112), (3.113), and (3.104)), we obtain a set of equations:

$$
\begin{bmatrix} \mathbf{c} \\ \mathbf{0} \end{bmatrix} = \begin{bmatrix} \mathbf{T}_G \mathbf{T}_{1e} & \mathbf{T}_G \mathbf{T}_1 \mathbf{T}_L \\ \mathbf{A}_{0e} & \mathbf{A}_0 \mathbf{A}_L \end{bmatrix} \begin{bmatrix} \mathbf{b}_{eL} \\ \mathbf{b}_{OUT} \end{bmatrix} \qquad (3.114)
$$

the solution of which is a vector of outgoing waves $\mathbf{b}_{OUT}$ at the output ports of the circuit.

The incoming wave variables now may be computed by using (3.104). The coefficient matrix in (3.114) is a square and nonsingular matrix. It is equivalent to the connection scattering matrix **W** defined in Chapter 2.

Example 3.7

Figure 3.29 presents a circuit of the reflection type phase modulator with a ferrite circulator. The matrix $\mathbf{T}_c$ is the transfer scattering matrix of the ferrite circulator. The term Γ_2 is the reflection coefficient of the load, and Γ_3 is the reflection coefficient of the modulating element.

The matrix $\mathbf{T}_G$ is the coefficient matrix in the set of equations describing the signal generator:

$$\begin{bmatrix} c \\ 0 \end{bmatrix} = \begin{bmatrix} -S_G & 1 & 0 \\ 0 & 0 & 1 \end{bmatrix} \begin{bmatrix} a_1 \\ b_1 \\ 0 \end{bmatrix} = \mathbf{T}_G \begin{bmatrix} a_1 \\ b_1 \\ 0 \end{bmatrix}$$

The ferrite circulator describes the equation:

$$\begin{bmatrix} b_1 \\ a_1 \\ 0 \end{bmatrix} = \begin{bmatrix} t_{11} & t_{12} & t_{13} & 0 \\ t_{21} & t_{22} & t_{23} & 0 \\ t_{31} & t_{32} & t_{33} & 1 \end{bmatrix} \begin{bmatrix} a_2 \\ b_2 \\ a_3 \\ b_3 \end{bmatrix} = \mathbf{T}_c \begin{bmatrix} a_2 \\ b_2 \\ a_3 \\ b_3 \end{bmatrix}$$

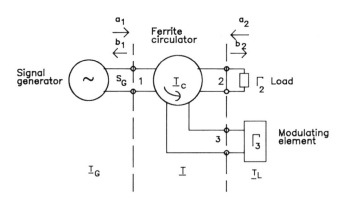

Figure 3.29 Circuit of the reflection type phase modulator with transmission layers.

The set of equations for output port loads of the circuit is

$$
\begin{bmatrix} a_2 \\ b_2 \\ a_3 \\ b_3 \end{bmatrix} = \begin{bmatrix} \Gamma_2 & 0 \\ 1 & 0 \\ 0 & \Gamma_3 \\ 0 & 1 \end{bmatrix} \begin{bmatrix} b_2 \\ b_e \end{bmatrix} = \mathbf{A}_L \begin{bmatrix} b_2 \\ b_e \end{bmatrix}
$$

By multiplying the transfer scattering matrices $\mathbf{T}_G \cdot \mathbf{T}_c \cdot \mathbf{T}_L$, we have

$$
\begin{bmatrix} c \\ 0 \end{bmatrix} = \begin{bmatrix} -S_G & 1 & 0 \\ 0 & 0 & 1 \end{bmatrix} \begin{bmatrix} t_{11} & t_{12} & t_{13} \\ t_{21} & t_{22} & t_{23} \\ t_{31} & t_{32} & t_{33} \end{bmatrix} \begin{bmatrix} \Gamma_2 & 0 \\ 1 & 0 \\ 0 & \Gamma_3 \\ 0 & 1 \end{bmatrix} \begin{bmatrix} b_2 \\ b_e \end{bmatrix}
$$

or

$$
\begin{bmatrix} c \\ 0 \end{bmatrix} = \begin{bmatrix} a_{22} - S_G a_{12} + \Gamma_2 a_{21} - S_G \Gamma_2 a_{11} & \Gamma_3 (a_{23} - S_G a_{13}) \\ a_{32} + \Gamma_2 a_{31} & 1 + \Gamma_3 a_{33} \end{bmatrix} \begin{bmatrix} b_2 \\ b_e \end{bmatrix}
$$

From this set of equations, we can compute the output port outgoing wave b_2.

REFERENCES

[1] M. Caulton, "Lumped Elements in Microwave Integrated Circuits," in *Advances in Microwaves*, Vol. 8, L. Young and H. Sobol, eds., Academic Press, New York, 1974.

[2] R.S. Pengelly and D.C. Rickard, "Design, Measurement and Application of Lumped Elements up to J-Band," *Proc. 7th European Microwave Conf.*, Copenhagen, 1977, pp. 460–464.

[3] E. Petenpaul, H. Kapusta, A. Weisegerber, H. Mampe, J. Luginsland, and I. Wolf, "CAD Models of Lumped Elements on GaAs up to 18 GHz," *IEEE Trans. Microwave Theory Tech.*, Vol. MTT-36, No. 2, February 1988, pp. 294–304.

[4] L.O. Chua and Pen-Min Lin, *Computer-Aided Analysis of Electronic Circuits*, Prentice-Hall, Englewood Cliffs, NJ, 1975.

[5] D.A. Calahan, *Computer-Aided Network Design*, McGraw-Hill, New York, 1972.

[6] V.A. Monaco and P. Tiberio, "Computer-Aided Analysis of Microwave Circuits," *IEEE Trans. Microwave Theory Tech.*, Vol. MTT-22, March 1974, pp. 249–263.

[7] K.C. Gupta, R. Garg, and R. Chadha, *Computer-Aided Design of Microwave Circuits*, Artech House, Dedham, MA, 1981.

[8] V.A. Monaco and P. Tiberio, "On the Transformation of a Lumped Element Linear Network into a Circuit Composed of Multiports," *Alta Frequenza*, Vol. 39, November 1970, pp. 1013–1014.

[9] M.A. Murray-Lasso, "Black-Box Models for Linear Integrated Circuits," *IEEE Trans. Educ.* (Special Issue on Educational Aspects of Circuit Design by Computer—I), Vol. E-12, September 1969, pp. 1170–1280.

[10] M.A. Murray-Lasso, "Analysis of Linear Integrated Circuits by Digital Computer Using Black-Box Techniques," in *Computer Aided Design*, G.H. Herskowitz, ed., McGraw-Hill, New York, 1968.

[11] W.N. Parker, "DIPNET, a General Distributed Parameter Network Analysis Program," *IEEE Trans. Microwave Theory Tech.* (Special Issue on Computer-Oriented Microwave Practices), Vol MTT-17, August 1969, pp. 495–505.

[12] P.L. Green, General Purpose Programs for the Frequency Domain Analysis of Microwave Circuits,' *IEEE Trans. Microwave Theory Tech.* (Special Issue on Computer-Oriented Microwave Practices) Vol. MTT-17, August 1969, pp. 506–514.

[13] T.N. Trick and J. Vlach, "Computer-Aided Design of Broadband Amplifiers with Complex Loads,' *IEEE Trans. Microwave Theory Tech.*, Vol. MTT-18, September 1979, pp. 541–547.

[14] V.G. Gelnovatch and I.L. Chase, "DEMON: An Optimal Seeking Computer Program for the Design of Microwave Circuits," *IEEE J. Solid-State Circuits* (Special Issue on Microwave Circuits), Vol SC-5, December 1970, pp. 303–309.

[15] P. Penfield, Jr., *MARTHA Users Manual*, Cambridge, MA, MIT Press, 1971.

[16] J.W. Bandler, S. Daijavad, and Q.J. Ahang, "Computer Aided Design of Branched Cascaded Networks," *Proc. IEEE Int. Symp. Circuits and Systems*, Kyoto, Japan, 1985, pp. 1579–1582.

[17] J.W. Bandler, M.R.M. Rizk, and H.L. Abdel-Malek, "New Results in Network Simulation Sensitivity, and Tolerance Analysis for Cascaded Structures," *IEEE Trans. Microwave Theory Tech.*, Vol. MTT-26, 1978, pp. 963–972.

[18] L. Weinberg, "Scattering Matrix and Transfer Scattering Matrix," in *Amplifiers*, R.F. Shea, ed. McGraw-Hill, New York, 1966.

[19] V.A. Monaco and P. Tiberio, "Automatic Scattering Matrix Computation of Microwave Circuits," *Alta Frequenza*, Vol. 39, February 1970, pp. 59–64.

[20] P. Miazga, "Computer Aided Analysis of Microwave Circuits by Means of Transfer Scattering Matrices," Ph.D. Dissertation, Warsaw University of Technology, Warsaw, Poland, 1989, [in Polish].

Chapter 4

Computer-Aided Sensitivity Analysis
of Microwave Circuits

ensitivities, or partial derivatives of certain network functions with respect to circuit parameters are very important in the design of microwave circuits. Computer-aided analysis allows us to simulate a network and check whether the network satisfies the imposed design requirements. Network simulation does not tell us how to improve the network performance and the result of the design procedure depends strongly on the professional experience, wisdom, and intuition of the designer. Sensitivity analysis in connection with the optimization methods allows us to make the design procedure more algorithmic and, in consequence, allows automation of the circuit design.

In the computer simulation we always use a network or system model, which is a simplified mathematical description of the actual network and computations are always done with a finite accuracy. Because of this, simulation results will always differ from the ideal measurement results of a real network. Sensitivity analysis lets us estimate simulation errors; and, by using the results of the sensitivity analysis, we can arrange network parameters according to increasing values of sensitivity measure. In consequence the procedure can lead to the identification of circuit parameters with excessive values of the sensitivity or help us simplify the model by eliminating elements for which sensitivities are negligible. Very often, we are interested in the behavior of a network in the presence of parasitic elements. In a network model these elements have no nominal values. Sensitivities are defined with respect to these elements and can be used to predict response variations when parasitics take on small values.

Real networks and systems usually must operate at changing environmental conditions (i.e., at changing temperature, humidity, atmospheric pressure, electromagnetic radiation *et cetera*). Circuit elements change their values due to natural aging. By taking into account these dependencies we can determine network or system sensitivities with respect to environment conditions parameters.

Information on network sensitivities may be used extensively in computer-aided optimal tolerance assignment, optimal centering, and postproduction tuning [1–4].

4.1 SENSITIVITY DEFINITION

Sensitivity is the derivative of a differentiable function f with respect to any parameter p:

$$D_p(f) = \frac{\partial f}{\partial p} \tag{4.1}$$

This definition is convenient and useful for computer applications, but is not independent of scale. The widely used definition is the normalized sensitivity:

$$S_p(f) = \frac{\partial (\ln f)}{\partial (\ln p)} = \frac{p}{f} \frac{\partial f}{\partial p} = \frac{p}{f} D_p(f) \tag{4.2}$$

which forms the basis for comparing various designs.

In microwave circuit design f can be a node voltage, an incoming or outgoing wave variable, a network function such as the reflection coefficient, transducer power gain, and so on. Parameter p can be the component value, the frequency variable f, the operating temperature or humidity, and so on.

In microwave circuit design the function of interest often is a complex function represented in the form

$$f = |f| e^{j\phi} \tag{4.3}$$

Taking natural logarithm

$$\ln f = \ln |f| + j\phi \tag{4.4}$$

and differentiating with respect to a parameter p yields

$$\frac{1}{f} \frac{\partial f}{\partial p} = \frac{1}{|f|} \frac{\partial |f|}{\partial p} + j \frac{\partial \phi}{\partial p} \tag{4.5}$$

This complex equation can be split into two equations

$$\frac{\partial |f|}{\partial p} = |f| \operatorname{Re} \left\{ \frac{1}{f} \frac{\partial f}{\partial p} \right\} \tag{4.6}$$

$$\frac{\partial \phi}{\partial p} = \operatorname{Im} \left\{ \frac{1}{f} \frac{\partial f}{\partial p} \right\} \tag{4.7}$$

where Re and Im denote real and imaginary parts. The sensitivity of the magnitude value of the complex function is given in linear units. The sensitivity of the argument ϕ of a complex function is measured in radians.

Very often the magnitude (the absolute value) is measured in decibels. To obtain the corresponding formula, we write (4.4) in the from

$$\ln f = A + j\phi \qquad (4.8)$$

where $A = \ln |f|$.

Differentiation with respect to p yields

$$\frac{1}{f}\frac{\partial f}{\partial p} = \frac{\partial A}{\partial p} + j\frac{\partial \phi}{\partial p} \qquad (4.9)$$

Sensitivity $\partial A / \partial p$ is expressed in nepers. To convert it to decibels we multiply the value of $\partial A / \partial p$ by $20 \cdot \log_{10} e = 8.686$:

$$\frac{\partial A}{\partial p}(dB) = 8.686\frac{\partial A}{\partial p} = 8.686\,\mathrm{Re}\left\{\frac{1}{f}\frac{\partial f}{\partial p}\right\} \qquad (4.10)$$

4.2 TELLEGEN'S THEOREM

Tellegen's theorem was published in 1952 [5, 6]. Thanks to the work of S. W. Director and R. A. Rohrer [7], it became a basic tool in the computer-aided sensitivity analysis of electronic and microwave circuits.

Consider a class of networks with identical topologies; that is, networks whose models are isomorfic graphs. Examples of networks with identical topology are shown in Figure 4.1.

Theorem: If N and $\hat{N}$ are any networks with identical topologies, then

$$\hat{\mathbf{I}}^T\mathbf{V} = 0 \qquad (4.11)$$

$$\hat{\mathbf{V}}^T\mathbf{I} = 0 \qquad (4.12)$$

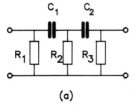

(a)

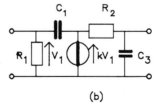

(b)

Figure 4.1 Networks with identical topologies.

where

$\mathbf{V}$ = a vector of branch voltages in the network N,
$\mathbf{I}$ = a vector of branch currents in the network N,
$\hat{\mathbf{V}}$ = a vector of branch voltages in the network $\hat{N}$,
$\hat{\mathbf{I}}$ = a vector of branch currents in the network $\hat{N}$,

The vector of branch voltages $\mathbf{V}$ may be given as

$$\mathbf{V} = \mathbf{B}^T \mathbf{V}_N \tag{4.13}$$

where $\mathbf{V}_N$ is a vector of node voltages, and $\mathbf{B}$ is the incidence matrix of the network graph [5]. From (4.11) and (4.13) we have

$$\hat{\mathbf{I}}^T \mathbf{V} = \hat{\mathbf{I}}^T \mathbf{B}^T \mathbf{V}_N = \left(\mathbf{B}\hat{\mathbf{I}}\right)^T \mathbf{V}_N = 0 \tag{4.14}$$

The last equality in (4.14) comes from the current Kirchhoff's law for the network $\hat{N}$. In the same way, using voltage Kirchhoff's law, we can prove the equality (4.2).

Network N is the original network. Corresponding to it, network $\hat{N}$ with an identical structure is called the *adjoint network*.

Subtracting both sides of (4.11) and (4.12), we obtain the "difference" form of Tellegen's equation:

$$\hat{\mathbf{I}}^T \mathbf{V} - \hat{\mathbf{V}}^T \mathbf{I} = 0 \tag{4.15}$$

Let us assume that in the original network N was a small deviation in one of the network parameters, causing a distortion in vectors of branch voltages and branch currents in network N. The vectors of branch voltages and branch currents of the distorted original network are $\mathbf{V} + \Delta\mathbf{V}$ and $\mathbf{I} + \Delta\mathbf{I}$, respectively. Because the distorted original network N and the adjoint network $\hat{N}$ have the same structure, they satisfy Tellegen's equation:

$$\hat{\mathbf{I}}^T(\mathbf{V} + \Delta\mathbf{V}) - \hat{\mathbf{V}}^T(\mathbf{I} + \Delta\mathbf{I}) = 0 \tag{4.16}$$

By using (4.15) and (4.16), we have

$$\hat{\mathbf{I}}^T \Delta\mathbf{V} - \hat{\mathbf{V}}^T \Delta\mathbf{I} = 0 \tag{4.17}$$

Equation (4.17) is widely used in the computer-aided sensitivity analysis of electronic and microwave circuits. The equation relates deviations of branch voltage and branch currents of network N that originate from parameter deviations.

In the case of microwave circuits described by the scattering matrix $\mathbf{S}$, which is equivalent to the form (4.15) of Tellegen's equation,

$$\boldsymbol{\beta}^T \mathbf{a} - \boldsymbol{\alpha}^T \mathbf{b} = 0 \tag{4.18}$$

where

$\mathbf{a}$ = a vector of incoming wave variables at the ports of network N,
$\mathbf{b}$ = a vector of outgoing wave variables at the ports of network N,
$\boldsymbol{\alpha}$ = a vector of incoming wave variables at the ports of network $\hat{N}$,
$\boldsymbol{\beta}$ = a vector of outgoing wave variables at the ports of network $\hat{N}$.

Equation (4.18) is applicable to two multiport networks with the same topologies in which the reference impedances of ports in one network are equal to the reference impedances of corresponding ports in the second network. In Figure 4.2 we show two networks that satisfy Tellegen's theorem.

We can prove the equality of (4.18). Because the outgoing waves at the ports of the original network N and the second network $\hat{N}$ satisfy the equations

$$\mathbf{b} = \boldsymbol{\Gamma} \mathbf{a} \quad \text{and} \quad \boldsymbol{\beta} = \hat{\boldsymbol{\Gamma}} \boldsymbol{\alpha} \tag{4.19}$$

where

$\boldsymbol{\Gamma}$ = the connection matrix of the first network N,
$\hat{\boldsymbol{\Gamma}}$ = the connection matrix of the second network $\hat{N}$,

we have

$$\boldsymbol{\beta}^T \mathbf{a} - \boldsymbol{\alpha}^T \mathbf{b} = \left(\hat{\boldsymbol{\Gamma}} \boldsymbol{\alpha} \right)^T - \boldsymbol{\alpha} \boldsymbol{\Gamma} \mathbf{a} = \boldsymbol{\alpha}^T (\hat{\boldsymbol{\Gamma}}^T - \boldsymbol{\Gamma}) \mathbf{a} \tag{4.20}$$

We see that expression (4.20) equals zero if

$$\hat{\boldsymbol{\Gamma}}^T = \boldsymbol{\Gamma} \tag{4.21}$$

The method for obtaining the connection matrix $\boldsymbol{\Gamma}$ was discussed in Section 3.2.1. It follows from (3.65), defining the connection matrix $\boldsymbol{\Gamma}$ for a pair of connected ports, that (4.21) is satisfied when the reference impedances of ports in the first network are equal to the reference impedances of the corresponding ports in the second network. In particular, when port reference impedances in each connected pair of ports in a circuit are real and equal to each other, the elements of the connection matrix $\boldsymbol{\Gamma}$ are only zeros and ones (see equation (3.65)). In such a case the connection matrices of both networks satisfy (4.21), and Tellegen's equation (4.18) is satisfied, too.

Hence, from the above considerations, Tellegen's equation (4.18) is satisfied

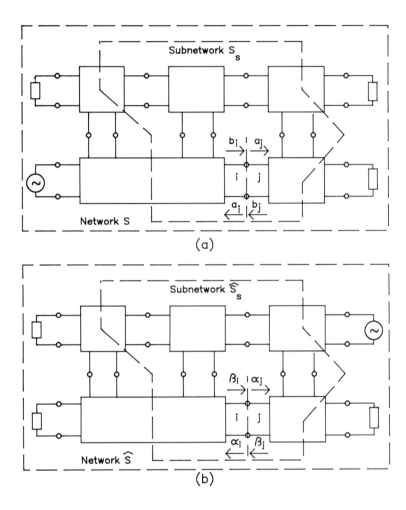

Figure 4.2 Two networks that satisfy Tellegen's theorem.

when the original network and the adjoint network fulfill two conditions:

1. the topologies of both circuits are the same,
2. connection matrix $\hat{\boldsymbol{\Gamma}}$ of the adjoint network is equal to transposed connection matrix $\boldsymbol{\Gamma}^T$ of the original network.

If we introduce a small distortion in one of circuit parameters, then the vectors of incoming and outgoing waves at the ports of the original network will change to $\mathbf{a} + \Delta\mathbf{a}$ and $\mathbf{b} + \Delta\mathbf{b}$, respectively. Because the topologies of the distorted original network N

and the adjoint network $\hat{N}$ are the same, they satisfy Tellegen's equation:

$$\boldsymbol{\beta}^T(\mathbf{a} + \Delta\mathbf{a}) - \boldsymbol{\alpha}^T(\mathbf{b} + \Delta\mathbf{b}) = 0 \qquad (4.22)$$

Subtracting both sides of (4.18) and (4.22), we have

$$\boldsymbol{\beta}^T\Delta\mathbf{a} - \boldsymbol{\alpha}^T \Delta\mathbf{b} = 0 \qquad (4.23)$$

Equation (4.23) is the basic relation of the adjoint network method for the computation of microwave network sensitivities. Equations (4.18) and (4.23) are valid for any two subnetworks or complete networks of the same topology. For complete networks, the external ports related to generators and loads have to be taken into account in equations (4.18) and (4.23).

4.3 SENSITIVITY ANALYSIS OF MICROWAVE NETWORKS DESCRIBED BY THE NODAL ADMITTANCE MATRIX

4.3.1 The Transposed Matrix Method

Consider a general network described by the nodal admittance matrix equation

$$\mathbf{Y}\mathbf{V}_N = \mathbf{I}_{N0} \qquad (4.24)$$

Let us assume a small deviation in parameter p of the network. Then differentiating the equation (4.24) with respect to the variable parameter p, we obtain

$$\mathbf{Y}\frac{\partial \mathbf{V}_N}{\partial p} + \frac{\partial \mathbf{Y}}{\partial p}\mathbf{V}_N = 0 \qquad (4.25)$$

or

$$\frac{\partial \mathbf{V}_N}{\partial p} = -\mathbf{Y}^{-1}\frac{\partial \mathbf{Y}}{\partial p}\mathbf{V}_N = 0 \qquad (4.26)$$

The sensitivity $\partial V_{Nr}/\partial p$ of the rth node voltage V_{Nr} with respect to the circuit parameter p may be computed by multiplying the left-hand side of the equation (4.26) by a row vector $\mathbf{J}_r^T$, called the *adjoint vector*, all of whose elements are null except the rth element, which equals 1:

$$\mathbf{J}_r^T = [0, 0, \dots, 0, 1, 0, \dots, 0] \qquad (4.27)$$

This operation yields

$$\frac{\partial V_{Nr}}{\partial p} = \mathbf{J}_r^T \frac{\partial \mathbf{V}_N}{\partial p} = -\mathbf{J}_r^T \mathbf{Y}^{-1} \frac{\partial \mathbf{Y}}{\partial p} \mathbf{V}_N$$

$$= -\left[(\mathbf{Y}^T)^{-1}\mathbf{J}_r\right]^T \frac{\partial \mathbf{Y}}{\partial p} \mathbf{V}_N = -\hat{\mathbf{V}}_N^T \frac{\partial \mathbf{Y}}{\partial p} \mathbf{V}_N \qquad (4.28)$$

In (4.28), we have

$$\hat{\mathbf{V}}_N = (\mathbf{Y}^T)^{-1}\mathbf{J}_r = \hat{\mathbf{Y}}^{-1}\mathbf{J}_r \qquad (4.29)$$

which is a solution vector of a matrix equation in the form:

$$\mathbf{Y}^T \hat{\mathbf{V}}_N = \mathbf{J}_r \qquad (4.30)$$

called the *transposed* or *adjoint matrix equation*. The transposed matrix method in some aspects is similar to the adjoint network method. In fact, (4.30) may be interpreted as a nodal admittance matrix equation of the adjoint network.

According to (4.30) the nodal admittance matrix $\hat{\mathbf{Y}}$ of the adjoint network equals the transposed admittance matrix of the original network:

$$\hat{\mathbf{Y}} = \mathbf{Y}^T \qquad (4.31)$$

The adjoint network is excited by a current source connected between the reference node and the rth node of the circuit with nominal current 1 A. The equivalence between this method and the adjoint network method is then evident.

In a typical case, a deviation in parameter p influences the characterization of one element only, leaving the characterization of other elements of the circuit unchanged. For such a case the nonzero elements of $\partial \mathbf{Y}/\partial p$ correspond to the affected circuit element.

Sensitivities of the rth nodal voltage with respect to other circuit parameters can be found by computing the right-hand side of the relation:

$$\frac{\partial V_{Nr}}{\partial p_k} = -\hat{\mathbf{V}}_N^T \frac{\partial \mathbf{Y}}{\partial p_k} \mathbf{V}_N \qquad (4.32)$$

for the different $\partial \mathbf{Y}/\partial p_k$ corresponding to different circuit parameters p_k. In general, we may write

$$\mathbf{G} = \nabla \mathbf{V}_{Nr} = \begin{bmatrix} \dfrac{\partial V_{Nr}}{\partial p_1} \\[2mm] \dfrac{\partial V_{Nr}}{\partial p_2} \\[1mm] \vdots \\[1mm] \dfrac{\partial V_{Nr}}{\partial p_n} \end{bmatrix} = \begin{bmatrix} -\hat{\mathbf{V}}_N^T \dfrac{\partial \mathbf{Y}}{\partial p_1} \mathbf{V}_N \\[2mm] -\hat{\mathbf{V}}_N^T \dfrac{\partial \mathbf{Y}}{\partial p_2} \mathbf{V}_N \\[1mm] \vdots \\[1mm] -\hat{\mathbf{V}}_N^T \dfrac{\partial \mathbf{Y}}{\partial p_n} \mathbf{V}_N \end{bmatrix} \qquad (4.33)$$

This requires the partial derivatives of the nodal admittance matrices of circuit elements.

According to these considerations, the computation of the sensitivities of the rth node voltage V_{Nr} with respect to any number of variable parameters requires solution of two matrix equations, the matrix equation of the analyzed circuit and the transposed matrix equation. For a circuit composed of reciprocal elements $\mathbf{Y} = \mathbf{Y}^T$; the adjoint network is the same as the original network; and the sensitivity analysis requires solution of two sets of linear equations with the same coefficient matrices, but with different right-hand vectors only.

4.3.2 The Direct Method

Computer-aided sensitivity analysis by means of the direct method is based on the application of the partial derivative relation for the vector of the nodal voltage $\mathbf{V}_N$. Because $\mathbf{V}_N = \mathbf{Y}^{-1}\mathbf{I}_{N0}$, we have

$$\frac{\partial \mathbf{V}_N}{\partial p} = -\mathbf{Y}^{-1}\frac{\partial \mathbf{Y}}{\partial p}\mathbf{Y}^{-1}\mathbf{I}_{N0} = -\mathbf{Y}^{-1}\frac{\partial \mathbf{Y}}{\partial p}\mathbf{V}_N \qquad (4.34)$$

or

$$\frac{\partial \mathbf{V}_N}{\partial p} = \mathbf{Y}^{-1}\boldsymbol{\xi} \qquad (4.35)$$

where

$$\boldsymbol{\xi} = -\frac{\partial \mathbf{Y}}{\partial p}\mathbf{V}_N \qquad (4.36)$$

As we can see, the sensitivities of all voltages in the circuit with respect to one parameter p are a solution vector of a matrix equation:

$$\mathbf{Y}\frac{\partial \mathbf{V}_N}{\partial \mathbf{p}} = \boldsymbol{\xi} \qquad (4.37)$$

in which the coefficient matrix is the nodal admittance matrix $\mathbf{Y}$ and the right-hand vector $\boldsymbol{\xi}$ is defined by (4.36).

Computation of network sensitivities by the direct method requires two solutions of network equations. First, we have to solve the nodal admittance matrix equation (4.24) for $\mathbf{V}_N$ and then, after determining the $\boldsymbol{\xi}$ vector by using (4.36), we must solve the matrix equation (4.37) for $\partial \mathbf{V}_N/\partial p$. We should note here that the coefficient matrices in both matrix equations to be solved are the same and only the right-hand vectors are different. The direct method permits determination of the sensitivities of all node voltages in the circuit with respect to a single parameter p. If the sensitivities of all node voltages are required with respect to several circuit parameters, as many solutions of the equation (4.37) are required as there are parameters.

4.3.3 Derivation of Sensitivities of Microwave Circuits Described by the Admittance Matrices

Table 4.1 presents the results of applying (4.32) to a number of commonly used elements of microwave circuits. These relations should be treated as a "library" of formulas that will be applicable in computer programs used for the solution of certain design problems.

4.3.4 Gradient Vector Computation of Circuit Functions

There are a number of ways in which the adjoint network method can be used effectively in gradient computations of such circuit functions as voltage gain, effective voltage gain, power gain, transducer power gain, input and output reflection coefficient, *et cetera*. Let us consider gradient computation of such functions as the transducer power gain, effective voltage gain, and the reflection coefficient.

Transducer Power Gain

According to the description given in Figure 4.3, transducer power gain G_T equals

$$G_T = \frac{P_L}{P_{SA}} = \frac{|V_L|^2 G_L}{\frac{|I_S|^2}{4G_S}} = 4G_S G_L \frac{V_L V_L^*}{|I_S|^2} \qquad (4.38)$$

A gradient vector ∇G_T is equal to

$$\nabla G_T = \frac{4G_S G_L}{|I_s|^2} \nabla(V_L V_L^*)$$

$$= \frac{4G_S G_L}{|I_s|^2} 2\,\mathrm{Re}\{V_L^* \nabla V_L\} \qquad (4.39)$$

Table 4.1
Sensitivity Expressions for Some Lumped and Distributed Elements of Microwave
Circuits Described by the Admittance Matrices

Circuit Element	Sensitivity (*component of* **G**, *see* (4.33))	Variable Parameter p
Lumped G	$-\begin{bmatrix} \hat{V}_{Ni} \\ \hat{V}_{Nj} \end{bmatrix}^{T} \begin{bmatrix} 1 & -1 \\ -1 & 1 \end{bmatrix} \begin{bmatrix} V_{Ni} \\ V_{Nj} \end{bmatrix}$	G
Lumped L	$\dfrac{1}{j\omega L^{2}} \begin{bmatrix} \hat{V}_{Ni} \\ \hat{V}_{Nj} \end{bmatrix}^{T} \begin{bmatrix} 1 & -1 \\ -1 & 1 \end{bmatrix} \begin{bmatrix} V_{Ni} \\ V_{Nj} \end{bmatrix}$	L
Lumped C	$-j\omega \begin{bmatrix} \hat{V}_{Ni} \\ \hat{V}_{Nj} \end{bmatrix}^{T} \begin{bmatrix} 1 & -1 \\ -1 & 1 \end{bmatrix} \begin{bmatrix} V_{Ni} \\ V_{Nj} \end{bmatrix}$	C
Voltage controlled current source	$-\hat{V}_{2}V_{1}$	g_{m}
Short-circuited transmission line	$-\coth(\gamma l)\hat{V}V$	Y_{0}
	$Y_{0}\gamma \, \mathrm{csch}^{2}(\gamma l)\hat{V}V$	l

Table 4.1 (*Continued*)

Circuit Element	Sensitivity (*component of* **G**, *see* (4.33))	Variable Parameter *p*
	$-\tanh(\gamma l)\hat{V}V$	Y_0
Open-circuited transmission line	$-Y_0\gamma\,\mathrm{sech}^2(\gamma l)\hat{V}V$	l
$\gamma_e = \alpha_e + j\beta_e$ $\gamma_0 = \alpha_0 + j\beta_0$ $\beta_e = \dfrac{2\pi}{\lambda_e};\quad \beta_0 = \dfrac{2\pi}{\lambda_0}$ A section of coupled transmission lines	$-\hat{V}_N^T\begin{bmatrix} A_e & A_e & B_e & B_e \\ A_e & A_e & B_e & B_e \\ B_e & B_e & A_e & A_e \\ B_e & B_e & A_e & A_e \end{bmatrix}\mathbf{V}_N$ $A_{e,0} = \dfrac{1}{2}\coth(\gamma_{e,0}l)$ $B_{e,0} = \dfrac{-1}{2}\,\mathrm{csch}(\gamma_{e,0}l)$ $-\hat{V}_N^T\begin{bmatrix} A_0 & -A_0 & -B_0 & B_0 \\ -A_0 & A_0 & B_0 & -B_0 \\ -B_0 & B_0 & A_0 & -A_0 \\ B_0 & -B_0 & -A_0 & A_0 \end{bmatrix}\mathbf{V}_N$	Y_{0e} Y_{00}
	$-\hat{\mathbf{V}}_N^T\begin{bmatrix} C & D & E & F \\ D & C & F & E \\ E & F & C & D \\ F & E & D & C \end{bmatrix}\mathbf{V}_N$ $C = \dfrac{-1}{2}\left[Y_{0e}\dfrac{\gamma_e}{\sinh^2(\gamma_e l)} + Y_{00}\dfrac{\gamma_0}{\sinh^2(\gamma_0 l)}\right]$ $D = \dfrac{-1}{2}\left[Y_{0e}\dfrac{\gamma_e}{\sinh^2(\gamma_e l)} - Y_{00}\dfrac{\gamma_0}{\sinh^2(\gamma_0 l)}\right]$ $E = \dfrac{1}{2}\left[Y_{0e}\dfrac{\gamma_e\cosh(\gamma_e l)}{\sinh^2(\gamma_e l)} - Y_{00}\dfrac{\gamma_0\cosh(\gamma_0 l)}{\sinh^2(\gamma_0 l)}\right]$ $F = \dfrac{1}{2}\left[Y_{0e}\dfrac{\gamma_e\cosh(\gamma_e l)}{\sinh^2(\gamma_e l)} + Y_{00}\dfrac{\gamma_0\cosh(\gamma_0 l)}{\sinh^2(\gamma_0 l)}\right]$	1

Table 4.1 (*Continued*)

Circuit Element	Sensitivity (*component of* **G**, *see* (4.33))	Variable Parameter p
Transmission line section	$-\begin{bmatrix} \hat{V}_1 \\ \hat{V}_2 \end{bmatrix}^T \begin{bmatrix} \coth(\gamma l) & -\operatorname{csch}(\gamma l) \\ -\operatorname{csch}(\gamma l) & \coth(\gamma l) \end{bmatrix} \begin{bmatrix} V_1 \\ V_2 \end{bmatrix}$	Y_0
	$\dfrac{Y_0 \gamma}{\sinh^2(\gamma l)}$ $\cdot \begin{bmatrix} \hat{V}_1 \\ \hat{V}_2 \end{bmatrix}^T \begin{bmatrix} 1 & \cosh(\gamma l) \\ \cosh(\gamma l) & 1 \end{bmatrix} \begin{bmatrix} V_1 \\ V_2 \end{bmatrix}$	l
	$\dfrac{Y_0}{2R} \begin{bmatrix} \hat{V}_1 \\ \hat{V}_2 \end{bmatrix} \begin{bmatrix} \coth\theta & -\operatorname{csch}\theta \\ -\operatorname{csch}\theta & \coth\theta \end{bmatrix}$ $\cdot \begin{bmatrix} 1 & -\dfrac{\theta}{\sinh\theta} \\ -\dfrac{\theta}{\sinh\theta} & 1 \end{bmatrix} \begin{bmatrix} V_1 \\ V_2 \end{bmatrix}$	R
$Y_0 = \sqrt{\dfrac{j\omega C}{R}}$ $\theta = \sqrt{j\omega RC}$ $R = r \cdot l = $ total resistance $C = c \cdot l = $ total capacitance Distributed RC line section	$\dfrac{-Y_0}{2C} \begin{bmatrix} \hat{V}_1 \\ \hat{V}_2 \end{bmatrix} \begin{bmatrix} \coth\theta & -\operatorname{csch}\theta \\ -\operatorname{csch}\theta & \coth\theta \end{bmatrix}$ $\cdot \begin{bmatrix} 1 & -\dfrac{\theta}{\sinh\theta} \\ -\dfrac{\theta}{\sinh\theta} & 1 \end{bmatrix} \begin{bmatrix} V_1 \\ V_2 \end{bmatrix}$	C

In (4.39), ∇ is the gradient operator given as

$$\nabla = \begin{bmatrix} \dfrac{\partial}{\partial p_1}, \dfrac{\partial}{\partial p_2}, \dots, \dfrac{\partial}{\partial p_n} \end{bmatrix}^T \qquad (4.40)$$

From relation (4.39), we can see that the computation of the gradient vector of the transducer power gain requires evaluation of the sensitivities of the output voltage V_L. This implies the analysis of the original network excited at the input port and the

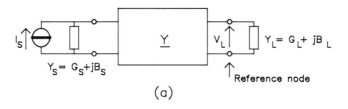

(a)

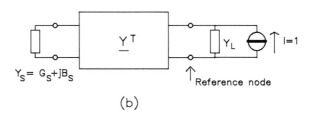

(b)

Figure 4.3 (a) Original network with the current source connected to the input port, and (b) adjoint network with the current source connected to the output port.

analysis of the adjoint network excited by the current source $I = -1$ A connected to the output port of the circuit.

Effective Voltage Gain

In accordance with notation presented in Figure 4.3, effective voltage gain of the network equals

$$K_v = \frac{V_L}{E_s} = \frac{Y_s}{I_s} V_L \tag{4.41}$$

and the gradient vector of K_v is

$$\nabla K_v = \frac{Y_s}{I_s} \nabla V_L \tag{4.42}$$

Gradient computation of the effective voltage gain of the circuit presented in Figure 4.3 (a) requires evaluation of the sensitivities of the output voltage V_L. The adjoint network and the computational effort are exactly the same as in the case of transducer power gain gradient computation.

The gradient vector of the magnitude of the effective voltage gain may be found from

$$\nabla \left(20 \log_{10} | K_v | \right) = 8.6458 \, \nabla \left(\ln | K_v | \right)$$
$$= 8.6458 \, \nabla \left(\text{Re} \{ \ln K_v \} \right)$$
$$= 8.6458 \, \text{Re} \left\{ \frac{1}{K_v} \nabla K_v \right\} \qquad (4.43)$$

and the gradient vector of the phase of K_v

$$\nabla \arg \{ K_v \} = \nabla \left(\ln K_v \right) = \text{Im} \left\{ \frac{1}{K_v} \nabla K_v \right\} \qquad (4.44)$$

Reflection Coefficient

Figure 4.4 presents the original and adjoint networks appropriate for the reflection coefficient sensitivity computation. According to the descriptions given in Figure 4.4 the reflection coefficient between Z_0 and the one-port network describes this relation:

$$\Gamma = \frac{Y_0 - Y_{in}}{Y_0 + Y_{in}} = \frac{2 Y_0}{Y_0 + Y_{in}} - 1 = \frac{2 Y_0 I_s}{V_{in}} - 1 \qquad (4.45)$$

from which we have

$$\nabla \Gamma = - \frac{2 Y_0 I_s}{V_{in}^2} \cdot \nabla V_{in} \qquad (4.46)$$

Evaluation of the reflection coefficient gradient vector $\nabla \Gamma$ requires computation of the sensitivities of the voltage V_{in}. Two network analyses yield the information required for the evaluation of ∇V_{in}. Figure 4.4 (b) presents the adjoint network excited by the current source $I = 1$ A at the port of interest.

Example 4.1

In Figure 4.5, we show a three-section quarter-wave transformer. The circuit function of interest is the reflection coefficient at the input port. To keep the example simple, the discontinuity reactances between transformer sections are neglected.

The reflection coefficient at the input of the transformer circuit is

$$\Gamma_{in} = \frac{Y_0 - Y_{in}}{Y_0 + Y_{in}} = \frac{2 Y_0 I_s}{V_{N1}} - 1$$

According to this equation, to evaluate Γ_{in} we have to find the nodal voltage V_{N1}.

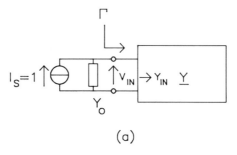

(a)

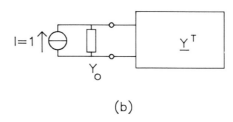

(b)

Figure 4.4 (a) Original network and (b) adjoint network for the reflection coefficient sensitivity computation.

The nodal admittance matrix equation of the original circuit is

$$
\begin{array}{c}
\begin{array}{cccc} 1 & \quad 2 & \quad 3 & \quad 4 \end{array} \\
\begin{array}{c} 1 \\ 2 \\ 3 \\ 4 \end{array}
\begin{bmatrix}
Y_0 + Y_{11}^{(1)} & -Y_{12}^{(1)} & 0 & 0 \\
-Y_{21}^{(1)} & Y_{22}^{(1)} + Y_{11}^{(2)} & -Y_{12}^{(2)} & 0 \\
0 & -Y_{21}^{(2)} & Y_{22}^{(2)} + Y_{11}^{(3)} & -Y_{12}^{(3)} \\
0 & 0 & -Y_{21}^{(3)} & Y_{22}^{(3)} + Y_L
\end{bmatrix}
\begin{bmatrix}
V_{N1} \\ V_{N2} \\ V_{N3} \\ V_{N4}
\end{bmatrix}
=
\begin{bmatrix}
I_s \\ 0 \\ 0 \\ 0
\end{bmatrix}
\end{array}
$$

where

$$
Y^{(i)} =
\begin{bmatrix}
Y_{11}^{(i)} & Y_{12}^{(i)} \\
Y_{21}^{(i)} & Y_{22}^{(i)}
\end{bmatrix}
= Y_{0i}
\begin{bmatrix}
\coth(\gamma l_i) & -\operatorname{csch}(\gamma l_i) \\
-\operatorname{csch}(\gamma l_i) & \coth(\gamma l_i)
\end{bmatrix}
$$

$i = 1, 2, 3$, are the admittance matrices of transmission line sections composing the transformer circuit.

The solution of the nodal admittance matrix equation delivers required value of V_{N1}. As we will see later it is very convenient to use $I_s = 1$ A.

To obtain the sensitivities of the input port reflection coefficient given by (4.46)

$$
\nabla \Gamma_{\text{in}} = -\frac{2 Y_0 I_s}{V_{\text{in}}^2} \cdot \nabla V_{\text{in}} = -\frac{2 Y_0 I_s}{V_{N1}} \nabla V_{N1}
$$

we must evaluate sensitivities of the V_{N1} voltage, which means that the adjoint network must be excited by a 1 A current source connected between the first node and the reference node of the circuit. All elements in the circuit are reciprocal, and so the adjoint network is the same as the original network. Because the original network and the adjoint network are identical, one analysis will suffice to evaluate sensitivities of the V_{N1} nodal voltage. Because $\hat{\mathbf{V}}_N = \mathbf{V}_N$, and assuming that the network parameters of interest are Y_{01}, Y_{02}, Y_{03}, l_1, l_2, and l_3, we have ($I_s = 1$ A)

$$\nabla \Gamma_{\text{in}} = -\frac{2Y_0}{V_{N1}^2} \nabla V_{N1} = -\frac{2Y_0}{V_{N1}^2} \begin{bmatrix} \dfrac{\partial V_{N1}}{\partial Y_{01}} \\[2mm] \dfrac{\partial V_{N1}}{\partial Y_{02}} \\[2mm] \dfrac{\partial V_{N1}}{\partial Y_{03}} \\[2mm] \dfrac{\partial V_{N1}}{\partial l_1} \\[2mm] \dfrac{\partial V_{N1}}{\partial l_2} \\[2mm] \dfrac{\partial V_{N1}}{\partial l_3} \end{bmatrix}$$

$$= \begin{bmatrix} -\begin{bmatrix} V_{N1} \\ V_{N2} \end{bmatrix}^T \begin{bmatrix} \coth(\gamma l_1) & -\operatorname{csch}(\gamma l_1) \\ -\operatorname{csch}(\gamma l_1) & \coth(\gamma l_1) \end{bmatrix} \begin{bmatrix} V_{N1} \\ V_{N2} \end{bmatrix} \\[4mm] -\begin{bmatrix} V_{N2} \\ V_{N3} \end{bmatrix}^T \begin{bmatrix} \coth(\gamma l_2) & -\operatorname{csch}(\gamma l_2) \\ -\operatorname{csch}(\gamma l_2) & \coth(\gamma l_2) \end{bmatrix} \begin{bmatrix} V_{N2} \\ V_{N3} \end{bmatrix} \\[4mm] -\begin{bmatrix} V_{N3} \\ V_{N4} \end{bmatrix}^T \begin{bmatrix} \coth(\gamma l_3) & -\operatorname{csch}(\gamma l_3) \\ -\operatorname{csch}(\gamma l_3) & \coth(\gamma l_3) \end{bmatrix} \begin{bmatrix} V_{N3} \\ V_{N4} \end{bmatrix} \\[4mm] -\dfrac{Y_{01}\gamma}{\sinh^2(\gamma l_1)} \begin{bmatrix} V_{N1} \\ V_{N2} \end{bmatrix}^T \begin{bmatrix} 1 & \cosh(\gamma l_1) \\ \cosh(\gamma l_1) & 1 \end{bmatrix} \begin{bmatrix} V_{N1} \\ V_{N2} \end{bmatrix} \\[4mm] -\dfrac{Y_{02}\gamma}{\sinh^2(\gamma l_2)} \begin{bmatrix} V_{N2} \\ V_{N3} \end{bmatrix}^T \begin{bmatrix} 1 & \cosh(\gamma l_2) \\ \cosh(\gamma l_2) & 1 \end{bmatrix} \begin{bmatrix} V_{N2} \\ V_{N3} \end{bmatrix} \\[4mm] -\dfrac{Y_{03}\gamma}{\sinh^2(\gamma l_3)} \begin{bmatrix} V_{N3} \\ V_{N4} \end{bmatrix}^T \begin{bmatrix} 1 & \cosh(\gamma l_3) \\ \cosh(\gamma l_3) & 1 \end{bmatrix} \begin{bmatrix} V_{N3} \\ V_{N4} \end{bmatrix} \end{bmatrix}$$

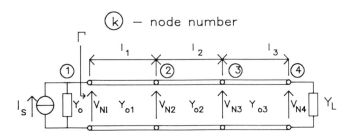

Figure 4.5 Three-section quarter-wave impedance transformer.

4.4 SENSITIVITY ANALYSIS OF MICROWAVE NETWORKS DESCRIBED BY THE SCATTERING MATRIX

4.4.1 The Adjoint Network Method

Consider a microwave network in which we assume a small deviation in parameter p. The deviation in p affects the characterization of a kth multiport corresponding to this parameter. If this element is characterized by the equation:

$$\mathbf{b}^{(k)} = \mathbf{S}^{(k)}\mathbf{a}^{(k)} \tag{4.47}$$

then

$$\frac{\partial \mathbf{b}^{(k)}}{\partial p} = \frac{\partial \mathbf{S}^{(k)}}{\partial p}\mathbf{a}^{(k)} + \mathbf{a}^{(k)}\frac{\partial \mathbf{a}^{(k)}}{\partial p} \tag{4.48}$$

The term related to the element under consideration in Tellegen's equation (4.23) (small changes $\Delta\mathbf{a}$ and $\Delta\mathbf{b}$ have been replaced by partial derivatives) can be written as

$$\boldsymbol{\beta}^{(k)T}\frac{\partial \mathbf{a}^{(k)}}{\partial p} - \boldsymbol{\alpha}^{(k)T}\frac{\partial \mathbf{b}^{(k)}}{\partial p} = \left(\boldsymbol{\beta}^{(k)} - \mathbf{S}^{(k)}\boldsymbol{\alpha}^{(k)}\right)^{T}\frac{\partial \mathbf{a}^{(k)}}{\partial p} - \boldsymbol{\alpha}^{(k)T}\frac{\partial S^{(k)}}{\partial p}\mathbf{a}^{(k)} \tag{4.49}$$

In this expression, $\boldsymbol{\alpha}^{(k)}$ and $\boldsymbol{\beta}^{(k)}$ are incoming and outgoing wave variables at ports of the kth element of the second network, which is topologically identical to the original network. If we assume that for the element under consideration in the second network:

$$\beta^{(k)} = \mathbf{S}^{(k)T}\alpha^{(k)} \tag{4.50}$$

the expression (4.49) reduces to

$$\beta^{(k)T}\frac{\partial \mathbf{a}^{(k)}}{\partial p} - \alpha^{(k)T}\frac{\partial \mathbf{b}^{(k)}}{\partial p} = -\alpha^{(k)T}\frac{\partial \mathbf{S}^{(k)}}{\partial p}\mathbf{a}^{(k)} \tag{4.51}$$

If all components in the second network are defined as in (4.50), this network described by α and β wave variables is called the *adjoint network*.

Considering the entire microwave network and using (4.51), Tellegen's equation (4.23) can be rewritten in the form

$$\beta_e^T\frac{\partial \mathbf{a}_e}{\partial p} - \alpha_e^T\frac{\partial \mathbf{b}_e}{\partial p} = +\sum_{i=1}^{m}\alpha^{(i)T}\frac{\partial \mathbf{S}^{(i)}}{\partial p}\mathbf{a}^{(i)} \tag{4.52}$$

where $\mathbf{a}_e$, $\mathbf{b}_e$, α_e, and β_e are vectors of incoming and outgoing wave variables at the load and generator ports, respectively, in the original and in the adjoint networks; and m is the number of elements of the network.

Normally, parameter p affects the characterization of no more than one element of the circuit. For other elements $\partial \mathbf{S}/\partial p$ equal zero. For such a case, we have

$$\beta_e^T\frac{\partial \mathbf{a}_e}{\partial p} - \alpha_e^T\frac{\partial \mathbf{b}_e}{\partial p} = \alpha^{(k)T}\frac{\partial \mathbf{S}^{(k)}}{\partial p}\mathbf{a}^{(k)} \tag{4.53}$$

Remember that if the change in p affects the characterization of more than one element, the sum in (4.52) must include all such circuit elements. In the group delay computation discussed in Section 4.4.7 sensitivities with respect to frequency are considered and the sum in the equation similar to (4.52) has to be taken over all frequency dependent elements in the network.

If we assume that the original network is matched at all the external ports, $\partial \mathbf{b}_e/\partial p = 0$ and (4.52) simplify to

$$G = \beta_e^T\frac{\partial \mathbf{a}_e}{\partial p} = \sum_{i=1}^{m}\alpha^{(i)T}\frac{\partial \mathbf{S}^{(i)}}{\partial p}\mathbf{a}^{(i)} \tag{4.54}$$

If the adjoint network is excited in such a way that all entries of the β_e vector are ones, then $\mathbf{G}$ in (4.54) is the sum of the incoming wave sensitivities at all output ports of the network.

In practice we are interested in the sensitivity of the incoming wave variable at the rth output port of the network. In such a case all output ports of the adjoint network have to be matched, and at the same time, the rth port of the adjoint network has to be excited to have $\beta_{er} = 1$. Figure 4.6 presents the original and adjoint networks for $\partial a_{er} / \partial p$ sensitivity computation. Equation (4.54) is now in the form

$$\frac{\partial a_{er}}{\partial p} = \sum_{i=1}^{m} \boldsymbol{\alpha}^{(i)T} \frac{\partial \mathbf{S}^{(i)}}{\partial p} \mathbf{a}^{(i)} \tag{4.55}$$

This equation can be written in the general form describing sensitivities of the a_{er} wave variable with respect to many circuit parameters. According to the earlier

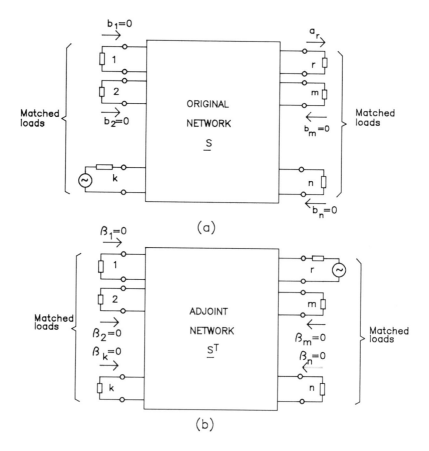

Figure 4.6 (a) Original network and (b) adjoint network for the first-order and second-order sensitivity computations.

discussion a vector of sensitivities of the wave variable a_{er} is given by

$$
\mathbf{G} = \nabla a_{er} =
\begin{bmatrix}
\dfrac{\partial a_{er}}{\partial p_1} \\[2mm]
\dfrac{\partial a_{er}}{\partial p_2} \\[2mm]
\vdots \\[2mm]
\dfrac{\partial a_{er}}{\partial p_n}
\end{bmatrix}
=
\begin{bmatrix}
\displaystyle\sum_{i=1}^{m} \alpha^{(i)} \dfrac{\partial \mathbf{S}^{(i)}}{\partial p_1} \mathbf{a}^{(i)} \\[4mm]
\displaystyle\sum_{i=1}^{m} \alpha^{(i)} \dfrac{\partial \mathbf{S}^{(i)}}{\partial p_2} \mathbf{a}^{(i)} \\[4mm]
\vdots \\[4mm]
\displaystyle\sum_{i=1}^{m} \alpha^{(i)} \dfrac{\partial \mathbf{S}^{(i)}}{\partial p_n} \mathbf{a}^{(i)}
\end{bmatrix}
\qquad (4.56)
$$

The sensitivities with respect to all circuit parameters can be determined from two analyses: one of the original network and the other of the adjoint network all of whose elements have scattering matrices equal to the transposed scattering matrices of the corresponding elements of the original network. According to (4.55) and (4.56) to evaluate the sensitivities, we need to know the partial derivatives of the element scattering matrices.

4.4.2 Sensitivity Invariants of Scattering Matrices and Their Use for Evaluation of Differential Scattering Matrices

As has been proved, some sensitivity invariants of $\mathbf{S}$ matrices exist for networks and elements composed of lumped resistors, inductors and capacitors, gyrators, uniform transmission lines, uniformly distributed RC lines, and either current controlled voltage sources or voltage controlled current sources [8]. The sensitivity invariants for $\mathbf{S}$ matrices have been obtained by first finding the sensitivity invariants for $\mathbf{Z}$ matrices with respect to the impedance-based parameters of a network. For individual elements of a network these parameters are defined as follows: R for lumped resistors, $L = 1/\Gamma$ for lumped inductors, $D = 1/C$ for lumped capacitors, gyrator ratio $\alpha = -1/\gamma$ for gyrators, characteristic impedances $Z_0 = 1/Y_0$ for uniform transmission lines, total series resistances $R_{RC} = 1/G_{RC}$, and total shunt capacitances $D_{RC} = 1/C_{RC}$ for uniformly distributed RC lines, and transfer resistances r_m for current-controlled voltage sources. Thus, the set of impedance-based parameters describing the elements of the network is

$$
\{z\} = \{R, L, D, Z_0, R_{RC}, D_{RC}, r_m\} \qquad (4.57)
$$

For a network composed of the elements just defined and for the adjoint network in which all elements have impedance matrices equal to the transpositions of the impedance matrices of the original elements, the Tellegen's equation is

$$\mathbf{V}_c^T \hat{\mathbf{I}}_c = \mathbf{V}_e^T \hat{\mathbf{I}}_e \tag{4.58}$$

where the $\mathbf{V}_c$ and $\mathbf{V}_e$ are the vectors of voltages across, respectively, the internal and external ports in the original network; and $\hat{\mathbf{I}}_c$ and $\hat{\mathbf{I}}_e$ are the currents flowing into the corresponding ports for the adjoint network. From the results presented in [9] we know that the sensitivity of the (i, j) element of the overall impedance matrix $\mathbf{Z}$ with respect to parameter p_k contained in an element with the impedance matrix $\mathbf{Z}^e$ is

$$\frac{\partial Z_{ij}}{\partial p_k} = \hat{\mathbf{I}}_B^T \frac{\partial \mathbf{Z}^e}{\partial p_k} \mathbf{I}_B \tag{4.59}$$

where $\mathbf{I}_B$ and $\hat{\mathbf{I}}_B$ contain currents flowing into the element ports in the original and adjoint networks, respectively. Equation (4.59) assumes that the original network is excited by a unit current source at port j, and the adjoint network is excited by a unit current source at port i, with other ports left open-circuited in both networks.

If p_k is one of impedance parameters listed in (4.57) (except R_{RC} or D_{RC} because in this case two parameters describe an element), and if p_k belongs to $\mathbf{Z}^e$,

$$p_k \frac{\partial \mathbf{Z}^e}{\partial p_k} = \mathbf{Z}^e \tag{4.60}$$

or if p_k does not belong to $\mathbf{Z}^e$,

$$p_k \frac{\partial \mathbf{Z}^e}{\partial p_k} = 0 \tag{4.61}$$

For the RC line equivalent to (4.60), the equation has the form:

$$R_{RC} \frac{\partial \mathbf{Z}_{RC}}{\partial R_{RC}} + D_{RC} \frac{\partial \mathbf{Z}_{RC}}{\partial D_{RC}} = \mathbf{Z}_{RC} \tag{4.62}$$

Equations (4.60) and (4.62) indicate that the impedance matrices are homogeneous of degree 1.

From (4.59), (4.60), and (4.61), if p_k is contained in $\mathbf{Z}^e$,

$$p_k \frac{\partial Z_{ij}}{\partial p} = p_k \hat{\mathbf{I}}_B^T \frac{\partial \mathbf{Z}^e}{\partial p_k} \mathbf{I}_B = \hat{\mathbf{I}}_B^T \mathbf{Z}^e \mathbf{I}_B = \hat{\mathbf{I}}_B^T \mathbf{V}_B \tag{4.63}$$

and for the RC lines:

$$R_{RC} \frac{\partial Z_{ij}}{\partial R_{RC}} + D_{RC} \frac{\partial Z_{ij}}{\partial D_{RC}} = R_{RC} \hat{\mathbf{I}}_B^T \frac{\partial \mathbf{Z}_{RC}}{\partial R_{RC}} \mathbf{I}_B + D_{RC} \hat{\mathbf{I}}_B^T \frac{\partial \mathbf{Z}_{RC}}{\partial D_{RC}} \mathbf{I}_B$$

$$= \hat{\mathbf{I}}_B^T \mathbf{Z}_{RC} \mathbf{I}_B = \hat{\mathbf{I}}_B^T \mathbf{V}_B \tag{4.64}$$

We can see from (4.63) and (4.64) that the sum over all the parameters p_k in the set $\{z\}$ given by (4.57) is equal to

$$\sum_{\{p_k\}} p_k \frac{\partial Z_{ij}}{\partial p_k} = \sum_c V_c \hat{I}_c \tag{4.65}$$

or, using (4.58),

$$\sum_{\{p_k\}} p_k \frac{\partial Z_{ij}}{\partial p_k} = \sum_e V_e \hat{I}_e \tag{4.66}$$

Because the adjoint network is excited by a unit current source at the external port i and all other external ports are left open, the right-hand side of (4.66) is equal to V_i. Because

$$\sum_e V_e \hat{I}_e = V_i = Z_{ij} I_j = Z_{ij} \tag{4.67}$$

and I_j is unity in the original network, we find that

$$\sum_{\{p_k\}} p_k \frac{\partial Z_{ij}}{\partial p_k} = Z_{ij} \tag{4.68}$$

Because the choice of (i, j) indices here is arbitrary, we can conclude that

$$\sum_{\{p_k\}} p_k \frac{\partial Z_e}{\partial p_k} = \mathbf{Z} \tag{4.69}$$

From (4.68), we thus have

$$\sum_{\{p_k\}} \frac{p_k}{Z_{ij}} \frac{\partial Z_{ij}}{\partial p_k} = 1 \tag{4.70}$$

The sum of normalized sensitivities of Z_{ij} is unity, independent of Z_{ij}. As the preceding derivation shows, if the circuit element impedance matrices are homogeneous of degree 1 in impedance parameters, then the overall impedance matrix exhibits the same property.

To obtain the sensitivity invariants of the $\mathbf{S}$ matrix elements with respect to the impedance parameters $\{z\}$, we have to differentiate the formula (1.90) defining the $\mathbf{S}$

matrix in terms of the $\mathbf{Z}$ matrix. This yields

$$\frac{\partial \mathbf{S}}{\partial p_k} = \mathbf{F}(\mathbf{G} + \mathbf{G}^+)(\mathbf{Z} + \mathbf{G})^{-1} \frac{\partial \mathbf{Z}}{\partial p_k}(\mathbf{Z} + \mathbf{G})^{-1}\mathbf{F}^{-1} \qquad (4.71)$$

Taking the sum over all p_k in $\{z\}$, we obtain

$$\sum_{\{p_k\}} p_k \frac{\partial \mathbf{S}}{\partial p_k} = \mathbf{F}(\mathbf{G} + \mathbf{G}^+)(\mathbf{Z} + \mathbf{G})^{-1} \sum_{\{p_k\}} p_k \frac{\partial \mathbf{S}}{\partial p_k}(\mathbf{Z} + \mathbf{G})^{-1}\mathbf{F}^{-1} \qquad (4.72)$$

and using (4.60)

$$\sum_{\{p_k\}} p_k \frac{\partial \mathbf{S}}{\partial p_k} = \mathbf{F}(\mathbf{G} + \mathbf{G}^+)(\mathbf{Z} + \mathbf{Z})^{-1}\mathbf{Z}(\mathbf{Z} + \mathbf{G})^{-1}\mathbf{F}^{-1}$$

$$= \frac{1}{2}(\mathbf{I} - \mathbf{S} \cdot \mathbf{S}) \qquad (4.73)$$

where $\mathbf{I}$ is the unity matrix. We should mention here that the normalizing impedances at the ports of $\mathbf{S}$ are not included in $\{p_k\}$. Parameter p_k could be the corresponding admittance parameter defined as

$$\{y\} = \{G, 1/L, C, \gamma, Y_0, G_{RC}, C_{RC}, g_m\} \qquad (4.74)$$

In this case,

$$\sum_{\{p_k\}} p_k \frac{\partial \mathbf{S}}{\partial p_k} = -\frac{1}{2}(\mathbf{I} - \mathbf{S} \cdot \mathbf{S}) \qquad (4.75)$$

The invariants just derived also can be used for individual elements of microwave circuits. In the case of circuit elements containing one impedance, or its dual admittance parameter, from (4.73) and (4.75), we have

$$\frac{\partial \mathbf{S}}{\partial p} = \frac{1}{2p}(\mathbf{I} - \mathbf{S} \cdot \mathbf{S}) \qquad (4.76)$$

if p is a parameter from $\{z\}$, and

$$\frac{\partial \mathbf{S}}{\partial p} = -\frac{1}{2p}(\mathbf{I} - \mathbf{S} \cdot \mathbf{S}) \qquad (4.77)$$

if p is a parameter from $\{y\}$.

By using (4.76) and (4.77), we can simplify sensitivity computation of circuit elements containing only one parameter of $\{z\}$ or its dual $\{y\}$.

For this kind of element, we have

$$\alpha^T \frac{\partial S}{\partial p} \mathbf{a} = \frac{1}{2p} (\alpha^T \mathbf{a} - \beta^T \mathbf{b}) \tag{4.78}$$

if p is a parameter from $\{z\}$, and

$$\alpha^T \frac{\partial S}{\partial p} \mathbf{a} = -\frac{1}{2p} (\alpha^T \mathbf{a} - \beta^T \mathbf{b}) \tag{4.79}$$

if p is a parameter from $\{y\}$.

The right-hand sides of these expressions do not contain derivatives of S matrices, and they are simple to compute. Of course, the relations (4.76) and (4.77) can be used directly for sensitivity computation by (4.56).

Example 4.2

Consider the circuit element presented in Figure 4.7. It is a series *RLC* network interposed in series between the input and the output port. The S matrix of this element, taken from Table 1.2, is

$$S = \frac{1}{Z + 2Z_N} \begin{bmatrix} Z & 2Z_N \\ 2Z_N & Z \end{bmatrix}$$

where $Z = R + j[\omega L - (1/\omega C)]$.

Its sensitivity with respect to Z impedance (which is a single impedance parameter of this element) can be obtained as

$$\frac{\partial S}{\partial Z} = \frac{1}{2Z} (\mathbf{I} - S \cdot S) = \frac{2Z_N}{(Z + 2Z_N)^2} \begin{bmatrix} 1 & -1 \\ -1 & 1 \end{bmatrix}$$

We can obtain the same result through a direct differentiation. By using the above

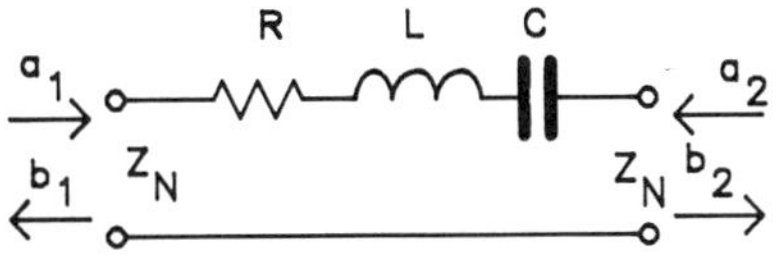

Figure 4.7 Series *RLC* network as a two-port with the same reference impedances Z_N at both ports.

relation, we also have

$$\frac{\partial S}{\partial R} = \frac{\partial S}{\partial Z}\frac{\partial Z}{\partial R} = \frac{\partial S}{\partial Z} \cdot 1 = \frac{2Z_N}{(Z + 2Z_N)^2}\begin{bmatrix} 1 & -1 \\ -1 & 1 \end{bmatrix}$$

$$\frac{\partial S}{\partial L} = \frac{\partial S}{\partial Z}\frac{\partial Z}{\partial L} = \frac{\partial S}{\partial Z} \cdot j\omega = \frac{j\omega 2 Z_N}{(Z + 2Z_N)^2}\begin{bmatrix} 1 & -1 \\ -1 & 1 \end{bmatrix}$$

$$\frac{\partial S}{\partial C} = \frac{\partial S}{\partial Z}\frac{\partial Z}{\partial C} = \frac{\partial S}{\partial Z} \cdot \frac{j}{\omega C^2} = \frac{j 2 Z_N}{\omega C^2(Z + 2Z_N)^2}\begin{bmatrix} 1 & -1 \\ -1 & 1 \end{bmatrix}$$

These relations may be used for sensitivity computation of wave variables. The elements of **G** in (4.56) related to the discussed element are

$$\frac{\partial b_e}{\partial Z} = \alpha\frac{\partial S}{\partial Z}\mathbf{a} = \frac{2Z_N}{(Z + 2Z_N)^2}\begin{bmatrix} \alpha_1 \\ \alpha_2 \end{bmatrix}^T\begin{bmatrix} 1 & -1 \\ -1 & 1 \end{bmatrix}\begin{bmatrix} a_1 \\ a_2 \end{bmatrix}$$

$$= \frac{2Z_N}{(Z + 2Z_N)^2}(\alpha_1 - \alpha_2)(a_1 - a_2)$$

$$\frac{\partial b_e}{\partial R} = \frac{\partial b_e}{\partial Z}\frac{\partial Z}{\partial R} = \frac{2Z_N}{(Z + 2Z_N)^2}(\alpha_1 - \alpha_2)(a_1 - a_2)$$

$$\frac{\partial b_e}{\partial L} = \frac{\partial b_e}{\partial Z}\frac{\partial Z}{\partial L} = \frac{j 2 \omega Z_N}{(Z + 2Z_N)^2}(\alpha_1 - \alpha_2)(a_1 - a_2)$$

$$\frac{\partial b_e}{\partial C} = \frac{\partial b_e}{\partial Z}\frac{\partial Z}{\partial C} = \frac{j 2 Z_N}{\omega C^2(Z + 2Z_N)^2}(\alpha_1 - \alpha_2)(a_1 - a_2)$$

We obtain equivalent results by using relation (4.78). We now have

$$\frac{\partial b_e}{\partial Z} = \frac{1}{2Z}(\alpha^T\mathbf{a} - \beta^T\mathbf{b}) = \frac{1}{2Z}\left(\begin{bmatrix} \alpha_1 \\ \alpha_2 \end{bmatrix}^T\begin{bmatrix} a_1 \\ a_2 \end{bmatrix} - \begin{bmatrix} \beta_1 \\ \beta_2 \end{bmatrix}^T\begin{bmatrix} b_1 \\ b_2 \end{bmatrix}\right)$$

$$\frac{\partial b_e}{\partial R} = \frac{\partial b_e}{\partial Z}\frac{\partial Z}{\partial R} = \frac{1}{2Z}\left(\begin{bmatrix} \alpha_1 \\ \alpha_2 \end{bmatrix}^T\begin{bmatrix} a_1 \\ a_2 \end{bmatrix} - \begin{bmatrix} \beta_1 \\ \beta_2 \end{bmatrix}^T\begin{bmatrix} b_1 \\ b_2 \end{bmatrix}\right)$$

$$\frac{\partial b_e}{\partial L} = \frac{\partial b_e}{\partial Z}\frac{\partial Z}{\partial L} = \frac{j\omega}{2Z}\left(\begin{bmatrix} \alpha_1 \\ \alpha_2 \end{bmatrix}^T\begin{bmatrix} a_1 \\ a_2 \end{bmatrix} - \begin{bmatrix} \beta_1 \\ \beta_2 \end{bmatrix}^T\begin{bmatrix} b_1 \\ b_2 \end{bmatrix}\right)$$

$$\frac{\partial b_e}{\partial C} = \frac{\partial b_e}{\partial Z}\frac{\partial Z}{\partial C} = \frac{j}{2Z\omega C^2}\left(\begin{bmatrix} \alpha_1 \\ \alpha_2 \end{bmatrix}^T\begin{bmatrix} a_1 \\ a_2 \end{bmatrix} - \begin{bmatrix} \beta_1 \\ \beta_2 \end{bmatrix}^T\begin{bmatrix} b_1 \\ b_2 \end{bmatrix}\right)$$

In these relations, a_1, a_2, b_1, b_2, and α_1, α_2, β_1, β_2, are respectively incoming and outgoing wave variables at both ports of the element discussed in the original and adjoint networks.

4.4.3 The Transposed Matrix Method for Networks Described by the Connection Scattering Matrix

A computational method for determining sensitivities of microwave networks described by the connection scattering matrix is based on considerations in [10–12]. By differentiating the equation $\mathbf{W}\mathbf{a} = \mathbf{c}$ (3.60) with respect to a parameter p, we have

$$\frac{\partial \mathbf{a}}{\partial p} = -\mathbf{W}^{-1}\frac{\partial \mathbf{W}}{\partial p}\mathbf{W}^{-1}\mathbf{c} = \mathbf{W}^{-1}\frac{\partial \mathbf{S}}{\partial p}\mathbf{a} \tag{4.80}$$

In (3.60), we assume $\partial \mathbf{c}/\partial p = 0$.

The sensitivity of the wave variable incoming at the rth port of the network may be computed by the left-hand-side multiplication of (4.80) by a row vector $\boldsymbol{\gamma}_r^T$, where all elements are equal to zero except for the rth element that equals one:

$$\boldsymbol{\gamma}_r^T = [0, 0, \ldots, 0, 1, 0, \ldots, 0] \tag{4.81}$$

By performing this operation, we have

$$\frac{\partial a_r}{\partial p} = \boldsymbol{\gamma}_r^T \frac{\partial \mathbf{a}}{\partial p} = \boldsymbol{\gamma}_r^T \mathbf{W}^{-1}\frac{\partial \mathbf{S}}{\partial p}\mathbf{a} = \left[(\mathbf{W}^T)^{-1}\boldsymbol{\gamma}_r\right]^T\frac{\partial \mathbf{S}}{\partial p}\mathbf{a}$$

$$= \boldsymbol{\alpha}^T\frac{\partial \mathbf{S}}{\partial p}\mathbf{a} \tag{4.82}$$

where

$$\boldsymbol{\alpha} = (\mathbf{W}^T)^{-1}\boldsymbol{\gamma}_r = \hat{\mathbf{W}}^{-1}\boldsymbol{\gamma}_r \tag{4.83}$$

is a solution vector of the transposed or adjoint matrix equation:

$$\mathbf{W}^T\boldsymbol{\alpha} = \boldsymbol{\gamma}_r \tag{4.84}$$

The preceding matrix equation may be treated as the equation of the adjoint network excited at the rth port by a generator with an impressed wave equal to 1. According to (4.84) the connection scattering matrix $\hat{\mathbf{W}}$ of the adjoint network must be equal to the transposed connection scattering matrix $\mathbf{W}$ of the original network.

Because

$$\hat{\mathbf{W}} = \mathbf{W}^T = (\boldsymbol{\Gamma} - \mathbf{S})^T = \boldsymbol{\Gamma} - \mathbf{S}^T \tag{4.85}$$

then topologies of the original and adjoint networks are the same, whereas the elements of the adjoint network have the scattering matrices $\hat{S}^{(1)}, \hat{S}^{(2)}, \ldots, \hat{S}^{(m)}$ equal to the transposed scattering matrices of the corresponding elements of the original network. Equation (4.84) is equivalent to (4.50). The equivalence between the transposed matrix method and the adjoint network method then is evident.

The general relation (4.82) defining sensitivities of the wave variable incoming at the rth port with respect to many network parameters has a form analogous to (4.34):

$$
G = \nabla a_r =
\begin{bmatrix}
\alpha^T \dfrac{\partial S}{\partial p_1} a \\[2mm]
\alpha^T \dfrac{\partial S}{\partial p_2} a \\[2mm]
\vdots \\[2mm]
\alpha^T \dfrac{\partial S}{\partial p_n} a
\end{bmatrix}
=
\begin{bmatrix}
\sum\limits_{k \in E_1} \alpha^{(k)T} \dfrac{\partial S^{(k)}}{\partial p_1} a^{(k)} \\[2mm]
\sum\limits_{k \in E_2} \alpha^{(k)T} \dfrac{\partial S^{(k)}}{\partial p_2} a^{(k)} \\[2mm]
\vdots \\[2mm]
\sum\limits_{k \in E_n} \alpha^{(k)T} \dfrac{\partial S^{(k)}}{\partial p_n} a^{(k)}
\end{bmatrix}
\tag{4.86}
$$

where $E_1, E_2, \ldots, E_n$ represent sets of circuit elements that depend respectively on parameters p_1, p_2, et cetera.

For each parameter p_i the matrix $\partial S / \partial p_i$ will be formed, and the products indicated on the right-hand side of (4.86) evaluated. As the vectors a and α are independent of the parameter index i, we see that the application of (4.86) requires solving only two sets of linear equations, (3.60) and (4.84), irrespective of the number of parameters p_i.

The method discussed may be used effectively for the computation of S parameter sensitivities with respect to any independent variable parameter p.

Let us assume that we analyze a network for which we wish to compute the transfer function $S_{ij} = |S_{ij}| e(j\phi_{ij})$ and then evaluate its sensitivities. If the jth port of the circuit is excited by a matched generator with impressed wave $c_g = 1$ and all external ports are terminated in matched loads, then for $i \neq j$, the transfer function S_{ij} is given by

$$
S_{ij} = b_i |_{a_j = 1} = a_l
\tag{4.87}
$$

In (4.87), a_l is the incoming wave variable at the lth port of the matched load terminating the ith port of the network. The condition $a_j = 1$ is imposed by the generator connected to port j.

Now, if the adjoint network with $\hat{W} = W^T$ is excited at port l by a matched generator with impressed wave $\gamma_l = 1$ ($\alpha_i = 1$) and because $\partial S_{ij} / \partial p = \partial a_l / \partial p$, we obtain

$$
\frac{\partial S_{ij}}{\partial p} = \alpha^T \frac{\partial S}{\partial p} a = \sum_{k \in E} \alpha^{(k)T} \frac{\partial S^{(k)}}{\partial p} a^{(k)}
\tag{4.88}
$$

The sensitivities of any S_{ij} are determined by exciting suitably ($a_j = \alpha_i = 1$) the original and the adjoint network. In (4.88), the summation is taken over all circuit elements that are dependent on the parameter p.

4.4.4 Derivation of Sensitivities of Microwave Circuits Described by the Scattering Matrices

Table 4.2 presents the results of applying the formula (4.82) to a number of commonly used elements of microwave circuits. The table forms a library of relations that may be applied for the solution of certain computer-aided design problems.

4.4.5 The Sensitivity Analysis Direct Method for Networks Described by the Connection Scattering Matrix

To evaluate the sensitivities of incoming wave variables at all network ports with respect to a single variable parameter p, we differentiate the equation $\mathbf{Wa} = \mathbf{c}$ (3.60) with respect to parameter p, obtaining

$$\frac{\partial \mathbf{W}}{\partial p} \mathbf{a} + \mathbf{W} \frac{\partial \mathbf{a}}{\partial p} = 0 \tag{4.89}$$

and rewriting as follows

$$\frac{\partial \mathbf{a}}{\partial p} = -\mathbf{W}^{-1} \frac{\partial \mathbf{W}}{\partial p} \mathbf{a} = \mathbf{W}^{-1} \frac{\partial \mathbf{S}}{\partial p} \mathbf{a} \tag{4.90}$$

or

$$\mathbf{W} \frac{\partial \mathbf{a}}{\partial p} = \delta \tag{4.91}$$

where

$$\delta = \frac{\partial \mathbf{S}}{\partial p} \mathbf{a} = \begin{bmatrix} \dfrac{\partial \mathbf{S}^{(1)}}{\partial p} \mathbf{a}^{(1)} \\[2ex] \dfrac{\partial \mathbf{S}^{(2)}}{\partial p} \mathbf{a}^{(2)} \\[1ex] \vdots \\[1ex] \dfrac{\partial \mathbf{S}^{(m)}}{\partial p} \mathbf{a}^{(m)} \end{bmatrix} \tag{4.92}$$

In (4.92), $\mathbf{S}^{(1)}, \mathbf{S}^{(2)}, \ldots, \mathbf{S}^{(m)}$ are scattering matrices of individual circuit elements, and $\mathbf{a}^{(1)}, \mathbf{a}^{(2)}, \ldots, \mathbf{a}^{(m)}$ are vectors of incoming wave variables related to these circuit elements.

Table 4.2
Sensitivity Expressions for Some Lumped and Distributed Elements of Microwave Circuits

Circuit Element	Sensitivity (component of **G**, see (4.56))	Variable Parameter p
Z_N — Z	$\dfrac{1}{2Z}(\alpha a - \beta b)$	Z
Z_N, C, L, R $Z = R + j\left(\omega L - \dfrac{1}{\omega C}\right)$	$\dfrac{1}{2Z}(\alpha a - \beta b)$	R
	$\dfrac{j\omega}{2Z}(\alpha a - \beta b)$	L
	$\dfrac{j}{2\omega C^2 Z}(\alpha a - \beta b)$	C
Z_N — Y	$-\dfrac{1}{2Y}(\alpha a - \beta b)$	Y
G, L, C $Y = G + j\left(\omega C - \dfrac{1}{\omega L}\right)$	$\dfrac{-1}{2Y}(\alpha a - \beta b)$	G
	$-\dfrac{j\omega}{2Y}(\alpha a - \beta b)$	C
	$-\dfrac{j}{2\omega L^2 Y}(\alpha a - \beta b)$	L
Z, Z_N, Z_N	$\dfrac{1}{2Z}(\boldsymbol{\alpha}^T \mathbf{a} - \boldsymbol{\beta}\mathbf{b})$	Z
R L C, Z_N, Z_N $Z = R + j\left(\omega L - \dfrac{1}{\omega C}\right)$	$\dfrac{1}{2Z}(\boldsymbol{\alpha}^T \mathbf{a} - \boldsymbol{\beta}^T \mathbf{b})$	R
	$\dfrac{j\omega}{2Z}(\boldsymbol{\alpha}^T \mathbf{a} - \boldsymbol{\beta}^T \mathbf{b})$	L
	$\dfrac{j}{2\omega C^2 Z}(\boldsymbol{\alpha}^T \mathbf{a} - \boldsymbol{\beta}^T \mathbf{b})$	C

Table 4.2 (*Continued*)

Circuit Element	Sensitivity (*component of* **G**, *see* (4.56))	Variable Parameter p
Z_N — (parallel C, L, G) — Z_N $Y = G + j\left(\omega C - \dfrac{1}{\omega L}\right)$	$-\dfrac{1}{2Y}(\boldsymbol{\alpha}^T\mathbf{a} - \boldsymbol{\beta}^T\mathbf{b})$	G
	$i - \dfrac{j\omega}{2Y}(\boldsymbol{\alpha}^T\mathbf{a} - \boldsymbol{\beta}^T\mathbf{b})$	C
	$-\dfrac{j}{2\omega L^2 Y}(\boldsymbol{\alpha}^T\mathbf{a} - \boldsymbol{\beta}^T\mathbf{b})$	L
Z_N — Y — Z_N	$\dfrac{-1}{2Y}(\boldsymbol{\alpha}^T\mathbf{a} - \boldsymbol{\beta}^T\mathbf{b})$	Y
Z_N — (G, L, C) — Z_N $Y = G + j\left(\omega L - \dfrac{1}{\omega C}\right)$	$-\dfrac{1}{2Y}(\boldsymbol{\alpha}^T\mathbf{a} - \boldsymbol{\beta}^T\mathbf{b})$	G
	$-\dfrac{j\omega}{2Y}(\boldsymbol{\alpha}^T\mathbf{a} - \boldsymbol{\beta}^T\mathbf{b})$	C
	$-\dfrac{j}{2\omega L^2 Y}(\boldsymbol{\alpha}^T\mathbf{a} - \boldsymbol{\beta}^T\mathbf{b})$	L
Z_N — (C, L, R) — Z_N $Z = R + j\left(\omega L - \dfrac{1}{\omega C}\right)$	$\dfrac{1}{2Z}(\boldsymbol{\alpha}^T\mathbf{a} - \boldsymbol{\beta}^T\mathbf{b})$	R
	$\dfrac{j\omega}{2Z}(\boldsymbol{\alpha}^T\mathbf{a} - \boldsymbol{\beta}^T\mathbf{b})$	L
	$\dfrac{j}{2\omega C^2 Z}(\boldsymbol{\alpha}^T\mathbf{a} - \boldsymbol{\beta}^T\mathbf{b})$	C
Z_N — Z_0 — Z_N $\gamma = \alpha + j\beta$ A transmission line section	$\dfrac{1}{2Z_0}(\boldsymbol{\alpha}^T\mathbf{a} - \boldsymbol{\beta}^T\mathbf{b})$	Z_0
	$-\dfrac{Z_0\gamma}{2Z_N\sinh^2(\gamma l)}(\boldsymbol{\alpha} - \boldsymbol{\beta})^T \cdot \begin{bmatrix} 1 & \cosh(\gamma l) \\ \cosh(\gamma l) & 1 \end{bmatrix}(\mathbf{a} - \mathbf{b})$	l
	$-\dfrac{1}{2Y_0}(\boldsymbol{\alpha}^T\mathbf{a} - \boldsymbol{\beta}^T\mathbf{b})$	Y_0

Table 4.2 (*Continued*)

Circuit Element	Sensitivity (*component of* **G**, *see* (4.56))	Variable Parameter p
  Open-circuited parallel stub $\gamma = \alpha + j\beta$	$\dfrac{1}{2Z_0}(\boldsymbol{\alpha}^T\mathbf{a} - \boldsymbol{\beta}^T\mathbf{b})$	Z_0
	$-\dfrac{Z_0\gamma}{2Z_N \sinh^2(\gamma l)}(\boldsymbol{\alpha}-\boldsymbol{\beta})^T$ $\cdot\begin{bmatrix}1 & 1\\1 & 1\end{bmatrix}(\mathbf{a}-\mathbf{b})$	l
	$-\dfrac{1}{2Y_0}(\boldsymbol{\alpha}^T\mathbf{a}-\boldsymbol{\beta}^T\mathbf{b})$	Y_0
  $\gamma = j\beta$ Lossless open-circuited parallel stub	$\dfrac{1}{2Z_0}(\boldsymbol{\alpha}^T\mathbf{a}-\boldsymbol{\beta}^T\mathbf{b})$	Z_0
	$\dfrac{jZ_0\beta}{2Z_N \sin^2(\beta l)}(\boldsymbol{\alpha}-\boldsymbol{\beta})^T$ $\cdot\begin{bmatrix}1 & 1\\1 & 1\end{bmatrix}(\mathbf{a}-\mathbf{b})$	l
	$-\dfrac{1}{2Y_0}(\boldsymbol{\alpha}^T\mathbf{a}-\boldsymbol{\beta}^T\mathbf{b})$	Y_0
  $\gamma = j\beta$ Lossless transmission line section	$\dfrac{1}{2Z_0}(\boldsymbol{\alpha}\mathbf{a}-\boldsymbol{\beta}\mathbf{b})$	Z_0
	$\dfrac{jZ_0\beta}{2Z_N \sin^2(\beta l)}(\boldsymbol{\alpha}-\boldsymbol{\beta})^T$ $\cdot\begin{bmatrix}1 & \cos(\beta l)\\\cos(\beta l) & 1\end{bmatrix}(\mathbf{a}-\mathbf{b})$	l
	$-\dfrac{1}{2Y_0}(\boldsymbol{\alpha}^T\mathbf{a}-\boldsymbol{\beta}^T\mathbf{b})$	Y_0

Table 4.2 (*Continued*)

Circuit Element	Sensitivity (*component of* $\mathbf{G}$, *see* (4.56))	Variable Parameter p
Z_N Z_N Z_0 l Short-circuited parallel stub $\gamma = \alpha + j\beta$	$-\dfrac{1}{2Z_0}(\boldsymbol{\alpha}^T\mathbf{a} - \boldsymbol{\beta}^T\mathbf{b})$	Z_0
	$-\dfrac{Z_0\gamma}{2Z_N\cosh^2(\gamma l)}(\boldsymbol{\alpha} - \boldsymbol{\beta})^T$ $\cdot\begin{bmatrix}1 & 1\\1 & 1\end{bmatrix}(\mathbf{a} - \mathbf{b})$	l
	$-\dfrac{1}{2Y_0}(\boldsymbol{\alpha}^T\mathbf{a} - \boldsymbol{\beta}^T\mathbf{b})$	Y_0
Z_N Z_N Z_0 l $Z_0, \gamma = j\beta$ Lossless short-circuited parallel stub	$\dfrac{1}{2Z_0}(\boldsymbol{\alpha}^T\mathbf{a} - \boldsymbol{\beta}^T\mathbf{b})$	Z_0
	$\dfrac{jZ_0\beta}{2Z_N\cos^2(\beta l)}(\boldsymbol{\alpha} - \boldsymbol{\beta})^T$ $\cdot\begin{bmatrix}1 & 1\\1 & 1\end{bmatrix}(\mathbf{a} - \mathbf{b})$	l
	$-\dfrac{1}{2Y_0}(\boldsymbol{\alpha}^T\mathbf{a} - \boldsymbol{\beta}^T\mathbf{b})$	Y_0
l Z_N Z_0 $\gamma = \alpha + j\beta$ Short-circuited transmission line section	$\dfrac{1}{2Z_0}(\alpha a - \beta b)$	Z_0
	$\dfrac{Z_0\gamma}{2Z_N\cosh^2(\gamma l)}(\alpha - \beta)(a - b)$	l
	$-\dfrac{1}{2Y_0}(\alpha a - \beta b)$	Y_0

Table 4.2 (*Continued*)

Circuit Element	Sensitivity (*component of* **G**, *see* (4.56))	Variable Parameter p
$\gamma = j\beta$ A lossless short-circuited transmission line section	$\dfrac{1}{2Z_0}(\alpha a - \beta b)$	Z_0
	$\dfrac{jZ_0\beta}{2Z_N\cos^2(\beta l)}(\alpha - \beta)(a - b)$	l
	$-\dfrac{1}{2Y_0}(\alpha a - \beta b)$	Y_0
$\gamma = \alpha + j\beta$ Open-circuited transmission line section	$\dfrac{1}{2Z_0}(\alpha a - \beta b)$	Z_0
	$\dfrac{-Z_0\gamma}{2Z_N\sinh^2(\gamma l)}(\alpha - \beta)(a - b)$	l
	$-\dfrac{1}{2Y_0}(\alpha a - \beta b)$	Y_0
$\gamma = j\beta$ Lossless open-circuited transmission line section	$\dfrac{1}{2Z_0}(\alpha a - \beta b)$	Z_0
	$\dfrac{jZ_0\beta}{2Z_N\sin^2(\beta l)}(\alpha - \beta)(a - b)$	l
	$-\dfrac{1}{2Y_0}(\alpha a - \beta b)$	Y_0

Table 4.2 (*Continued*)

Circuit Element	Sensitivity (*component of* **G**, *see* (4.56))	Variable Parameter p
 Voltage controlled current source	$-2Z_N\alpha_2 a_1$	g_m
 Voltage controlled voltage source	$2\alpha_2 a_1$	k
 Current controlled current source	$-2\alpha_2 a_1$	β
 Current controlled voltage source	$\dfrac{2}{Z_N}\alpha_2 a_1$	r_m
 Ideal transformer	$\dfrac{2}{1+n^2}\alpha^T\begin{bmatrix} 0 & 1 \\ -1 & 0 \end{bmatrix}\mathbf{b}$	n
 Gyrator	$\dfrac{2Z_N}{g^2+Z_N^2}\mathbf{b}^T\begin{bmatrix} 0 & -1 \\ 1 & 0 \end{bmatrix}\alpha$	g

Table 4.2 (*Continued*)

Circuit Element	Sensitivity (*component of* **G**, *see* (4.56))	Variable Parameter p
Ideal directional coupler $c = \dfrac{b_3}{a_2}$	$\dfrac{1}{\sqrt{1 - c^2}}\left(\mathbf{b}^T \mathbf{1}'\alpha + \beta \mathbf{1}'\mathbf{a}\right)$ where $\mathbf{1}' = \begin{bmatrix} 0 & 0 & 1 \\ 0 & 0 & 0 \\ -1 & 0 & 0 \end{bmatrix}$	c
$c = \dfrac{b_4}{a_1}$ Ideal directional coupler	$\dfrac{-j}{\sqrt{1 - c^2}}\mathbf{b}^T \begin{bmatrix} 0 & 0 & 1 & 0 \\ 0 & 0 & 0 & 1 \\ 1 & 0 & 0 & 0 \\ 0 & 1 & 0 & 0 \end{bmatrix}\alpha$	c

According to (4.91) and (4.92) to compute $\partial\mathbf{a}/\partial p$ we first must solve (3.60) $\mathbf{Wa} = \mathbf{c}$ and find vector $\mathbf{a}$. Thus the right-hand vector $\boldsymbol{\delta}$ of the matrix equation (4.91) may be generated using (4.92). The solution of (4.91) is the sensitivity of the whole vector $\mathbf{a}$ with respect to a single variable parameter p. If sensitivities with respect to many parameters p_i are required, the right-hand vector $\boldsymbol{\delta}$ must be formed and (4.91) is solved for each p_i in turn.

4.4.6 Gradient Vector Computation of Circuit Functions

The knowledge of sensitivities of wave variables at circuit ports allows us to evaluate gradients of circuit functions such as reflection coefficient, transducer power gain, insertion loss, *et cetera*.

Reflection Coefficient

Let us assume that we are interested in computing the gradient vector of the input port reflection coefficient in the circuit presented in Figure 4.8. The input reflection coefficient is given by

$$\Gamma_{in} = \frac{b_{in}}{a_{in}} \tag{4.93}$$

and its gradient:

$$\nabla \Gamma_{in} = \left[\frac{\partial \Gamma_{in}}{\partial p_1}, \frac{\partial \Gamma_{in}}{\partial p_2}, \ldots, \frac{\partial \Gamma_{in}}{\partial p_n} \right]^T \tag{4.94}$$

Because

$$b_g = \Gamma_g a_g + c_g \tag{4.95}$$

then, under matching condition ($\Gamma_g = 0$) at the generator port:

$$b_g = c_g$$

$$\nabla \Gamma_{in} = \frac{\nabla b_{in}}{a_{in}} = \frac{\nabla a_g}{c_g} = \frac{1}{c_g} \mathbf{G} = \Big|_{c_g = 1} \mathbf{G} \tag{4.96}$$

where $\mathbf{G}$ is the vector of sensitivities of wave variable incoming at the generator port. The vector $\mathbf{G}$ is given by (4.86) in which α is a vector of wave variables of the adjoint network. According to (4.84), the adjoint network will be excited at the same port as

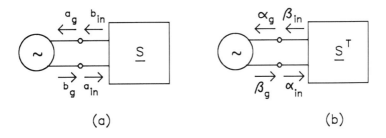

(a)   (b)

Figure 4.8 (a) Original network and (b) adjoint network for gradient computation of the input reflection coefficient.

the original network. If the elements in the analyzed circuit are reciprocal, $S = S^T$, then both networks are identical. In such case only one analysis is required for computing the gradient vector of the input port reflection coefficient.

Because the reflection coefficient is a complex function $\Gamma = |\Gamma| \cdot e^{j\phi}$, we compute the amplitude gradient vector:

$$\nabla |\Gamma_{in}| = |\Gamma_{in}| \operatorname{Re}\left\{ \frac{1}{\Gamma_{in}} \nabla\Gamma_{in} \right\} \tag{4.97}$$

and the phase gradient vector:

$$\nabla\phi = \operatorname{Im}\left\{ \frac{1}{\Gamma_{in}} \nabla\Gamma_{in} \right\} \tag{4.98}$$

In both cases, the gradient vector $\nabla\Gamma_{in}$ is given by (4.96).

Transducer Power Gain

The transducer power gain is defined as

$$G_T = \frac{P_L}{P_{SA}} \tag{4.99}$$

where P_L is the active power dissipated in the load impedance, and P_{SA} is the available power of the signal generator.

According to the notation presented in Figure 4.9:

$$P_L = |a_L|^2 \left(1 - |\Gamma_L|^2\right) \tag{4.100}$$

where $\Gamma_L = b_L / a_L$, and

$$P_{SA} = \frac{|c_g|^2}{1 - |\Gamma_g|^2} \tag{4.101}$$

where $\Gamma_g = (b_g - c_g)/a_g$.

Thus, we have

$$G_T = \frac{|a_L|^2}{|c_g|^2} \left(1 - |\Gamma_L|^2\right)\left(1 - |\Gamma_g|^2\right) \tag{4.102}$$

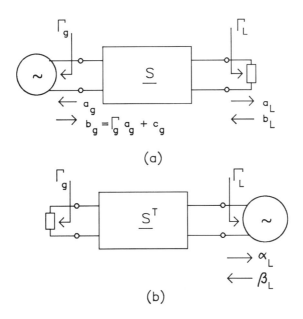

Figure 4.9 Networks for gradient computation of the transducer power gain: (a) original network excited at the input port; (b) adjoint network excited at the output port.

Then, the gradient vector of the transducer power gain is given as

$$\nabla G_T = \frac{1}{|c_g|^2}\left(1 - |\Gamma_L|^2\right)\left(1 - |\Gamma_L|^2\right) \cdot \nabla \, |a_L|^2$$

$$= \frac{1}{|c_g|^2}\left(1 - |\Gamma_L|^2\right)\left(1 - |\Gamma_L|^2\right) \cdot 2\,\mathrm{Re}\{a_L^* \nabla a_L\}$$

$$= \frac{1}{|c_g|^2}\left(1 - |\Gamma_L|^2\right)\left(1 - |\Gamma_L|^2\right) \cdot 2\,\mathrm{Re}\{a_L^* \mathbf{G}\} \qquad (4.103)$$

The elements of the vector **G** are the sensitivities of the wave variable a_L incoming at the load port. Vector **G** is defined by (4.86), in which $\boldsymbol{\alpha}$ is a vector of wave variables of the adjoint network. According (4.84) the adjoint network will be excited at the load port, as is shown in Figure 4.9 (b). Evaluation of the gradient vector of the transducer power gain requires performing two analyses: analysis of the original network excited at the input port and analysis of the adjoint network excited at the output port. In derivation of (4.103), we have used the relation:

$$\frac{\partial |f|^2}{\partial p} = 2 |f| \frac{\partial |f|}{\partial p} = 2 ff^* \text{Re}\left\{\frac{1}{f} \frac{\partial f}{\partial p}\right\} = 2 \text{Re}\left\{f^* \frac{\partial f}{\partial p}\right\} \quad (4.104)$$

Insertion Loss

Frequency performance of microwave filters is usually described by the insertion loss function defined as

$$\text{IL} = \frac{P_L}{P_{L0}} \quad (4.105)$$

where P_L is the active power dissipated in the load impedance Z_L of the filter, and P_{L0} is the active power dissipated in the load impedance connected directly to the generator port.

In accordance with the description in Figure 4.10:

$$P_{L0} = |c_g|^2 \frac{1 - |\Gamma_L|^2}{|1 - \Gamma_g \Gamma_L|^2} \quad (4.106)$$

$$P_L = |a_L|^2 \left(1 - |\Gamma_L|^2\right) \quad (4.107)$$

$$\text{IL} = \frac{|a_L|^2}{|c_g|^2} |1 - \Gamma_g \Gamma_L|^2 \quad (4.108)$$

The gradient vector of the insertion loss then is given as

$$\nabla(\text{IL}) = \frac{1}{|c_g|^2} |1 - \Gamma_g \Gamma_L|^2 \nabla |a_L|^2$$

$$= \frac{1}{|c_g|^2} |1 - \Gamma_g \Gamma_L|^2 2 \text{Re}\{a_L^* \nabla a_L\}$$

$$= \frac{1}{|c_g|^2} |1 - \Gamma_g \Gamma_L|^2 2 \text{Re}\{a_L^* \mathbf{G}\} \quad (4.109)$$

where vector $\mathbf{G}$ describes equation (4.86). In accordance with (4.84) the adjoint network will be excited at the load port, as shown in Figure 4.10(c).

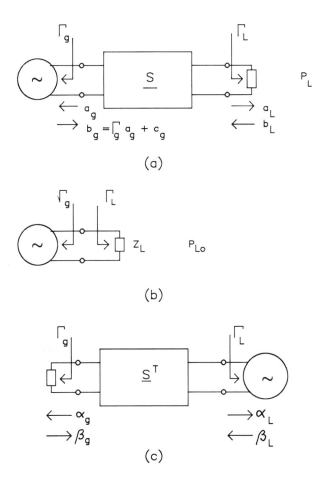

Figure 4.10 Networks for gradient computation of the insertion loss: (a) original network excited at the input port; (b) load impedance connected directly to the generator port; (c) adjoint network with the signal generator exciting the output port.

4.4.7 Evaluation of Group Delay of Microwave Network Transmission Functions

A characteristic of great importance for microwave networks such as linear amplifiers and filters is the group delay *versus* frequency of transmission functions.

In microwave circuits described by the scattering matrices the group delay is defined as

$$\tau_{ij} = -\frac{\partial \phi_{ij}}{\partial \omega} \tag{4.110}$$

where $\phi_{ij} = \arg\{S_{ij}\}$, and S_{ij} is the transmission scattering parameter between the output port i and input port j.

We may compute τ_{ij} by means of the incremental ratio:

$$\tau_{ij} = -\frac{\Delta\phi_{ij}}{\Delta\omega} \tag{4.111}$$

If the increment $\Delta\omega$ is not small enough, a great error can be made in computing τ_{ij}. The incremental values $\Delta\phi_{ij}/\Delta\omega$ tend to the true value of τ_{ij} only in the limit as $\Delta\omega \to 0$, but very small values of $\Delta\omega$ are precluded by roundoff errors.

Using the adjoint network method we can compute exact values of the group delay function. From (4.7) and $p = \omega$ we have

$$\tau_{ij} = -\frac{\partial\phi_{ij}}{\partial\omega} = -\mathrm{Im}\left\{\frac{1}{S_{ij}}\frac{\partial S_{ij}}{\partial\omega}\right\} \tag{4.112}$$

Exciting the jth port of the circuit by a matched generator with $c_g = 1$ and terminating the kth output port with a matched load, we may compute the transmission function S_{ij} from

$$S_{ij}\big|_{a_j=1} = b_i = a_l \tag{4.113}$$

where a_l is the incoming wave variable at the lth port of the matched load terminating the ith port of the network.

To compute the partial derivative of S_{ij} with respect to ω, we refer to (4.88) in which we set $p = \omega$:

$$\frac{\partial S_{ij}}{\partial\omega} = \frac{\partial a_l}{\partial\omega} = \alpha^T\frac{\partial \mathbf{S}}{\partial\omega}\mathbf{a} = \sum_{k\in E_\omega}\alpha^{(k)T}\frac{\partial \mathbf{S}^{(k)}}{\partial\omega}\mathbf{a}^{(k)} \tag{4.114}$$

In the preceding relation, E_ω represents a set of all frequency-dependent element of the analyzed circuit.

By inserting (4.114) into (4.112), we obtain the relation for exact computation o the group delay of any transmission function:

$$\tau_{ij} = -\mathrm{Im}\left\{\frac{1}{a_l}\sum_{k\in E_\omega}\alpha^{(k)T}\frac{\partial \mathbf{S}^{(k)}}{\partial\omega}\mathbf{a}^{(k)}\right\} \tag{4.115}$$

Formulas for the right-hand side of (4.115) have to be obtained for specifi elements. To do so, we may use formulas presented in Table 4.2. We only have t remember that, if some appropriate $\partial\mathbf{S}^{(k)}/\partial p_j$ are available, then

$$\frac{\partial\mathbf{S}^{(k)}}{\partial\omega} = \sum_j\frac{\partial\mathbf{S}^{(k)}}{\partial p_j}\frac{\partial p_j}{\partial\omega} \tag{4.116}$$

Thus, for example, given $Z = R + j[\omega L - (1/\omega C)]$, the ω sensitivity of a Z impedance as a one-port or two-port is $j[L + (1/\omega^2 C)]$ times the Z sensitivity given in Table 4.2.

Example 4.3

Figure 4.11 presents a one-stage microwave transistor amplifier. Seven elements form the network: a signal generator E1 and a load E7, two transmission line sections E3 and E5, two short-circuited shunt stubs E2 and E6, and a *field-effect transistor* (FET) E4. The transmission line sections together with short-circuited shunt stubs form matching networks at the input (E2 and E3) and at the output (E5 and E6) of the transistor.

The matrix equation $\mathbf{Wa} = \mathbf{c}$ of the circuit has the following form:

$$
\begin{bmatrix}
-S^{(1)} & 1 & 0 & 0 & 0 & 0 & 0 & 0 & 0 & 0 & 0 & 0 \\
1 & -S^{(2)} & 0 & 0 & 0 & 0 & 0 & 0 & 0 & 0 & 0 \\
0 & & 1 & 0 & 0 & 0 & 0 & 0 & 0 & 0 & 0 \\
0 & 0 & 1 & -S^{(3)} & 0 & 0 & 0 & 0 & 0 & 0 & 0 \\
0 & 0 & 0 & & 1 & 0 & 0 & 0 & 0 & 0 & 0 \\
0 & 0 & 0 & 0 & 1 & -S^{(4)} & 0 & 0 & 0 & 0 & 0 \\
0 & 0 & 0 & 0 & 0 & & 1 & 0 & 0 & 0 & 0 \\
0 & 0 & 0 & 0 & 0 & 0 & 1 & -S^{(5)} & 0 & 0 & 0 \\
0 & 0 & 0 & 0 & 0 & 0 & 0 & & 1 & 0 & 0 \\
0 & 0 & 0 & 0 & 0 & 0 & 0 & 0 & 1 & -S^{(6)} & 0 \\
0 & 0 & 0 & 0 & 0 & 0 & 0 & 0 & 0 & & 1 \\
0 & 0 & 0 & 0 & 0 & 0 & 0 & 0 & 0 & 1 & -S^{(7)}
\end{bmatrix}
\begin{bmatrix}
a_1 \\ a_2 \\ a_3 \\ a_4 \\ a_5 \\ a_6 \\ a_7 \\ a_8 \\ a_9 \\ a_{10} \\ a_{11} \\ a_{12}
\end{bmatrix}
=
\begin{bmatrix}
1 \\ 0 \\ 0 \\ 0 \\ 0 \\ 0 \\ 0 \\ 0 \\ 0 \\ 0 \\ 0 \\ 0
\end{bmatrix}
$$

where (see Table 2.2)

$$
S^{(1)} = \frac{Z_g - Z_N}{Z_g + Z_N}
$$

$$
S^{(2)} = \frac{1}{2Y_N + Y_{02}\coth(\gamma l_2)}
\begin{bmatrix}
-Y_{02}\coth(\gamma l_2) & 1 \\
1 & -Y_{02}\coth(\gamma l_2)
\end{bmatrix}
$$

$$
S^{(3)} = \frac{1}{Z_{03}^2 + Z_N^2 + 2Z_{03}Z_N\coth(\gamma l_3)}
\begin{bmatrix}
Z_{03}^2 - Z_N^2 & 2Z_{03}Z_N\operatorname{csch}(\gamma l_3) \\
2Z_{03}Z_N\operatorname{csch}(\gamma l_3) & Z_{03}^2 - Z_N^2
\end{bmatrix}
$$

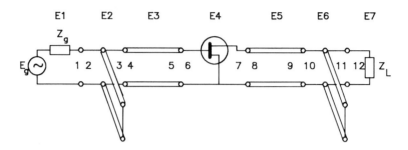

Figure 4.11 The network of a single-stage transistor amplifier.

The fourth element is the FET described by the scattering matrix $\mathbf{S}^{(4)}$. The scattering parameters of the transistor should be given as numerical values at a discrete set of frequency points at a given dc bias.

The scattering matrix $\mathbf{S}^{(5)}$ of the element E5 is similar to $\mathbf{S}^{(3)}$, only instead of Z_{03} and l_3 we should substitute Z_{05} and l_5. Similarly, the scattering matrix $\mathbf{S}^{(6)}$ of the element E6 is identical to $\mathbf{S}^{(2)}$, except that instead of Z_{02} and l_2 we should substitute Z_{06} and l_6, respectively.

The scattering element of the load impedance is

$$S^{(7)} = \frac{Z_L - Z_N}{Z_L + Z_N}$$

In the expressions defining the scattering matrices of particular circuit elements, Z_N is the reference impedance common to the all ports of the analyzed circuit. Because scattering parameters of FET are usually determined in the $Z_0 = 50\ \Omega$ measurement systems, it is very convenient to assume that $Z_N = Z_0 = 50\ \Omega$.

In the following discussion we assume that the transducer power gain G_T of the amplifier is the circuit function of interest. The transducer power gain will be computed from the relation ($a_L = a_{12}$, $c_g = 1$):

$$G_T = |a_{12}|^2 \left(1 - |\Gamma_L|^2\right)\left(1 - |\Gamma_g|^2\right)$$

$$= |a_{12}|^2 \left(1 - \left|\frac{a_{11}}{a_{12}}\right|^2\right)\left(1 - \left|\frac{b_1 - 1}{a_1}\right|^2\right)$$

$$= \left(|a_{12}|^2 - |a_{11}|^2\right)\left(1 - \left|\frac{b_1 - 1}{a_1}\right|^2\right)$$

To evaluate the gradient vector ∇G_T of the transducer power gain we should compute the sensitivities of the a_{12} wave variable, the incoming wave variable at the load port. To do this we solve a matrix equation $\mathbf{W}^T\boldsymbol{\alpha} = \boldsymbol{\gamma}_{12}$ of the adjoint network in the form given as (only the FET is a nonreciprocal two-port $[\mathbf{S}^{(4)T} \neq \mathbf{S}^{(4)}]$):

$$
\begin{array}{c}
1 \\ 2 \\ 3 \\ 4 \\ 5 \\ 6 \\ 7 \\ 8 \\ 9 \\ 10 \\ 11 \\ 12
\end{array}
\begin{bmatrix}
-\mathbf{S}^{(1)} & 1 & 0 & 0 & 0 & 0 & 0 & 0 & 0 & 0 & 0 & 0 \\
1 & -\mathbf{S}^{(2)} & 0 & 0 & 0 & 0 & 0 & 0 & 0 & 0 & 0 & 0 \\
0 & & 1 & 0 & 0 & 0 & 0 & 0 & 0 & 0 & 0 & 0 \\
0 & 0 & 1 & -\mathbf{S}^{(3)} & 0 & 0 & 0 & 0 & 0 & 0 & 0 & 0 \\
0 & 0 & 0 & & 1 & 0 & 0 & 0 & 0 & 0 & 0 & 0 \\
0 & 0 & 0 & 0 & 1 & -\mathbf{S}^{(4)T} & 0 & 0 & 0 & 0 & 0 & 0 \\
0 & 0 & 0 & 0 & 0 & & 1 & 0 & 0 & 0 & 0 & 0 \\
0 & 0 & 0 & 0 & 0 & 0 & 1 & -\mathbf{S}^{(5)} & 0 & 0 & 0 & 0 \\
0 & 0 & 0 & 0 & 0 & 0 & 0 & & 1 & 0 & 0 & 0 \\
0 & 0 & 0 & 0 & 0 & 0 & 0 & 0 & 1 & -\mathbf{S}^{(6)} & 0 & 0 \\
0 & 0 & 0 & 0 & 0 & 0 & 0 & 0 & 0 & & 1 & 0 \\
0 & 0 & 0 & 0 & 0 & 0 & 0 & 0 & 0 & 0 & 1 & -\mathbf{S}^{(7)}
\end{bmatrix}
\begin{bmatrix}
\alpha_1 \\ \alpha_2 \\ \alpha_3 \\ \alpha_4 \\ \alpha_5 \\ \alpha_6 \\ \alpha_7 \\ \alpha_8 \\ \alpha_9 \\ \alpha_{10} \\ \alpha_{11} \\ \alpha_{12}
\end{bmatrix}
=
\begin{bmatrix}
0 \\ 0 \\ 0 \\ 0 \\ 0 \\ 0 \\ 0 \\ 0 \\ 0 \\ 0 \\ 0 \\ 1
\end{bmatrix}
$$

The sensitivity of the wave variable a_{12} with respect to any variable network parameter should be computed using the formulas given in Table 4.2.

We now assume that the variable parameters of the circuit are Z_{02}, Z_{03}, Z_{05}, Z_{06}, l_2, l_3, l_5, and l_6. The gradient operator is then given as

$$
\boldsymbol{\nabla} = \left[\frac{\partial}{\partial Z_{02}}, \frac{\partial}{\partial Z_{03}}, \frac{\partial}{\partial Z_{05}}, \frac{\partial}{\partial Z_{06}}, \frac{\partial}{\partial l_2}, \frac{\partial}{\partial l_3}, \frac{\partial}{\partial l_5}, \frac{\partial}{\partial l_6} \right]^T
$$

The sensitivity of wave variable a_{12} with respect to the characteristic impedance Z_0 of a transmission line section (elements E3 and E5) is equal to

$$
\frac{\partial a_{12}}{\partial Z_0} = \frac{1}{2Z_0} \left(\boldsymbol{\alpha}^T \mathbf{a} - \boldsymbol{\beta}^T \mathbf{b} \right)
$$

Then, we have

$$
G_1 = \frac{\partial a_{12}}{\partial Z_{03}} = \frac{1}{2Z_{03}} \left(\begin{bmatrix} \alpha_4 \\ \alpha_5 \end{bmatrix}^T \begin{bmatrix} a_4 \\ a_5 \end{bmatrix} - \begin{bmatrix} \beta_4 \\ \beta_5 \end{bmatrix}^T \begin{bmatrix} b_4 \\ b_5 \end{bmatrix} \right)
$$

$$
= \frac{1}{2Z_{03}} \left(\begin{bmatrix} \alpha_4 \\ \alpha_5 \end{bmatrix}^T \begin{bmatrix} a_4 \\ a_5 \end{bmatrix} - \begin{bmatrix} \alpha_3 \\ \alpha_6 \end{bmatrix}^T \begin{bmatrix} a_3 \\ a_6 \end{bmatrix} \right)
$$

$$
G_2 = \frac{\partial a_{12}}{\partial Z_{05}} = \frac{1}{2Z_{05}} \left(\begin{bmatrix} \alpha_8 \\ \alpha_9 \end{bmatrix}^T \begin{bmatrix} a_8 \\ a_9 \end{bmatrix} - \begin{bmatrix} \beta_7 \\ \beta_{10} \end{bmatrix}^T \begin{bmatrix} b_7 \\ b_{10} \end{bmatrix} \right)
$$

The sensitivity of wave variable a_{12} with respect to the length l of a transmission line section (elements E3 and E5) is

$$\frac{\partial a_{12}}{\partial l} = \frac{\gamma Z_0}{2 Z_N \sinh^2(\gamma l)} \cdot (\alpha - \beta)^T \begin{bmatrix} \cosh(\gamma l) & 1 \\ 1 & \cosh(\gamma l) \end{bmatrix} (\mathbf{a} - \mathbf{b})$$

According to this relation, we have

$$G_3 = \frac{\partial a_{12}}{\partial l_3} = \frac{\gamma Z_{03}}{2 Z_N \sinh(\gamma l_3)} \begin{bmatrix} \alpha_4 - \beta_4 \\ \alpha_5 - \beta_5 \end{bmatrix}^T \begin{bmatrix} \cosh(\gamma l_3) & 1 \\ 1 & \cosh(\gamma l_3) \end{bmatrix}$$

$$\cdot \begin{bmatrix} a_4 - b_4 \\ a_5 - b_5 \end{bmatrix}$$

$$G_3 = \frac{\gamma Z_{03}}{2 Z_N \sinh(\gamma l_3)} \begin{bmatrix} \alpha_4 - \alpha_3 \\ \alpha_5 - \alpha_6 \end{bmatrix}^T \begin{bmatrix} \cosh(\gamma l_3) & 1 \\ 1 & \cosh(\gamma l_3) \end{bmatrix} \cdot \begin{bmatrix} a_4 - a_3 \\ a_5 - a_6 \end{bmatrix}$$

$$G_4 = \frac{\partial a_{12}}{\partial l_5} = \frac{\gamma Z_{05}}{2 Z_N \sinh(\gamma l_5)} \begin{bmatrix} \alpha_8 - \alpha_7 \\ \alpha_9 - \alpha_{10} \end{bmatrix}^T \begin{bmatrix} \cosh(\gamma l_5) & 1 \\ 1 & \cosh(\gamma l_5) \end{bmatrix}$$

$$\cdot \begin{bmatrix} a_8 - a_7 \\ a_9 - a_{10} \end{bmatrix}$$

Sensitivity of wave variable a_{12} with respect to the characteristic impedance of a short-circuited parallel stub is

$$\frac{\partial a_{12}}{\partial Z_0} = \frac{1}{2 Z_0} (\alpha^T \mathbf{a} - \beta^T \mathbf{b})$$

Sensitivity of wave variable a_{12} with respect to the length l of a parallel stub is

$$\frac{\partial a_{12}}{\partial l} = \frac{\gamma Z_0}{2 Z_N \cosh^2(\gamma l)} (\alpha - \beta)^T \begin{bmatrix} 1 & 1 \\ 1 & 1 \end{bmatrix} (\mathbf{a} - \mathbf{b})$$

Then, we have

$$G_5 = \frac{\partial a_{12}}{\partial Z_{02}} = \frac{1}{2 Z_{02}} \left(\begin{bmatrix} \alpha_2 \\ \alpha_3 \end{bmatrix}^T \begin{bmatrix} a_2 \\ a_3 \end{bmatrix} - \begin{bmatrix} \beta_2 \\ \beta_3 \end{bmatrix}^T \begin{bmatrix} b_2 \\ b_3 \end{bmatrix} \right)$$

$$= \frac{1}{2 Z_{02}} \left(\begin{bmatrix} \alpha_2 \\ \alpha_3 \end{bmatrix}^T \begin{bmatrix} a_2 \\ a_3 \end{bmatrix} - \begin{bmatrix} \alpha_1 \\ \alpha_4 \end{bmatrix}^T \begin{bmatrix} a_1 \\ a_4 \end{bmatrix} \right)$$

$$G_6 = \frac{\partial a_{12}}{\partial Z_{06}} = \frac{1}{2 Z_{06}} \left(\begin{bmatrix} \alpha_{10} \\ \alpha_3 \end{bmatrix}^T \begin{bmatrix} a_{10} \\ a_3 \end{bmatrix} - \begin{bmatrix} \alpha_9 \\ \alpha_{12} \end{bmatrix}^T \begin{bmatrix} a_9 \\ a_{12} \end{bmatrix} \right)$$

$$G_7 = \frac{\partial a_{12}}{\partial l_2} = \frac{\gamma Z_{02}}{2 Z_N \cosh^2 (\gamma l_2)} \begin{bmatrix} \alpha_2 - \beta_2 \\ \alpha_3 - \beta_3 \end{bmatrix}^T \begin{bmatrix} 1 & 1 \\ 1 & 1 \end{bmatrix} \begin{bmatrix} a_2 - b_2 \\ a_3 - b_3 \end{bmatrix}$$

$$= \frac{\gamma Z_{02}}{2 Z_N \cosh^2 (\gamma l_2)} \begin{bmatrix} \alpha_2 - \alpha_1 \\ \alpha_3 - \alpha_4 \end{bmatrix}^T \begin{bmatrix} 1 & 1 \\ 1 & 1 \end{bmatrix} \begin{bmatrix} a_2 - a_1 \\ a_3 - a_4 \end{bmatrix}$$

$$G_8 = \frac{\partial a_{12}}{\partial l_6} = \frac{\gamma Z_{06}}{2 Z_N \cosh^2 (\gamma l_6)} \begin{bmatrix} \alpha_{10} - \alpha_9 \\ \alpha_{11} - \alpha_{12} \end{bmatrix}^T \begin{bmatrix} 1 & 1 \\ 1 & 1 \end{bmatrix} \begin{bmatrix} a_{10} - a_9 \\ a_{11} - a_{12} \end{bmatrix}$$

The gradient vector of the transducer power gain ∇G_T (see (4.103)) is equal to

$$\nabla G_T = \frac{1}{|c_g|^2} \left(1 - |\Gamma_L|^2 \right) \left(1 - |\Gamma_g|^2 \right) 2 \operatorname{Re} \{ a_L^* \mathbf{G} \}$$

$$= \left(1 - \frac{|a_{11}|^2}{|a_{12}|^2} \right) \left(1 - \frac{|a_2 - 1|^2}{|a_1|^2} \right) 2 \operatorname{Re} \{ a_{12}^* \mathbf{G} \}$$

where

$$\mathbf{G} = \begin{bmatrix} G_1, G_2, G_3, G_4, G_5, G_6 \end{bmatrix}^T$$

4.5 SECOND-ORDER SENSITIVITIES OF MICROWAVE NETWORKS

As mentioned earlier, in many situations we are interested in the effect on circuit function F caused by an infinitesimally small change in a circuit parameter p.

Provided that parameter changes are infinitesimal and circuit function F is analytical at $\mathbf{p} = \mathbf{p}^0$ (nominal values of circuit parameters), we have linear relation between the parameter changes and the resulting change in circuit function F. For n parameters, we have

$$\Delta F = \sum_{i=1}^{n} \frac{\partial F}{\partial p_i} \Delta p_i \qquad (4.117)$$

or, in matrix notation:

$$\Delta F = \begin{bmatrix} \dfrac{\partial E}{\partial p_1} \\[2mm] \dfrac{\partial F}{\partial p_2} \\[2mm] \vdots \\[2mm] \dfrac{\partial F}{\partial p_n} \end{bmatrix}^T \begin{bmatrix} \Delta p_1 \\[2mm] \Delta p_2 \\[2mm] \vdots \\[2mm] \Delta p_n \end{bmatrix} = \mathbf{G}^T \, \Delta \mathbf{p} \qquad (4.118)$$

If the parameter changes Δp_i are incremental, then the resulting change in circuit function F should be expressed as the Taylor series expansion:

$$F \approx F_0 + \sum_{i=1}^{n} \frac{\partial F}{\partial p_i} \Delta p_i + \frac{1}{2} \sum_{i=1}^{n} \sum_{j=1}^{n} \frac{\partial^2 F}{\partial p_i \, \partial p_j} \Delta p_i \, \Delta p_j \qquad (4.119)$$

In some applications, the second term in the series will be adequate and only the first-order sensitivities $\partial F / \partial p_i$ have to be evaluated. However, for a better approximation of $\Delta F = F - F_0$ we must include the term involving second-order sensitivities $\partial^2 F / \partial p_i \, \partial p_j$. In the matrix notation this term in the truncated Taylor series expansion is

$$\frac{1}{2} \, \Delta \mathbf{p}^T \, \mathbf{H} \, \Delta \mathbf{p} \qquad (4.120)$$

where $\mathbf{H}$ is the square and symmetrical Hessian matrix whose (i, j) entry is the second-order sensitivity:

$$(\mathbf{H})_{ij} = \frac{\partial^2 F}{\partial p_i \, \partial p_j} \qquad (4.121)$$

A concise expression for F containing the first three terms of its Taylor expansion is

$$F \cong F_0 + \mathbf{G}^T \, \Delta \mathbf{p} + \frac{1}{2} \, \Delta \mathbf{p}^T \, \mathbf{H} \, \Delta \mathbf{p} \qquad (4.122)$$

4.5.1 Second-order Sensitivity Analysis by the Adjoint Network Method

The adjoint network concept also may be used to determine the second-order sensitivities of the wave variables in a microwave circuit.

Consider an original network with two distinct parameters p_1 and p_2, and assume that they are contained in a multiport element with the scattering matrix $\mathbf{S}^{(k)}$. Differentiating (4.48) with respect to p_2 we have

$$\frac{\partial^2 \mathbf{b}^{(k)}}{\partial p_1 \, \partial p_2} = \frac{\partial^2 \mathbf{S}^{(k)}}{\partial p_1 \, \partial p_2} \mathbf{a}^{(k)} + \frac{\partial \mathbf{S}^{(k)}}{\partial p_1} \frac{\partial \mathbf{a}^{(k)}}{\partial p_2} + \frac{\partial \mathbf{S}^{(k)}}{\partial p_2} \frac{\partial \mathbf{a}^{(k)}}{\partial p_1}$$

$$+ \mathbf{S}^{(k)} \frac{\partial^2 \mathbf{a}}{\partial p_1 \, \partial p_2} \tag{4.123}$$

Then, using this relation, Tellegen's expression (4.23) valid for the entire network may be written in the form

$$\boldsymbol{\beta}_e^T \frac{\partial^2 \mathbf{a}_e}{\partial p_1 \, \partial p_2} - \boldsymbol{\alpha}_e^T \frac{\partial^2 \mathbf{b}_e}{\partial p_1 \, \partial p_2}$$

$$= \sum_{i=1}^{m} \boldsymbol{\alpha}^{(i)T} \left(\frac{\partial^2 \mathbf{b}^{(i)}}{\partial p_1 \, \partial p_2} - \boldsymbol{\beta}^{(i)T} \frac{\partial^2 \mathbf{a}^{(k)}}{\partial p_1 \, \partial p_2} \right)$$

$$= \sum_{i=1}^{m} \left\{ \boldsymbol{\alpha}^{(i)T} \left(\frac{\partial^2 \mathbf{S}^{(i)}}{\partial p_1 \, \partial p_2} \mathbf{a}^{(i)} + \frac{\partial \mathbf{S}^{(i)}}{\partial p_1} \frac{\partial \mathbf{a}^{(i)}}{\partial p_2} + \frac{\partial \mathbf{S}^{(i)}}{\partial p_2} \frac{\partial \mathbf{a}^{(i)}}{\partial p_1} \right) \right.$$

$$\left. + \left(\boldsymbol{\alpha}^{(i)} \mathbf{S}^{(i)T} - \boldsymbol{\beta}^{(i)} \right)^T \frac{\partial^2 \mathbf{a}^{(i)}}{\partial p_1 \, \partial p_2} \right\} \tag{4.124}$$

where $\mathbf{a}_e$, $\mathbf{b}_e$, $\boldsymbol{\alpha}_e$, and $\boldsymbol{\beta}_e$ are vectors of incoming and outgoing wave variables at the load and generator ports, respectively, in the original and adjoint networks, and m is the number of elements in the network.

If all elements in the adjoint network satisfy the relation:

$$\boldsymbol{\beta}^{(i)} = \mathbf{S}^{(i)T} \boldsymbol{\alpha}^{(i)} \tag{4.125}$$

then (4.118) may be reduced to

$$\boldsymbol{\beta}_e^T \frac{\partial^2 \mathbf{a}_e}{\partial p_1 \, \partial p_2} - \boldsymbol{\alpha}_e^T \frac{\partial^2 \mathbf{b}_e}{\partial p_1 \, \partial p_2}$$

$$= \sum_{i=1}^{m} \boldsymbol{\alpha}^{(i)T} \left(\frac{\partial^2 \mathbf{S}^{(i)}}{\partial p_1 \, \partial p_2} \mathbf{a}^{(i)} + \frac{\partial \mathbf{S}^{(i)}}{\partial p_1} \frac{\partial \mathbf{a}^{(i)}}{\partial p_2} + \frac{\partial \mathbf{S}^{(i)}}{\partial p_2} \frac{\partial \mathbf{a}^{(i)}}{\partial p_1} \right) \tag{4.126}$$

Assuming matched terminations at all the external ports of the original network, $\partial^2 \mathbf{b}_e / \partial p_1 \partial p_2 = 0$, and the left-hand side of (4.126) reduces to

$$h = \boldsymbol{\beta}_e^T \frac{\partial^2 \mathbf{a}_e}{\partial p_1 \partial p_2} \tag{4.127}$$

If at the same time the adjoint network is excited so that all entries of the $\boldsymbol{\beta}_e$ vector are ones, then h in (4.127) is the sum of the incoming wave second-order sensitivities at all output ports of the network.

To evaluate the second-order sensitivity of the incoming wave variable a_r at the rth output port of the network, all output ports of the adjoint network have to be matched and at the same time the rth port has to be excited to have $\beta_{er} = 1$. Figure 4.6 shows the original and the adjoint network for the second-order sensitivity computation. Equation (4.126) now has the form:

$$\frac{\partial^2 a_{er}}{\partial p_1 \partial p_2} = \sum_{i=1}^{m} \boldsymbol{\alpha}^{(i)T} \left(\frac{\partial^2 \mathbf{S}^{(i)}}{\partial p_1 \partial p_2} \mathbf{a}^{(i)} + \frac{\partial \mathbf{S}^{(i)}}{\partial p_1} \frac{\partial \mathbf{a}^{(i)}}{\partial p_2} + \frac{\partial \mathbf{S}^{(i)}}{\partial p_2} \frac{\partial \mathbf{a}^{(i)}}{\partial p_1} \right)$$

$$\tag{4.128}$$

If the variable parameters p_1 and p_2 are identical, we have

$$\frac{\partial^2 a_{er}}{\partial p^2} = \sum_{i=1}^{m} \boldsymbol{\alpha}^{(i)T} \left(\frac{\partial^2 \mathbf{S}^{(i)}}{\partial p^2} \mathbf{a}^{(i)} + 2 \frac{\partial \mathbf{S}^{(i)}}{\partial p} \frac{\partial \mathbf{a}^{(i)}}{\partial p} \right) \tag{4.129}$$

When a parameter p_1 affects the characterization of one element with the scattering matrix $\mathbf{S}_{p1}$, while parameter p_2 affects another element described by the $\mathbf{S}_{p2}$ scattering matrix, the expression for the second-order sensitivity of a_{er} is

$$\frac{\partial a_{er}}{\partial p_1 \partial p_2} = \boldsymbol{\alpha}_{p1}^T \frac{\partial \mathbf{S}_{p1}}{\partial p_1} \frac{\partial \mathbf{a}_{p1}}{\partial p_2} + \boldsymbol{\alpha}_{p2}^T \frac{\partial \mathbf{S}_{p2}}{\partial p_2} \frac{\partial \mathbf{a}_{p2}}{\partial p_1} \tag{4.130}$$

where subscripts p_1 and p_2 denote quantities related to the appropriate elements.

To summarize the discussion thus far, we can write the matrix $\mathbf{H}$ of the second-order derivatives of a_r (the Hessian matrix) as follows:

$$\mathbf{H} = \nabla \nabla^T a_{er} = \mathbf{B}_r \tag{4.131}$$

where $\mathbf{B}_r$ is a matrix of the second-order sensitivities containing expressions of the form of (4.128), (4.129), or (4.130), as appropriate. In (4.131) ∇ is a gradient operator given by (4.40).

According to (4.128) computation of the second-order sensitivities (entries of the $\mathbf{B}_r$ matrix) requires

1. solution of the original network equations (the $\mathbf{a}$ vector),
2. solution of the appropriately excited adjoint network (the α vector),
3. as many analyses of the adjoint network as there are ports of elements containing variable parameters in the analyzed circuit.

The reason for the further analyses of the adjoint network is the requirement of the first-order sensitivities ($\partial \mathbf{a}/\partial p_1$ and $\partial \mathbf{a}/\partial p_2$) of the incident wave variables at ports of these elements.

4.5.2 Transposed Matrix Method for the Second-Order Sensitivity Analysis for Networks Described by the Connection Scattering Matrix

Obtaining the second-order sensitivities of microwave circuits described by the connection scattering matrix, in principle, is as simple as evaluating the first-order sensitivities, only more expensive computationally. Differentiate the equation (3.60) $\mathbf{Wa} = \mathbf{c}$ with respect to a parameter p_1 and assume for simplicity that the exciting generators do not depend on the parameters. We have

$$\frac{\partial \mathbf{W}}{\partial p_1} \mathbf{a} + \mathbf{W} \frac{\partial \mathbf{a}}{\partial p_1} = 0$$

Differentiate again with respect to the parameter p_2

$$\frac{\partial^2 \mathbf{W}}{\partial p_1 \partial p_2} \mathbf{a} + \frac{\partial \mathbf{W}}{\partial p_1} \frac{\partial \mathbf{a}}{\partial p_2} + \frac{\partial \mathbf{W}}{\partial p_2} \frac{\partial \mathbf{a}}{\partial p_1} + \mathbf{W} \frac{\partial^2 \mathbf{a}}{\partial p_1 \partial p_2} = 0$$

from which

$$\frac{\partial^2 \mathbf{a}}{\partial p_1 \partial p_2} = -\mathbf{W}^{-1} \left(\frac{\partial^2 \mathbf{W}}{\partial p_1 \partial p_2} a + \frac{\partial \mathbf{W}}{\partial p_1} \frac{\partial \mathbf{a}}{\partial p_2} + \frac{\partial \mathbf{W}}{\partial p_2} \frac{\partial \mathbf{a}}{\partial p_1} \right) \quad (4.132)$$

Because $\mathbf{W} = \boldsymbol{\Gamma} - \mathbf{S}$ and $\partial \boldsymbol{\Gamma} / \partial p_i = 0$, equation (4.132) becomes

$$\frac{\partial^2 \mathbf{a}}{\partial p_1 \partial p_2} = \mathbf{W}^{-1} \left(\frac{\partial^2 \mathbf{S}}{\partial p_1 \partial p_2} \mathbf{a} + \frac{\partial \mathbf{S}}{\partial p_1} \frac{\partial \mathbf{a}}{\partial p_2} + \frac{\partial \mathbf{S}}{\partial p_2} \frac{\partial \mathbf{a}}{\partial p_1} \right) \quad (4.133)$$

As in the first-order sensitivities, premultiplying (4.133) by the vector $\boldsymbol{\gamma}_r^T$ defined by (4.81), we introduce the adjoint equation:

$$\frac{\partial^2 a}{\partial p_1 \partial p_2} = \alpha^T \frac{\partial^2 \mathbf{S}}{\partial p_1 \partial p_2} \mathbf{a} + \alpha^T \left(\frac{\partial \mathbf{S}}{\partial p_1} \frac{\partial \mathbf{a}}{\partial p_2} + \frac{\partial \mathbf{S}}{\partial p_2} \frac{\partial \mathbf{a}}{\partial p_1} \right) \quad (4.134)$$

where α is the solution vector of the adjoint (transposed) system

$$\mathbf{W}^T \alpha = \gamma_r \tag{4.135}$$

Relation (4.134) allows computation of the second-order sensitivities of the wave variable a_r with respect to parameters p_1 and p_2. The computation of second-order sensitivities requires

1. the solution of the original network (the vector $\mathbf{a}$),
2. the solution of the adjoint system (the vector α),
3. the vectors of derivatives of the $\mathbf{a}$ vector with respect to p_1 and p_2.

These can be obtained in two ways. If the number n of variable parameters for which second-order derivatives are required is large (larger than the dimension of the $\mathbf{W}$ matrix), they are computed by a repeated application of the transposed matrix approach by considering each component of the $\mathbf{a}$ vector as an output. In other words, we must find the solution to the transposed (adjoint) matrix equation $\mathbf{W}^T \alpha = \gamma_r$ n times, for $\gamma_1, \gamma_2, \ldots, \gamma_n$. However, if the number of variable parameters is small, the vectors of the derivatives $\partial \mathbf{a} / \partial p_i$ may be computed by the direct method, in which systems of the form (4.89) are solved to obtain the vectors $\partial \mathbf{a} / \partial p_i$:

$$\mathbf{W} \frac{\partial \mathbf{a}}{\partial p_i} = \frac{\partial \mathbf{S}}{\partial p_i} \mathbf{a} \tag{4.136}$$

The number of repeated solutions of (4.136) in this case equals the number of parameters.

Formula (4.134) simplifies when the considered parameters belong to different elements of the circuit. In this case $\partial^2 \mathbf{S} / \partial p_1 \partial p_2$ becomes zero, and $\partial^2 a_r / \partial p_1 \partial p_2$ can be written as

$$\frac{\partial^2 a_r}{\partial p_1 \partial p_2} = \alpha^T \left(\frac{\partial \mathbf{S}}{\partial p_1} \frac{\partial \mathbf{a}}{\partial p_2} + \frac{\partial \mathbf{S}}{\partial p_2} \frac{\partial \mathbf{a}}{\partial p_1} \right) \tag{4.137}$$

4.6 SENSITIVITY ANALYSIS OF CASCADED TWO-PORT NETWORKS DESCRIBED BY THE CHAIN MATRICES

Figure 4.12 presents a network composed of cascaded two-ports. The overall chain matrix for the cascaded network in Figure 4.12 is expressed as

$$\mathbf{T} = \mathbf{T}^{(1)} \cdot \mathbf{T}^{(2)} \cdot \ldots \cdot \mathbf{T}^{(k-1)} \cdot \mathbf{T}^{(k)} \cdot \mathbf{T}^{(k+1)} \cdot \ldots \cdot \mathbf{T}^{(m)} \tag{4.138}$$

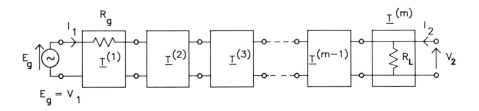

Figure 4.12 Cascaded connection of two-ports.

where $\mathbf{T}^{(k)}$ is the chain matrix of the kth subnetwork in the cascaded network. Let us assume that the overall chain matrix $\mathbf{T}$ depends on n variable parameters p_i, $i = 1, 2, \ldots, n$. If p_i belongs only to subnetwork l, p_j only to subnetwork k, then the first-order and second-order sensitivities of the overall $\mathbf{T}$ matrix are ($r = 1, 2$) [13]:

$$\frac{\partial^r \mathbf{T}}{\partial p_i^r} = \mathbf{T}^{(1)} \cdot \mathbf{T}^{(2)} \cdot \ldots \cdot \mathbf{T}^{(l-1)} \frac{\partial^r \mathbf{T}^{(l)}}{\partial p_i^r} \mathbf{T}^{(l+1)} \cdots \mathbf{T}^{(m)} \quad (4.139)$$

$$\frac{\partial^r \mathbf{T}}{\partial p_j^r} = \mathbf{T}^{(1)} \cdot \mathbf{T}^{(2)} \cdot \ldots \cdot \mathbf{T}^{(k-1)} \frac{\partial^r \mathbf{T}^{(k)}}{\partial p_j^r} \mathbf{T}^{(k+1)} \cdots \mathbf{T}^{(m)} \quad (4.140)$$

$$\frac{\partial^2 \mathbf{T}}{\partial p_i \, \partial p_j} = \mathbf{T}^{(1)} \cdot \mathbf{T}^{(2)} \cdot \ldots \cdot \mathbf{T}^{(k-1)} \frac{\partial \mathbf{T}^{(k)}}{\partial p_j} \mathbf{T}^{(k+1)}$$

$$\ldots \cdot \mathbf{T}^{(l-1)} \frac{\partial \mathbf{T}^{(l)}}{\partial p_i} \mathbf{T}^{(l+1)} \cdots \mathbf{T}^{(m)} \quad (4.141)$$

Table 4.3 presents first-order and second-order sensitivities of the chain matrices of some commonly used elements of microwave circuits. These may be used to very effectively compute gradient vectors and Hessian matrices of the objective functions in the optimization algorithms used in CAD of microwave circuits.

The same approach may be used to compute sensitivities of any interconnected two-port networks with cascaded, parallel, and series two-port connections [9].

Table 4.3
Chain Matrix Sensitivities of Some Lumped and Distributed Elements of Microwave Circuits

Circuit Element	p_i	$\dfrac{\partial T}{\partial p_i}$	$\dfrac{\partial^2 T}{\partial p_i^2}$	p_j	$\dfrac{\partial^2 T}{\partial p_i\,\partial p_j}$
R L C series $Z = R + j\left(\omega L - \dfrac{1}{\omega C}\right)$	$\ln R$	$\begin{bmatrix} 0 & R \\ 0 & 0 \end{bmatrix}$	$\begin{bmatrix} 0 & R \\ 0 & 0 \end{bmatrix}$	$\ln L$ or $\ln C$	$\begin{bmatrix} 0 & 0 \\ 0 & 0 \end{bmatrix}$
	$\ln L$	$\begin{bmatrix} 0 & j\omega L \\ 0 & 0 \end{bmatrix}$	$\begin{bmatrix} 0 & j\omega L \\ 0 & 0 \end{bmatrix}$	$\ln C$	$\begin{bmatrix} 0 & 0 \\ 0 & 0 \end{bmatrix}$
	$\ln C$	$\begin{bmatrix} 0 & \dfrac{-1}{\omega C} \\ 0 & 0 \end{bmatrix}$	$\begin{bmatrix} 0 & \dfrac{1}{j\omega C} \\ 0 & 0 \end{bmatrix}$		
G L C parallel $Y = \dfrac{1}{R} + \dfrac{1}{j\omega L} + j\omega C$	$\ln R$	$\begin{bmatrix} 0 & 0 \\ \dfrac{-1}{R} & 0 \end{bmatrix}$	$\begin{bmatrix} 0 & 0 \\ \dfrac{1}{R} & 0 \end{bmatrix}$	$\ln L$ or $\ln C$	$\begin{bmatrix} 0 & 0 \\ 0 & 0 \end{bmatrix}$
	$\ln L$	$\begin{bmatrix} 0 & 0 \\ \dfrac{1}{\omega L} & 0 \end{bmatrix}$	$\begin{bmatrix} 0 & 0 \\ \dfrac{1}{j\omega L} & 0 \end{bmatrix}$	$\ln C$	$\begin{bmatrix} 0 & 0 \\ 0 & 0 \end{bmatrix}$
	$\ln C$	$\begin{bmatrix} 0 & 0 \\ j\omega C & 0 \end{bmatrix}$	$\begin{bmatrix} 0 & 0 \\ j\omega C & 0 \end{bmatrix}$		

Circuit: C — L — R in series

$$Y = \frac{1}{R + j\left(\omega L - \dfrac{1}{\omega C}\right)}$$

ln R	$\begin{bmatrix} 0 & 0 \\ -RY^3 & 0 \end{bmatrix}$	$\begin{bmatrix} 0 & 0 \\ 2R^3Y^3 - RY^2 & 0 \end{bmatrix}$
ln L	$\begin{bmatrix} 0 & 0 \\ -j\omega LY^2 & 0 \end{bmatrix}$	$\begin{bmatrix} 0 & 0 \\ \dfrac{j\omega L(j\omega 2L - Z)}{Z^3} & 0 \end{bmatrix}$
ln C	$\begin{bmatrix} 0 & 0 \\ \dfrac{j2RY^3}{\omega C} & 0 \end{bmatrix}$	$\begin{bmatrix} 0 & 0 \\ \dfrac{Y^2(2Y - j\omega C)}{-\omega^2 C^2} & 0 \end{bmatrix}$
ln R		
ln L	$\begin{bmatrix} 0 & 0 \\ j\omega 2LRY^3 & 0 \end{bmatrix}$	
ln C	$\begin{bmatrix} 0 & 0 \\ \dfrac{j2RY}{\omega C} & 0 \end{bmatrix}$	
ln C	$\begin{bmatrix} 0 & 0 \\ \dfrac{-2LY}{C} & 0 \end{bmatrix}$	

Table 4.3 (*Continued*)

Circuit Element	p_i	$\dfrac{\partial \mathbf{T}}{\partial p_i}$	$\dfrac{\partial^2 \mathbf{T}}{\partial p_i^2}$	p_j	$\dfrac{\partial^2 \mathbf{T}}{\partial p_i\,\partial p_j}$
$\dfrac{1}{Z} = \dfrac{1}{R} + \dfrac{1}{j\omega L} + j\omega C$	$\ln R$	$\begin{bmatrix} 0 & \dfrac{Z^2}{R} \\ 0 & 0 \end{bmatrix}$	$\begin{bmatrix} 0 & \dfrac{Z^2}{R}\left(2\dfrac{Z}{R} - 1\right) \\ 0 & 0 \end{bmatrix}$		
	$\ln L$	$\begin{bmatrix} 0 & \dfrac{Z^2}{j\omega L} \\ 0 & 0 \end{bmatrix}$	$\begin{bmatrix} 0 & \dfrac{Z^2}{j\omega L}\left(\dfrac{2Z}{j\omega L} - 1\right) \\ 0 & 0 \end{bmatrix}$		
	$\ln C$	$\begin{bmatrix} 0 & -j\omega C Z^2 \\ 0 & 0 \end{bmatrix}$	$\begin{bmatrix} 0 & j\omega C Z^2(j\omega 2CZ - 1) \\ 0 & 0 \end{bmatrix}$		
	$\ln R$			$\ln L$	$\begin{bmatrix} 0 & \dfrac{2Z^3}{j\omega LR} \\ 0 & 0 \end{bmatrix}$
	$\ln L$			$\ln C$	$\begin{bmatrix} 0 & \dfrac{j\omega 2CZ^3}{R} \\ 0 & 0 \end{bmatrix}$
	$\ln C$			$\ln C$	$\begin{bmatrix} 0 & \dfrac{-2CZ^3}{L} \\ 0 & 0 \end{bmatrix}$

				$\dfrac{\partial^2 \mathbf{T}}{\partial p_i\, \partial p_j}$
p:1 Ideal transformer	$\ln n$	$\begin{bmatrix} n & 0 \\ 0 & \dfrac{-1}{n} \end{bmatrix}$	$\begin{bmatrix} n & 0 \\ 0 & \dfrac{1}{n} \end{bmatrix}$	
	$\ln Z_0$	$\begin{bmatrix} 0 & jZ_0 \sin\theta \\ -jY_0 \sin\theta & 0 \end{bmatrix}$	$\begin{bmatrix} 0 & jZ_0 \sin\theta \\ jY_0 \sin\theta & 0 \end{bmatrix}$	
$\theta = \beta l$ Lossless transmission line section	$\ln\theta$	$\begin{bmatrix} -\theta\sin\theta & jZ_0\theta\cos\theta \\ jY_0\theta\cos\theta & -\theta\sin\theta \end{bmatrix}$	$\begin{bmatrix} -a & jZ_0 b \\ jY_0 b & -a \end{bmatrix}$ $a = \theta^2\cos\theta + \theta\sin\theta$ $b = \theta\cos\theta - \theta^2\sin\theta$	
	p_j			
	$\ln Z_0$	$\ln\theta$	$\begin{bmatrix} 0 & jZ_0\theta\cos\theta \\ -jY_0\theta\cos\theta & 0 \end{bmatrix}$	

Table 4.3 (*Continued*)

Circuit Element	p_i	$\dfrac{\partial T}{\partial p_i}$	$\dfrac{\partial^2 T}{\partial p_i^2}$	p_j	$\dfrac{\partial^2 T}{\partial p_i \, \partial p_j}$
Z_0	$\ln Z_0$	$\begin{bmatrix} 0 & Z \\ 0 & 0 \end{bmatrix}$	$\begin{bmatrix} 0 & Z \\ 0 & 0 \end{bmatrix}$		
	$\ln\theta$	$\begin{bmatrix} 0 & \dfrac{jZ_0\theta}{\cos^2\theta} \\ 0 & 0 \end{bmatrix}$	$\begin{bmatrix} 0 & jZ_0 c \\ 0 & 0 \end{bmatrix}$ $c = \dfrac{\theta\cos\theta + 2\theta^2\sin\theta}{\cos^3\theta}$		
		p_j	$\dfrac{\partial^2 T}{\partial p_i \, \partial p_j}$		
$\theta = \beta l$ Series short-circuited stub $Z = jZ_0\,tg\theta$	$\ln Z_0$	$\ln\theta$	$\begin{bmatrix} 0 & \dfrac{jZ_0\theta}{\cos^2\theta} \\ 0 & 0 \end{bmatrix}$		

$\theta = \beta l$ Series open-circuited stub $Z = jZ_0 \cot\theta$	$\ln Z_0$	$\begin{bmatrix} 0 & Z \\ 0 & 0 \end{bmatrix}$	$\ln Z_0$	$\begin{bmatrix} 0 & Z \\ 0 & 0 \end{bmatrix}$
	$\ln\theta$	$\begin{bmatrix} 0 & \dfrac{-jZ_0\theta}{\sin^2\theta} \\ 0 & 0 \end{bmatrix}$	$\ln\theta$	$\begin{bmatrix} 0 & jZ_0 d \\ 0 & 0 \end{bmatrix}$ $d = \dfrac{\theta\sin\theta - 2\theta^2\cos\theta}{\sin^3\theta}$
		p_j		$\dfrac{\partial^2 \mathbf{T}}{\partial p_i\, \partial p_j}$
	$\ln Z_0$	$\ln\theta$	$\ln Z_0$	$\begin{bmatrix} 0 & \dfrac{-jZ_0\theta}{\sin^2\theta} \\ 0 & 0 \end{bmatrix}$

Table 4.3 (*Continued*)

Circuit Element	p_i	$\dfrac{\partial T}{\partial p_i}$	$\dfrac{\partial^2 T}{\partial p_i^2}$	p_j	$\dfrac{\partial^2 T}{\partial p_i\,\partial p_j}$
	$\ln Z_0$	$\begin{bmatrix} 0 & 0 \\ -Y & 0 \end{bmatrix}$	$\begin{bmatrix} 0 & 0 \\ Y & 0 \end{bmatrix}$		
$\theta = \beta l$ A shunt short-circuited stub $Y = \dfrac{-j\cot\theta}{Z_0}$	$\ln\theta$	$\begin{bmatrix} 0 & 0 \\ \dfrac{-j\theta}{Z_0\sin^2\theta} & 0 \end{bmatrix}$	$\begin{bmatrix} 0 & 0 \\ \dfrac{jd}{Z_0} & 0 \end{bmatrix}$ $d = \dfrac{\theta\sin\theta - 2\theta^2\cos\theta}{\sin^3\theta}$		
		p_j	$\dfrac{\partial^2 T}{\partial p_i\,\partial p_j}$		
	$\ln Z_0$	$\ln\theta$	$\begin{bmatrix} 0 & 0 \\ \dfrac{j\theta}{Z_0\sin^2\theta} & 0 \end{bmatrix}$		

$\theta = \beta l$

Shunt short-circuited stub

$Y = \dfrac{j \cot \theta}{Z_0}$

	$\ln Z_0$		
	$\ln \theta$	$\begin{bmatrix} 0 & 0 \\ -Y & 0 \end{bmatrix}$	$\begin{bmatrix} 0 & 0 \\ Y & 0 \end{bmatrix}$
		$\begin{bmatrix} 0 & 0 \\ \dfrac{j\theta}{Z_0 \cos^2 \theta} & 0 \end{bmatrix}$	$\begin{bmatrix} 0 & 0 \\ \dfrac{jc}{Z_0} & 0 \end{bmatrix}$ $c = \dfrac{\theta \cos \theta + 2\theta^2 \sin \theta}{\cos^3 \theta}$
		p_j	$\dfrac{\partial^2 T}{\partial p_i\, \partial p_j}$
	$\ln Z_0$	$\ln \theta$	$\begin{bmatrix} 0 & 0 \\ \dfrac{-j\theta}{Z_0 \cos^2 \theta} & 0 \end{bmatrix}$

REFERENCES

[1] J.W. Bandler and H.L. Abdel-Malek, "Optimal Centering, Tolerancing and Yield Determination via Updated Approximations and Cuts," *IEEE Trans. Circuits and Systems*, Vol. CAS-25, 1978, pp. 853–870.

[2] J.W. Bandler, P.C. Liu, and H. Tromp, "A Nonlinear Programming Approach to Optimal Design, Centering, Tolerancing and Tuning," *IEEE Trans. Circuits and Systems*, Vol. CAS-23, 1976, pp. 155–165.

[3] J.W. Bandler and P.C. Liu, "Automated Network Design with Optimal Tolerances," *IEEE Trans. Circuits and Systems*, Vol. CAS-21, 1974, pp. 219–222.

[4] B.J. Karafin, "Statistical Circuit Design: The Optimum Assignment of Component Tolerances for Electrical Networks," *BSTJ*, Vol. 50, 1971, pp. 1225–1242.

[5] B.D.H. Tellegen, "A General Network Theorem with Applications," *Philips Research Reports*, Vol. 7, 1952, pp. 259–269.

[6] B.D.H. Tellegen, "A General Network Theorem with Applications," *Proc. Inst. Radio Engineers*, Australia, Vol. 14, 1953, pp. 265–270.

[7] S.W. Director and R.A. Rohrer, "Automated Network Design—The Frequency Domain Case," *IEEE Trans. Circuit Theory*, Vol. CT-16, 1969, pp. 330–337.

[8] M. Sablatash and R.E. Seviora, "Sensitivity Invariants for Scattering Matrices," *IEEE Trans. Circuit Theory*, Vol. CT-18, 1971, pp. 288–290.

[9] K.W. Iobst, "A Direct Approach to the Frequency Domain Analysis and Optimization of Lumped-Distributed Active Two-Ports Using the Hessian Matrix," Ph.D. Dissertation, University of Maryland, 1981.

[10] G. Iuculano, V.A. Monaco, and P. Tiberio, "Network Sensitivities in Terms of Scattering Parameters," *Electronics Letters*, Vol. 8, 1972, pp. 53–54.

[11] V.A. Monaco and P. Tiberio, Two Properties for Circuit Sensitivities in Terms of Scattering Parameters," *Electronics Letters*, Vol. 8, 1972, pp. 382–283.

[12] V.A. Monaco and P. Tiberio, "On Linear Network Scattering Matrix Sensitivities," *Alta Frequenza*, Vol. 39, 1970, pp. 193–195.

[13] K.W. Iobst and K.A. Zaki, "An Optimization Technique for Lumped-Distributed Two-Ports," *IEEE Trans. Microwave Theory Tech.*, Vol. MTT-30, No. 12, 1982, pp. 2167–2171.

Chapter 5

Computer-Aided Noise Analysis
of Microwave Circuits

Before starting discussions directly connected to computer-aided analysis of microwave circuits let us recall some important definitions and descriptions for noise signals in microwave circuits and systems.

From the stochastic processes theory point of view, high-frequency noise is a normal, stationary stochastic process $n(t)$ with its statistical expectation equal to zero $E[n(t)] = 0$ [1–3].

Two stationary stochastic processes $n_1(t)$ and $n_2(t)$ are characterized by the following quantities.

a) Autocorrelation function:

$$R_{n_1 n_1}(\tau) = E[n_1(t)n_1(t + \tau)] \qquad (5.1)$$

$$R_{n_2 n_2}(\tau) = E[n_2(t)n_2(t + \tau)] \qquad (5.2)$$

b) Cross-correlation function:

$$R_{n_1 n_2}(\tau) = E[n_1(t)n_2(t + \tau)] \qquad (5.3)$$

In the frequency domain these quantities are replaced by

a) Power spectral density:

$$S_{n_1}(f) = \int_{-\infty}^{+\infty} R_{n_1 n_1}(\tau)\,e(-j2\pi f\tau)\,d\tau \qquad (5.4)$$

$$S_{n_2}(f) = \int_{-\infty}^{+\infty} R_{n_2 n_2}(\tau)\exp(-j2\pi f\tau)\,d\tau \qquad (5.5)$$

(b) Mutual power spectral density:

$$S_{n_1 n_2}(f) = \int_{-\infty}^{+\infty} R_{n_1 n_2}(\tau)\, e(-j2\pi f\tau)\, d\tau \qquad (5.6)$$

Spectral representation of noise plays a very important role in the analysis of narrow band circuits with center frequency f_0. We can assume that the power noise spectral densities are constant around f_0. In such a case two noise signals are described completely by four numbers: two real numbers $S_{n_1}(f_0)$ and $S_{n_2}(f_0)$ and one complex $S_{n_1 n_2}(f_0)$ [2–5].

In methods for noise analysis from classical circuit theory, noise signals are described by their complex amplitude densities. The definition of the spectral density is based on the Fourier transform [4, 5]. There is an interrelation between these two noise representations. Denoting by $N_1(f)$ and $N_2(f)$ complex amplitudes for noise signals $n_1(t)$ and $n_2(t)$, respectively, these relations are given by [3–5]

$$\overline{N_1(f_0)N_1^*(f_0)} = S_{n_1}(f_0) \qquad (5.7)$$

$$\overline{N_2(f_0)N_2^*(f_0)} = S_{n_2}(f_0) \qquad (5.8)$$

$$\overline{N_1(f_0)N_2^*(f_0)} = S_{n_1 n_2}(f_0) \qquad (5.9)$$

5.1 NOISE REPRESENTATION OF NOISY CIRCUITS

Due to the spectral representation of noise sources, noisy two-ports may be described by small-signal equations, as is very well known, for example, for transistor equivalent circuits. The circuit theory of linear noisy networks shows that any noisy two-port can be replaced by its equivalent circuit, which consists of the original two-port (now assumed to be noiseless) and two additional noise sources. There are many equivalent representations for noisy two-ports. The admittance form of the spectral representation of a noisy two-port is

$$\begin{bmatrix} I_1 \\ I_2 \end{bmatrix} = \begin{bmatrix} Y_{11} & Y_{12} \\ Y_{21} & Y_{22} \end{bmatrix} \begin{bmatrix} V_1 \\ V_2 \end{bmatrix} + \begin{bmatrix} I_{N1} \\ I_{N2} \end{bmatrix} \qquad (5.10)$$

The equivalent circuit for the admittance representation is given in Figure 5.1

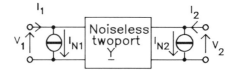

Figure 5.1 Admittance representation of a noisy two-port.

where I_{N1}, I_{N2} represent the port noise currents with input and output short-circuited simultaneously. These two current sources do not actually exist in the positions marked in Figure 5.1; they are merely concentrated equivalent representations of the effect of all noise-current and noise-voltage sources inside the two-port. These primary physical noise sources contribute to I_{N1} and I_{N2}. In general these two deduced quantities are not statistically independent. The arithmetic of noise quantities takes this into account in the following way. Products $\overline{N_1 \cdot N_2^*}$ of two noise quantities N_1 and N_2 represent the Fourier transform of the cross-correlation function $E[n_1(t) \cdot n_2(t + \tau)]$ of the corresponding time functions $n_1(t)$ and $n_2(t)$. These products have to be taken as zero if $n_1(t)$ and $n_2(t)$ are not correlated at all or if there is no correlation at the particular frequency of interest. Otherwise,

$$\overline{N_1 \cdot N_2^*} = c_{12} \sqrt{S_{n_1}(f) \cdot S_{n_2}(f)} \tag{5.11}$$

where S_{n_1} and S_{n_2} are spectral power densities of $n_1(t)$ and $n_2(t)$, and c_{12} is the so-called cross-correlation coefficient.

The noise parameters corresponding to the admittance matrix representation of a two-port are [6]

$$G_1 = \frac{\overline{|I_{N1}|^2}}{4kT_0\,df}, \quad G_2 = \frac{\overline{|I_{N2}|^2}}{4kT_0\,df} \tag{5.12}$$

$$\rho_c = \frac{\overline{I_{N1}^* I_{N2}}}{\sqrt{\overline{|I_{N1}|^2}\,\overline{|I_{N2}|^2}}} \tag{5.13}$$

G_1 and G_2 are equivalent noise conductances, and ρ_c is the correlation coefficient ($\rho_c = |\rho_c| e^{\phi c}$), df is an increment of frequency, T_0 is a standard temperature (290 K), and k the Boltzmann's constant.

The impedance representation of a noisy two-port (see Figure 5.2) is

$$\begin{bmatrix} V_1 \\ V_2 \end{bmatrix} = \begin{bmatrix} Z_{11} & Z_{12} \\ Z_{21} & Z_{22} \end{bmatrix} \begin{bmatrix} I_1 \\ I_2 \end{bmatrix} + \begin{bmatrix} V_{N1} \\ V_{N2} \end{bmatrix} \tag{5.14}$$

Equation (5.14) expresses the fact that a noisy two-port develops noise voltages V_{N1} and V_{N2} across both of its ports if they are simultaneously open-circuited. The representation in Figure 5.2, natural for the impedance matrix representation,

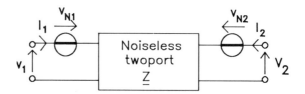

Figure 5.2 Impedance representation of a noisy two-port.

has the following noise parameters:

$$R_1 = \frac{\overline{|V_{N1}|^2}}{4kT_0\, df}, \quad R_2 = \frac{\overline{|V_{N2}|^2}}{4kT_0\, df} \tag{5.15}$$

$$\rho_v = \frac{\overline{V_{N1}^* V_{N2}}}{\sqrt{\overline{|V_{N1}|^2}\,\overline{|V_{N2}|^2}}} \tag{5.16}$$

where R_1 and R_2 are equivalent noise resistances, and ρ_v is the correlation coefficient ($\rho_v = |\rho_v|\, e^{\phi_v}$).

A third, equivalent form of the two-port noise representation (Rothe and Dahlke model [6]) formally uses two input noise sources, V_N and I_N (see Figure 5.3). This the chain matrix representation. The equivalence of the admittance matrix and chain matrix representations require

$$I_N = I_{N1} - \frac{Y_{11}}{Y_{21}} I_{N2}, \quad V_N = -\frac{I_{N2}}{Y_{21}} \tag{5.17}$$

So, the chain representation of a noisy two-port is

$$\begin{bmatrix} V_1 \\ I_1 \end{bmatrix} = \begin{bmatrix} A & B \\ C & D \end{bmatrix} \begin{bmatrix} V_2 \\ -I_2 \end{bmatrix} + \begin{bmatrix} V_N \\ I_N \end{bmatrix} \tag{5.1}$$

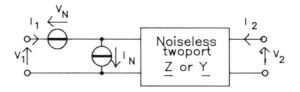

Figure 5.3 Equivalent representation with two noise sources at the input of a noisy two-port.

Noise parameters for the chain matrix representation of a noisy two-port are R_N, g_N, and $\gamma = |\rho| e^{\phi}$, where

$$R_N = \frac{\overline{|V_N|^2}}{4kT_0\,df}, \quad g_N = \frac{\overline{|I_N|^2}}{4kT_0\,df} \tag{5.19}$$

$$\rho = \frac{\overline{V_N^* I_N}}{\sqrt{\overline{|V_N|^2}\,\overline{|I_N|^2}}} \tag{5.20}$$

Note the lower case letter g_N for noise conductance, as in [6]. A noise conductance G_n is used there to denote the portion of g_N that corresponds to the part of the noise source $\overline{|I_N|^2}$ that is not correlated with noise source $\overline{|V_N|^2}$.

The wave representation of noise in a linear two-port originally was from Bauer and Rothe [7]. The complex noise waves are defined as

$$A = \frac{V + ZI}{2\sqrt{\mathrm{Re}\,Z}}, \quad B = \frac{V - Z^*I}{2\sqrt{\mathrm{Re}\,Z}} \tag{5.21}$$

and

$$|A|^2 - |B|^2 = \mathrm{Re}\{VI^*\} \tag{5.22}$$

where Z is a port normalizing impedance, which may be the same as that used to define the scattering parameters of the circuit. If we assume Z to be the characteristic impedance of a port transmission line, A and B are the complex amplitudes of the incoming and outgoing noise waves, respectively, and $|A|^2$ and $|B|^2$ are the corresponding active noise powers carried by these waves.

For a two-port network, the noise wave complex amplitudes at the ports are related by a linear matrix equation:

$$\begin{bmatrix} B_1 \\ B_2 \end{bmatrix} = \begin{bmatrix} S_{11} & S_{12} \\ S_{21} & S_{22} \end{bmatrix} \begin{bmatrix} A_1 \\ A_2 \end{bmatrix} + \begin{bmatrix} B_{N1} \\ B_{N2} \end{bmatrix} \tag{5.23}$$

in which S_{ij}, $i, j = 1, 2$ are the scattering parameters of the two-port. This equation is illustrated in Figure 5.4.

The noise parameters corresponding to scattering matrix representation of a noisy two-port are equivalent noise temperatures T_{B1}, T_{B2} and correlation coefficient $\rho_B = \rho_B| e^{\phi_B}$, where

$$T_{B1} = \frac{\overline{|B_{N1}|^2}}{k\,df}, \quad T_{B2} = \frac{\overline{|B_{N2}|^2}}{k\,df} \tag{5.24}$$

$$\rho_B = \frac{\overline{B_{N1}^* B_{N2}}}{\sqrt{\overline{|B_{N1}|^2}\,\overline{|B_{N2}|^2}}} \tag{5.25}$$

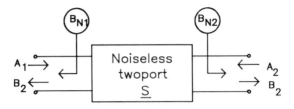

Figure 5.4 Wave representation of a noisy two-port with input and output port noise waves.

The wave representation of a noisy two-port with input and output noise wave sources may be formally replaced by the equivalent wave representation with two input noise wave sources, as it is shown in Figure 5.5. The two noise waves B_{N1} and B_{N2} in Figure 5.4 are formally replaced by two noise waves

$$A_N = \frac{B_{N2}}{S_{21}}, \quad B_N = -\frac{S_{11}}{S_{21}} B_{N2} + B_{N1} \tag{5.26}$$

The matrix equation for this representation is

$$\begin{bmatrix} A_1 \\ B_1 \end{bmatrix} = \begin{bmatrix} T_{11} & T_{12} \\ T_{21} & T_{22} \end{bmatrix} \begin{bmatrix} B_2 \\ A_2 \end{bmatrix} + \begin{bmatrix} A_N \\ B_N \end{bmatrix} \tag{5.27}$$

where T_{ij}, $i, j = 1, 2$ are the transfer scattering parameters of the two-port.

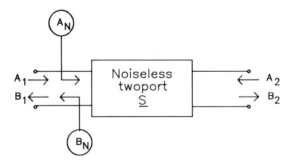

Figure 5.5 Wave representation of a noisy two-port with two input port noise wave sources.

The noise parameters natural for a **T** matrix are the equivalent noise temperatures T_A and T_B and the correlation coefficient $\rho_{AB} = |\rho_{AB}| e^{\phi_{AB}}$, where

$$T_A = \frac{\overline{|A_N|^2}}{k\,df}, \quad T_B = \frac{\overline{|B_N|^2}}{k\,df} \tag{5.28}$$

$$\rho_{AB} = \frac{\overline{A_N^* B_N}}{\sqrt{\overline{|A_N|^2}\,\overline{|B_N|^2}}} \tag{5.29}$$

Table 5.1 presents eight noise parameter sets for noisy two-ports. Anyone of these eight noise representations may be transformed into any other by means of the algebra presented in Table 5.2 [13].

In the general multiport case, the noisy network is represented by the noiseless equivalent of the original circuit with noise current sources connected across each port. Figure 5.6 shows this representation schematically. This is the admittance representation and the matrix equation for it is

$$\mathbf{I} = \mathbf{YV} + \mathbf{I}_N \tag{5.30}$$

where

$\mathbf{Y} =$ the admittance matrix of the multiport,
$\mathbf{V} = [V_1 \quad V_2 \quad \cdots \quad V_n]^T =$ a vector of port voltages,
$\mathbf{I} = [I_1 \quad I_2 \quad \cdots \quad I_n]^T =$ a vector of port currents,
$\mathbf{I}_N = [I_{N1} \quad I_{N2} \quad \cdots \quad I_{Nn}]^T =$ a vector of noise current sources.

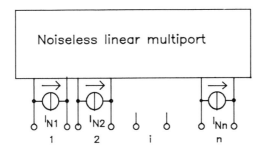

Figure 5.6 Equivalent circuit of a noisy linear multiport, admittance representation.

These noise sources are not statistically independent of each other, and so a complete description of the circuit must include information on their correlation.

As shown in Figure 5.7, noise generated in a multiport also can be represented by mutually correlated noise wave sources, one source at each port. The noiseless equivalent multiport with noise wave sources has the same scattering matrix S as the

Table 5.1

General Noise Parameter Sets for Noisy Two-Ports

		Noise Parameter Sets	Definitions
(1)		$G_1, G_2, \|\rho_c\|, \phi_c$	Figure 5.1; (5.12), (5.13); [6, 8]
(2)		$R_1, R_2, \|\rho_v\|, \phi_v$	Figure 5.2; (5.15), (5.16); [6, 8]
(3)		$R_N, g_N, \|\rho\|, \phi$	Figure 5.3; (5.19), (5.20); [6, 8]
(4)		$T_{B1}, T_{B2}, \|\rho_B\|, \phi_B$	Figure 5.4; (5.24), (5.25); [9]
(5)		$T_A, T_B, \|\rho_{AB}\|, \phi_{AB}$	Figure 5.5; (5.28), (5.29); [7]
(6)		$F_{min}, R_N, G_{opt}, B_{opt}$	$F_{min} = F_{e_{min}} + 1$ is minimum noise figure, R_N is as defined under (3), $G_{opt} + j B_{opt}$ is optimum signal source admittance at which noise figure is minimum, [6, 10, 11]
(7)		$F_{min}, g_N, R_{opt}, X_{opt}$	$F_{min} = F_{e_{min}} + 1$ is minimum noise figure, g_N is as defined under (3), $R_{opt} + j X_{opt}$ is optimum signal source impedance at which noise figure is minimum, [6, 10, 11]
(8)		$F_{min}, N, \|\Gamma_{opt}\|, \phi_{opt}$	$F_{min} = F_{e_{min}} + 1$ is minimum noise figure, $N = 4R_N G_{opt} = 4 g_N R_{opt}$ (R_N and g_N as defined under (3)), G_{opt} as defined under (6), X_{opt} as defined under (7)), $\Gamma_{opt} = \|\Gamma_{opt}\| e\phi_{opt}$ is optimum reflection coefficient of the signal source at which the noise figure is minimum, [11]–[13]

Table 5.2
Transformation Relations between Different Noise
Parameter Sets of Noisy Two-Ports

Transformation Type	Transformation Equations
(1) → (3)	$R_N = G_2 / \mid Y_{21} \mid^2$ $g_N = (\mid Y_{21} \mid^2 G_1 + \mid Y_{11} \mid^2 G_2 - 2\operatorname{Re}\{Y_{11}^* Y_{21} \rho^* \sqrt{G_1 G_2}\}] / \mid Y_{21} \mid^2$ $\rho \sqrt{R_N g_N} = (Y_{11} G_2 - Y_{21} \rho_c^* \sqrt{G_1 G_2}) / \mid Y_{21} \mid^2$
(3) → (1)	$G_1 = \mid Y_{11} \mid^2 R_N + g_N - 2\operatorname{Re}\{Y_{11} \rho^* \sqrt{R_N g_N}\}$ $G_2 = \mid Y_{21} \mid^2 R_N$ $\rho_c \sqrt{G_1 G_2} = Y_{21}(Y_{11}^* R_N - \rho^* \sqrt{R_N g_N})$
(3) → (2)	$R_1 = \mid Z_{11} \mid^2 g_N + R_N - 2\operatorname{Re}\{Z_{11} \rho \sqrt{R_N g_N}\}$ $R_2 = \mid Z_{21} \mid^2 g_N$ $\rho_v \sqrt{R_1 R_2} = Z_{21}(Z_{11}^* g_N - \rho \sqrt{R_N g_N})$
(2) → (3)	$g_N = R_2 / \mid Z_{21} \mid^2$ $R_N = (\mid Z_{21} \mid^2 R_1 + \mid Z_{11} \mid^2 R_2 - 2\operatorname{Re}\{Z_{11}^* Z_{21} \rho_v^* \sqrt{R_1 R_2}\}) / \mid Z_{21} \mid^2$ $\rho \sqrt{R_N g_N} = (Z_{11}^* R_2 - Z_{21}^* \rho_v \sqrt{R_1 R_2}) / \mid Z_{21} \mid^2$
(3) → (6)	$g_N = g_N$ $X_{\text{opt}} = \operatorname{Im}\{\rho \sqrt{R_N g_N}\} / g_N$ $R_{\text{opt}} = \sqrt{R_N / g_N - X_{\text{opt}}^2}$ $F_{e_{\min}} = 2(g_N R_{\text{opt}} + \operatorname{Re}\{\rho \sqrt{R_N g_N}\})$
(6) → (3)	$R_N = g_N(R_{\text{opt}}^2 + X_{\text{opt}}^2)$ $g_N = g_N$ $\operatorname{Re}\{\rho \sqrt{R_N g_N}\} = F_{e_{\min}} / 2 - g_N R_{\text{opt}}$ $\operatorname{Im}\{\rho \sqrt{R_N g_N}\} = g_N X_{\text{opt}}$
(3) → (5)	$T_A = \dfrac{T_0}{Z_0}(R_N + Z_0^2 g_N + 2Z_0 \operatorname{Re}\{\rho \sqrt{R_N g_N}\})$ $T_B = \dfrac{T_0}{Z_0}(R_N + Z_0^2 g_N - 2Z_0 \operatorname{Re}\{\rho \sqrt{R_N g_N}\})$ $\operatorname{Re}\{\rho_{AB}\} = \dfrac{T_0}{\sqrt{T_A T_B} Z_0}(-R_N + Z_0^2 g_N)$ $\operatorname{Im}\{\rho_{AB}\} = \dfrac{T_0}{\sqrt{T_A T_B}} 2\operatorname{Im}\{\rho \sqrt{R_N g_N}\}$

Table 5.2 (*Continued*)

Transformation Type	Transformation Equations
(5) → (3)	$$R_N = \frac{Z_0}{4T_0}(T_A + T_B - 2\sqrt{T_A T_B}\ \text{Re}\{\rho_{AB}\})$$ $$g_N = \frac{1}{4T_0 Z_0}(T_A + T_B + 2\sqrt{T_A T_B}\ \text{Re}\{\rho_{AB}\})$$ $$\text{Re}\{\rho\sqrt{R_N g_N}\} = (T_A - T_B)/4T_0$$ $$\text{Im}\{\rho\sqrt{R_N g_N}\} = \frac{\sqrt{T_A T_B}}{2T_0}\ \text{Im}\{\rho_{AB}\}$$
(6) → (8)	$$F_{e\min} = F_{e\min}$$ $$N = 4R_N G_{\text{opt}}$$ $$|\Gamma_{\text{opt}}| = \sqrt{\frac{(Y_0 - G_{\text{opt}})^2 + B_{\text{opt}}^2}{(Y_0 + G_{\text{opt}})^2 + B_{\text{opt}}^2}}$$ $$\phi_{\text{opt}} = \arctan\left\{\frac{B_{\text{opt}}}{Y_0 - G_{\text{opt}}}\right\} - \arctan\left\{\frac{B_{\text{opt}}}{Y_0 + G_{\text{opt}}}\right\}$$
(8) → (6)	$$F_{e\min} = F_{e\min}$$ $$R_N = \frac{N}{4G_{\text{opt}}}$$ $$G_{\text{opt}} = \text{Re}\left\{Y_0\frac{1 - |\Gamma_{\text{opt}}|e^{\phi_{\text{opt}}}}{1 + |\Gamma_{\text{opt}}|e^{\phi_{\text{opt}}}}\right\}$$ $$B_{\text{opt}} = \text{Im}\left\{Y_0\frac{1 - |\Gamma_{\text{opt}}|e^{\phi_{\text{opt}}}}{1 + |\Gamma_{\text{opt}}|e^{\phi_{\text{opt}}}}\right\}$$
(6) → (7)	$$F_{e\min} = F_{e\min}$$ $$g_N = R_N(G_{\text{opt}}^2 + B_{\text{opt}}^2)$$ $$R_{\text{opt}} = \frac{G_{\text{opt}}}{G_{\text{opt}}^2 + B_{\text{opt}}^2}$$ $$X_{\text{opt}} = -\frac{B_{\text{opt}}}{G_{\text{opt}}^2 + B_{\text{opt}}^2}$$
(7) → (6)	$$F_{e\min} = F_{e\min}$$ $$R_N = g_N(R_{\text{opt}}^2 + X_{\text{opt}}^2)$$ $$G_{\text{opt}} = \frac{R_{\text{opt}}}{R_{\text{opt}}^2 + X_{\text{opt}}^2}$$ $$B_{\text{opt}} = -\frac{X_{\text{opt}}}{R_{\text{opt}}^2 + X_{\text{opt}}^2}$$

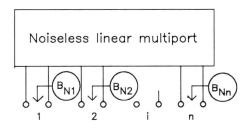

Figure 5.7 Wave representation of noise in a multiport.

original noisy network. The matrix equation for the noise wave representation of a linear multiport is

$$\mathbf{B} = \mathbf{SA} + \mathbf{B}_N \qquad (5.31)$$

where

$\mathbf{S}$ = the scattering matrix of the multiport,
$\mathbf{A} = [A_1 \quad A_2 \quad \cdots \quad A_n]^T$ = a vector of the port incoming noise waves,
$\mathbf{B} = [B_1 \quad B_2 \quad \cdots \quad B_n]^T$ = a vector of the port outgoing noise waves,
$\mathbf{B}_N = [B_{N1} \quad B_{N2} \quad \cdots \quad B_{Nn}]^T$ = a vector of the port equivalent noise wave
   sources.

5.2 CORRELATION MATRICES OF NOISY TWO-PORTS

A physically significant description of noise sources existing in the equivalent circuits of noisy two-ports is given by their self- and cross-power spectral densities, which are defined as the Fourier transform of their auto- and cross-correlation functions. Arranging these spectral densities in matrix form leads to the so-called correlation matrices [14, 15].

The so-called normalized noise correlation matrix for admittance representation is

$$\mathbf{C}_Y = \frac{1}{4kT_0\,df} \overline{\begin{bmatrix} I_{N1} \\ I_{N2} \end{bmatrix} \begin{bmatrix} I_{N1} \\ I_{N2} \end{bmatrix}^+} = \frac{1}{4kT_0\,df} \begin{bmatrix} \overline{I_{N1}I_{N1}^*} & \overline{I_{N1}I_{N2}^*} \\ \overline{I_{N2}I_{N1}^*} & \overline{I_{N2}I_{N2}^*} \end{bmatrix} \quad (5.32)$$

where the overbars denote statistical average, and

k = Boltzmann's constant,
T_0 = reference absolute temperature ($T_0 = 290$ K),
df = noise bandwidth,
$+$ = conjugate transpose,
$*$ = complex conjugate.

Noise correlation matrices for other noise representations of a noisy two-port are defined in the same way. In the case of wave representation normalization with respect to thermal noise available power $kT_0 \, df$ is used:

$$C_Y = \frac{1}{4kT_0 \, df} \mathscr{C}_Y = \frac{1}{4kT_0 \, df} \begin{bmatrix} \overline{I_{N1} I_{N1}^*} & \overline{I_{N1} I_{N2}^*} \\ \overline{I_{N2} I_{N1}^*} & \overline{I_{N2} I_{N2}^*} \end{bmatrix} \tag{5.33}$$

$$C_Z = \frac{1}{4kT_0 \, df} \mathscr{C}_Z = \frac{1}{4kT_0 \, df} \begin{bmatrix} \overline{V_{N1} V_{N1}^*} & \overline{V_{N1} V_{N2}^*} \\ \overline{V_{N2} V_{N1}^*} & \overline{V_{N2} V_{N2}^*} \end{bmatrix} \tag{5.34}$$

$$C_A = \frac{1}{4kT_0 \, df} \mathscr{C}_A = \frac{1}{4kT_0 \, df} \begin{bmatrix} \overline{V_N V_N^*} & \overline{V_N I_N^*} \\ \overline{I_N V_N^*} & \overline{I_N I_N^*} \end{bmatrix} \tag{5.35}$$

$$C_S = \frac{1}{kT_0 \, df} \mathscr{C}_S = \frac{1}{kT_0 \, df} \begin{bmatrix} \overline{B_{N1} B_{N1}^*} & \overline{B_{N1} B_{N2}^*} \\ \overline{B_{N2} B_{N1}^*} & \overline{B_{N2} B_{N2}^*} \end{bmatrix} \tag{5.36}$$

$$C_T = \frac{1}{kT_0 \, df} \mathscr{C}_T = \frac{1}{kT_0 \, df} \begin{bmatrix} \overline{A_N A_N^*} & \overline{-A_N B_N^*} \\ \overline{-B_N A_N^*} & \overline{B_N B_N^*} \end{bmatrix} \tag{5.37}$$

Noise correlation matrices are Hermitian matrices because

$$\text{Im } c_{11} = \text{Im } c_{22} = 0 \quad \text{and} \quad c_{12} = c_{21}^* \tag{5.38}$$

Thanks to this, noise properties of noisy linear two-ports are fully described by four real numbers:

$$c_{11}, \, c_{22}, \, \text{Re } c_{12}, \text{ and } \text{Im } c_{12} \tag{5.39}$$

Another very important aspect of these matrices is that they are positive semidefinite; that is,

$$c_{11} \geq 0 \quad (c_{22} \geq 0) \tag{5.40}$$

and

$$\det \mathbf{C} = c_{11} c_{22} - |c_{12}|^2 \geq 0 \tag{5.41}$$

Noise correlation matrices have been known for years [4, 16], but only in 197 did Reiss [16] and Hillbrandt and Russer [14] prove their importance for the analysis c linear noisy circuits.

5.3 RELATIONS BETWEEN DIFFERENT NOISE CORRELATION MATRICES OF NOISY TWO-PORTS

If, for a given two-port, two or more noise correlation matrices exist, these representations can be transformed into each other by simple transformation operations. The general transformation formula has the form [14]:

$$\mathbf{C}' = \mathbf{PCP}^+ \tag{5.42}$$

where $\mathbf{C}$ and $\mathbf{C}'$ denote respectively the correlation matrices of the original and resulting representations, and the plus sign indicates the conjugate transpose. Transformation matrix $\mathbf{P}$ can be obtained by establishing relations between the noise amplitudes of the original and resulting two-port and by expressing these relations in matrix form.

Transformation matrices for all possible combinations of the noise correlation matrices for noisy two-ports are given in Tables 5.3 and 5.4.

5.4 INTERCONNECTIONS OF NOISY TWO-PORTS

The correlation matrix of an interconnection of two noisy two-ports is a linear transformation of their individual correlation matrices [14]. A general form of this transformation is

$$\mathbf{C} = \mathbf{P}_1\mathbf{C}^{(1)}\mathbf{P}_1^+ + \mathbf{P}_2\mathbf{C}^{(2)}\mathbf{P}_2^+ \tag{5.43}$$

Table 5.3
Transformation Matrices $\mathbf{P}$ for Impedance, Admittance, and Chain Matrix Noise Representations

Resulting Noise Representation	Original Noise Representation		
	Y	**Z**	**A**
Y	$\begin{bmatrix} 1 & 0 \\ 0 & 1 \end{bmatrix}$	$\begin{bmatrix} Y_{11} & Y_{12} \\ Y_{21} & Y_{22} \end{bmatrix}$	$\begin{bmatrix} -Y_{11} & 1 \\ -Y_{21} & 0 \end{bmatrix}$
Z	$\begin{bmatrix} Z_{11} & Z_{12} \\ Z_{21} & Z_{22} \end{bmatrix}$	$\begin{bmatrix} 1 & 0 \\ 0 & 1 \end{bmatrix}$	$\begin{bmatrix} 1 & -Z_{11} \\ 0 & -Z_{21} \end{bmatrix}$
A	$\begin{bmatrix} 0 & A_{12} \\ 1 & A_{22} \end{bmatrix}$	$\begin{bmatrix} 1 & -A_{11} \\ 0 & -A_{21} \end{bmatrix}$	$\begin{bmatrix} 1 & 0 \\ 0 & 1 \end{bmatrix}$

Table 5.4
Transformation Matrices for Scattering Matrix
and Transfer Scattering Matrix Noise Representations

Resulting Noise Representation	Original Noise Representation	
	S	T
S	$\begin{bmatrix} 1 & 0 \\ 0 & 1 \end{bmatrix}$	$\begin{bmatrix} -S_{11} & 0 \\ -S_{21} & 0 \end{bmatrix}$
T	$\begin{bmatrix} 0 & -T_{11} \\ 1 & -T_{21} \end{bmatrix}$	$\begin{bmatrix} 1 & 0 \\ 0 & 1 \end{bmatrix}$

where $\mathbf{C}^{(1)}$ and $\mathbf{C}^{(2)}$ are correlation matrices of two-ports to be connected, $\mathbf{P}_1$ and $\mathbf{P}_2$ are transformation matrices, and $\mathbf{C}$ is the resulting correlation matrix of the interconnection. The transformation matrices corresponding to the various types of interconnections can be obtained by establishing relations between the noise amplitudes of the individual two-ports to be connected and the resulting two-port, and by expressing these relations in matrix form.

Of particular interest for the noise analysis of circuits composed of interconnected two-ports are series, parallel, and cascaded connections. The following formulas relate the resulting correlation matrix to the correlation matrices of the two-ports according to their connection.

Parallel connection:

$$\mathbf{C}_Y = \mathbf{C}_Y^{(1)} + \mathbf{C}_Y^{(2)} \tag{5.44}$$

Series connection:

$$\mathbf{C}_Z = \mathbf{C}_Z^{(1)} + \mathbf{C}_Z^{(2)} \tag{5.45}$$

Cascade connection:

$$\mathbf{C}_A = \mathbf{A}^{(1)}\mathbf{C}_A^{(2)}\mathbf{A}^{(1)\,+} + \mathbf{C}_A^{(1)} \tag{5.46}$$

Cascade connection:

$$\mathbf{C}_T = \mathbf{T}^{(1)}\mathbf{C}_T^{(2)}\mathbf{T}^{(1)\,+} + \mathbf{C}_T^{(1)} \tag{5.47}$$

In these equations the (1) and (2) superscripts refer to the connected two-ports and the plus superscript means the Hermitian complex conjugate. In the case of parallel and series connections of two-ports the resulting correlation matrix is a sum of respectively the correlation matrices in admittance and impedance representation of the original two-ports. For the cascade connection with an order indicated by the subscripts, the corresponding formulas are more complicated. Equation (5.46), which relates to chain matrix representation, contains the chain matrix $\mathbf{A}^{(1)}$ of the first two-port. In (5.47), applicable to the transfer scattering matrix representation, a matrix $\mathbf{T}^{(1)}$ is the transfer scattering matrix of the first two-port in the cascade.

5.5 CORRELATION MATRICES OF ACTIVE TWO-PORTS AND PASSIVE MULTIPORTS

In general, elements of microwave circuits can be divided into two groups: passive and active multiports. Lossy passive multiports generate only thermal noise. The correlation matrices of lossy passive multiports are

for admittance representation [17]:

$$\mathscr{C}_Y = 2kT\,df(\mathbf{Y} + \mathbf{Y}^+) \qquad (5.48)$$

for impedance representation [17]:

$$\mathscr{C}_Z = 2kT\,df(\mathbf{Z} + \mathbf{Z}^+) \qquad (5.49)$$

for scattering matrix representation [18]:

$$\mathscr{C}_S = kT\,df(\mathbf{I} + \mathbf{SS}^+) \qquad (5.50)$$

where

k = Boltzmann's constant,
T = physical absolute temperature of the two-port,
$\mathbf{I}$ = an identity matrix.

Of course, all three relations are applicable as well to lossy passive two-ports. Correlation matrices of lossy passive two-ports for chain and transfer scattering matrix representations may be obtained from these relations by using the transformation formulas given in Tables 5.3 and 5.4. Tables 5.5 and 5.6 present noise correlation matrices for some passive elements of microwave circuits represented by scattering matrices and admittance matrices, respectively.

For active two-ports, as for example bipolar or FET transistors, the correlation matrices can be obtained from their noise equivalent circuits [12, 19–22]. Where the correlation matrix cannot be derived from theoretical investigations, measurements of noise characteristics provide the required information. Proper measurement techniques

Table 5.5
Noise Wave Correlation Matrices for Some
Passive Elements Represented by Scattering Matrices

Circuit Element	Noise Wave Correlation Matrix $\mathscr{C}_s$
$Z = R + jX$ Z_N	$4kT_0\,df\,\dfrac{RZ_N}{\lvert Z + Z_N\rvert^2}$
$Z = R + jX$ Z_N Z_N	$\dfrac{4kT_0\,df\,RZ_N}{\lvert Z + 2Z_N\rvert^2}\begin{bmatrix} 1 & -1 \\ -1 & 1 \end{bmatrix}$
$Y = G + jB$ Y_N	$4kT_0\,df\,\dfrac{GY_N}{\lvert Y + Y_N\rvert^2}$
Y_N Y Y_N $Y = G + jB$	$\dfrac{4kT_0\,df\,GY_N}{\lvert Y + 2Y_N\rvert^2}\begin{bmatrix} 1 & 1 \\ 1 & 1 \end{bmatrix}$
Z_N Z_0 Z_N A transmission line section $\gamma = \alpha + j\beta, \quad \beta = \dfrac{2\pi}{\lambda_g}$ $\theta = \gamma l$	$kT_0\,df\begin{bmatrix} c_{11} & c_{12} \\ c_{12} & c_{11} \end{bmatrix}$ $c_{11} = \dfrac{4Z_0Z_N\{(Z_0^2 + Z_N^2)\,\text{Re}\{\coth\theta\} + Z_0Z_N\left[\lvert\coth\theta\rvert^2 - \lvert\operatorname{csch}\theta\rvert^2\right]\}}{(Z_0^2 + Z_N^2)\left[(Z_0^2 + Z_N^2) + 4Z_0Z_N\,\text{Re}\{\coth\theta\}\right]}$ $c_{12} = \dfrac{4Z_0Z_N(Z_0^2 - Z_N^2)\,\text{Re}\{\operatorname{csch}\theta\}}{(Z_0^2 + Z_N^2)\left[(Z_0^2 + Z_N^2) + 4Z_0Z_N\,\text{Re}\{\coth\theta\}\right]}$

$$kT_0\,\mathrm{d}f \begin{bmatrix} c_{11} & c_{12} & c_{13} & c_{14} \\ c_{12} & c_{11} & c_{14} & c_{13} \\ c_{13} & c_{14} & c_{11} & c_{12} \\ c_{14} & c_{13} & c_{12} & c_{11} \end{bmatrix}$$

$$c_{11} = 1 - \frac{1}{2}\left(|S_{11e}|^2 + |S_{110}|^2 + |S_{12e}|^2 + |S_{120}|^2 \right)$$

$$c_{12} = -\frac{1}{2}\left(|S_{11e}|^2 - |S_{110}|^2 + |S_{12e}|^2 - |S_{120}|^2 \right)$$

$$c_{13} = -\left[\operatorname{Re}\{S_{11e}S_{12e}^*\} - \operatorname{Re}\{S_{110}S_{120}^*\} \right]$$

$$c_{14} = -\left[\operatorname{Re}\{S_{11e}S_{12e}^*\} + \operatorname{Re}\{S_{110}S_{120}^*\} \right]$$

$$S_{11e,0} = \frac{Z_{0e,0}^2 - Z_N^2}{Z_{0e,0}^2 + Z_N^2 + 2Z_{0e,0}Z_N \operatorname{ctgh}(\gamma_{e,0}l)}$$

$$S_{12e,0} = \frac{2Z_{0e,0}Z_N \cosh(\gamma_{e,0}l)}{Z_{0e,0}^2 + Z_N^2 + 2Z_{0e,0}Z_N \operatorname{ctgh}(\gamma_{e,0}l)}$$

Z_{0e}, Z_{00}, l

$\gamma_e = \alpha_e + j\beta_e, \quad \beta_e = \dfrac{2\pi}{\lambda_{ge}}$

$\gamma_0 = \alpha_0 + j\beta_0, \quad \beta_0 = \dfrac{2\pi}{\lambda_{g0}}$

A section of coupled transmission lines

$$kT_0\,\mathrm{d}f \frac{4Z_0 Z_N \operatorname{Re}\{\tanh\theta\}}{|Z_0 \tanh\theta + 2Z_N|^2} \begin{bmatrix} 1 & -1 \\ -1 & 1 \end{bmatrix}$$

Short-ended series connected stub

$$kT_0\,\mathrm{d}f \frac{4Z_0 Z_N \operatorname{Re}\{\coth\theta\}}{|Z_0 \coth\theta + 2Z_N|^2} \begin{bmatrix} 1 & -1 \\ -1 & 1 \end{bmatrix}$$

Open-ended series connected stub

Table 5.5 (*Continued*)

Circuit Element	Noise Wave Correlation Matrix $\mathscr{C}_s$
  Short-ended shunt connected stub	$$kT_0 \, df \frac{4Y_0 Y_N \, \text{Re}\{\coth\theta\}}{|Y_0 \coth\theta + 2Y_N|^2} \begin{bmatrix} 1 & 1 \\ 1 & 1 \end{bmatrix}$$
  Open-ended shunt connected stub	$$kT_0 \, df \frac{4Y_0 Y_N \, \text{Re}\{\tanh\theta\}}{|Y_0 \tanh\theta + 2Y_N|^2} \begin{bmatrix} 1 & 1 \\ 1 & 1 \end{bmatrix}$$
  Distributed RC line section	As for a transmission line section with $Z_0 = \sqrt{\dfrac{R}{j\omega C}} = \sqrt{j\omega RC}$ and $\gamma l = \sqrt{j\omega RC}$ $R = r \cdot l$ is total resistance $C = c \cdot l$ is total capacitance

Table 5.6
Noise Correlation Matrices for Some Passive
Elements Represented by Admittance Matrices

Circuit Element	Noise Current Correlation Matrix $\mathscr{C}_Y$
Y $4kT_0 \, df \, \mathrm{Re}\{Y\} \begin{bmatrix} 1 & -1 \\ -1 & 1 \end{bmatrix}$ (circuit: element Y)	$4kT_0 \, df \, \mathrm{Re}\{Y\} \begin{bmatrix} 1 & -1 \\ -1 & 1 \end{bmatrix}$
l Y_0 $\gamma = \alpha + j\beta, \ \beta = \dfrac{2\pi}{\lambda_g}$ A transmission line section, $\theta = \gamma l$	$4kT_0 \, df \, Y_0 \begin{bmatrix} \mathrm{Re}\{\coth\theta\} & -\mathrm{Re}\{\mathrm{csch}\,\theta\} \\ -\mathrm{Re}\{\mathrm{csch}\,\theta\} & \mathrm{Re}\{\coth\theta\} \end{bmatrix}$
l Y_0 Short-ended series connected stub, $\theta = \gamma l$	$4kT_0 \, df \, Y_0 \, \mathrm{Re}\{\coth\theta\} \begin{bmatrix} 1 & -1 \\ -1 & 1 \end{bmatrix}$
l Y_0 Open-ended series connected stub, $\theta = \gamma l$	$4kT_0 \, df \, Y_0 \, \mathrm{Re}\{\tanh\theta\} \begin{bmatrix} 1 & -1 \\ -1 & 1 \end{bmatrix}$

Table 5.6 (*Continued*)

Circuit Element	Noise Current Correlation Matrix $\mathscr{C}_Y$
Π = network	$4kT_0\,df\begin{bmatrix} \mathrm{Re}\{Y_1 + Y_2\} & -\mathrm{Re}\{Y_2\} \\ -\mathrm{Re}\{Y_2\} & \mathrm{Re}\{Y_1 + Y_3\} \end{bmatrix}$
T = network	$\dfrac{4kT_0\,df}{\lvert Z_1Z_2 + Z_2Z_3 + Z_1Z_3\rvert^2}\begin{bmatrix} c_{11} & \mathrm{Re}\{Z_3\} \\ \mathrm{Re}\{Z_3\} & c_{11} \end{bmatrix}$ $c_{11} = \lvert Z_2\rvert^2[\mathrm{Re}\,Z_1 + \mathrm{Re}\,Z_3] + \lvert Z_3\rvert^2[\mathrm{Re}\,Z_1 + \mathrm{Re}\,Z_2]$ $+\,\mathrm{Re}\{Z_1Z_2^*Z_3\} + \mathrm{Re}\{Z_1Z_2Z_3^*\}$
Y_{0e}, Y_{0o}, l $\gamma_e = \alpha_e + j\beta_e,\ \beta_e = \dfrac{2\pi}{\lambda_{ge}}$ $\gamma_e = \alpha_e + j\beta_e,\ \beta_e = \dfrac{2\pi}{\lambda_{ge}}$ A section of coupled transmission lines	$2kT_0\,df\begin{bmatrix} c_{11} & c_{12} & c_{13} & c_{14} \\ c_{12} & c_{11} & c_{14} & c_{13} \\ c_{13} & c_{14} & c_{11} & c_{12} \\ c_{14} & c_{13} & c_{12} & c_{11} \end{bmatrix}$ $c_{11} = \mathrm{Re}\{Y_{0e}\coth(\gamma_e l) + Y_{0o}\coth(\gamma_0 l)\}$ $c_{12} = \mathrm{Re}\{Y_{0e}\coth(\gamma_e l) - Y_{0o}\coth(\gamma_0 l)\}$ $c_{13} = -\mathrm{Re}\{Y_{0e}\,\mathrm{csch}(\gamma_e l) - Y_{0o}\,\mathrm{csch}(\gamma_0 l)\}$ $c_{14} = -\mathrm{Re}\{Y_{0e}\,\mathrm{csch}(\gamma_e l) + Y_{0o}\,\mathrm{csch}(\gamma_0 l)\}$

are used for obtaining four device noise parameters: $F_{\min}$ = minimum noise figure, R_N = the equivalent noise resistance (as defined in (5.19)), and $Y_{\mathrm{opt}} = G_{\mathrm{opt}} + jB_{\mathrm{opt}}$ = the optimal source admittance at which the noise figure has a minimum value [10, 11, 23, 24]. The other equivalent four noise parameters are $F_{\min}$, G_N = the equivalent noise conductance (as defined in (5.19)), and $Z_{\mathrm{opt}} = R_{\mathrm{opt}} + jX_{\mathrm{opt}}$ = the optimum source impedance, or $F_{\min}$, R_N, and Γ_{opt} = optimum reflection coefficient of the signal source. With these parameters estimated, the correlation matrices of active two-ports are obtained as given below.

Admittance matrix representation:

$$\mathscr{C}_Y = 4kT\,\mathrm{d}f \cdot C_Y = 4kT\,\mathrm{d}f$$

$$\cdot \begin{bmatrix} \begin{matrix} \left(|\,Y_{\text{opt}}\,|^2 + |\,Y_{11}\,|^2\right)R_N \\[2pt] -2\,\mathrm{Re}\{Y_{11}\}\left[\dfrac{F_{e_{\min}}}{2} - R_N G_{\text{opt}}\right] \\[2pt] -2\,\mathrm{Im}\{Y_{11}\}R_N B_{\text{opt}} \end{matrix} & \vdots & \begin{matrix} Y_{11}Y_{21}^* R_N \\[4pt] -Y_{21}^*\left[\dfrac{F_{e_{\min}}}{2} - R_N Y_{\text{opt}}^*\right] \end{matrix} \\[4pt] \hline \\[-6pt] Y_{11}^* Y_{21} R_N - Y_{21}\left[\dfrac{F_{e_{\min}}}{2} - R_N Y_{\text{opt}}\right] & \vdots & |\,Y_{21}\,|^2 R_N \end{bmatrix}$$

$$(5.51)$$

Chain matrix representation:

$$\mathscr{C}_A = 4kT\,\mathrm{d}f \cdot C_A$$

$$= 4kT\,\mathrm{d}f \begin{bmatrix} R_N & \vdots & \dfrac{F_{e_{\min}}}{2} - R_N Y_{\text{opt}} \\[4pt] \hline \\[-6pt] \dfrac{F_{e_{\min}}}{2} - R_N Y_{\text{opt}}^* & \vdots & R_N\,|\,Y_{\text{opt}}\,|^2 \end{bmatrix} \qquad (5.52)$$

Transfer scattering matrix representation:

$$\mathscr{C}_T = kT\,\mathrm{d}f \cdot \mathbf{C}_T$$

$$= \frac{kT\,\mathrm{d}f}{1 - |\,\Gamma_{\text{opt}}\,|^2} \begin{bmatrix} F_{e_{\min}} + \left(N - F_{e_{\min}}\right)|\,\Gamma_{\text{opt}}\,|^2 & \vdots & N\Gamma_{\text{opt}} \\[4pt] \hline \\[-6pt] N\Gamma_{\text{opt}}^* & \vdots & N - F_{e_{\min}}\left(1 - |\,\Gamma_{\text{opt}}\,|^2\right) \end{bmatrix}$$

$$(5.53)$$

Scattering matrix representation:

$$\mathscr{C}_S = kT\,\mathrm{d}f \cdot C_S = \frac{kT\,\mathrm{d}f}{1 - |\,\Gamma_{\text{opt}}\,|^2}$$

$$\cdot \begin{bmatrix} \begin{matrix} \left[F_{e_{\min}} + \left(N - F_{e_{\min}}\right)|\,\Gamma_{\text{opt}}\,|^2\right]|\,S_{11}\,|^2 \\[2pt] +2N\,\mathrm{Re}\{\Gamma_{\text{opt}}S_{11}\} + N - F_{e_{\min}}\left(1 - |\,\Gamma_{\text{opt}}\,|^2\right) \end{matrix} & \vdots & \begin{matrix} \left[F_{e_{\min}} + \left(N - F_{e_{\min}}\right)|\,\Gamma_{\text{opt}}\,|^2\right]S_{11}S_{21}^* \\[2pt] +N\Gamma_{\text{opt}}^* S_{21}^* \end{matrix} \\[4pt] \hline \\[-6pt] \left[F_{e_{\min}} + \left(N - F_{e_{\min}}\right)|\,\Gamma_{\text{opt}}\,|^2\right]S_{11}^* S_{21} + N\Gamma_{\text{opt}}S_{21} & \vdots & \left[F_{e_{\min}} + \left(N - F_{e_{\min}}\right)|\,\Gamma_{\text{opt}}\,|^2\right]|\,S_{21}\,|^2 \end{bmatrix}$$

$$(5.54)$$

In the preceding representations, $F_{e\min} = F_{\min} - 1$ is the minimum excess noise figure, and

$$N = 4R_N G_{opt} = 4g_N R_{opt} \qquad (5.55)$$

The figure $F_{\min}$ and parameter N are known to be invariant under transformation through lossless reciprocal two-ports connected to the input of a noisy two-port [25]. Also for $F_{\min}$ and N to represent a physical two-port, the following inequality has to be satisfied [26]:

$$F_{e\min} \leq N \qquad (5.56)$$

This inequality follows directly from the property of the correlation matrices that have to be Hermitian and positive semidefinite.

The other restrictions satisfied by noise parameters of a physical two-port are [28]

$$R_N \geq \max(R_{N1}, R_{N2}) \qquad (5.57)$$

$$R_{N1} \geq \frac{\text{Re}\{y_{11}\}(F_{\min} - 1)}{|Y_{opt} + y_{11}|^2} \qquad (5.58)$$

$$R_{N2} \geq \frac{F_{\min} - 1}{4\,\text{Re}\{Y_{opt}\}} \qquad (5.59)$$

Example 5.1

Figure 5.8 shows a noise equivalent circuit of the FET, valid for high frequencies [20–22]. The important extrinsic thermal noise sources are those associated with the gate metallization resistance R_m and the source-gate resistance R_f. The noise generator i_g represents the induced gate noise of the intrinsic device. The intrinsic drain noise generator i_d is coupled with i_g, which is represented by correlation coefficient C:

$$jC = \frac{\overline{i_g^* i_d}}{\sqrt{|\overline{i_g^2}|\,|\overline{i_d^2}|}}$$

which approaches unity in magnitude for short-gate devices.

One need not include all of the equivalent circuit elements of the FET because some have little effect on noise performance. For simplicity we shall neglect the drain-gate capacitance C_{dg} and the drain resistance R_d. With these approximations, the noise equivalent circuit of the FET in common source configuration reduces to that shown in Figure 5.9.

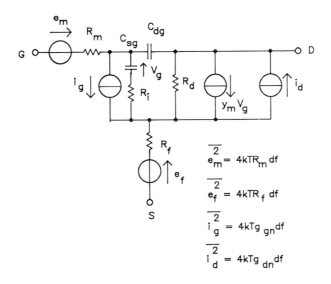

Figure 5.8 Noise equivalent circuit of FET; noise sources i_g and i_d are correlated.

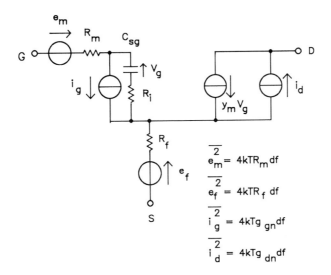

Figure 5.9 Simplified equivalent circuit used in noise analysis of FET.

Let us determine now noise currents I_{N1} and I_{N2} of the admittance representation of the two-port. By simultaneously short-circuiting the input and output of the circuit shown in Figure 5.9, we have

$$I_{N1} = \frac{\left(1 + y_m R_f + j\omega R_i C_{sg}\right)i_g + j\omega C_{sg}\left(e_m - e_f + R_f i_d\right)}{1 + y_m R_f + j\omega C_{sg}\left(R_f + R_i + R_m\right)}$$

$$I_{N2} = \frac{y_m}{1 + j\omega C_{sg}}\left(I_{N1} - i_g\right) - i_d$$

By using (5.32), we can compute now the noise correlation matrix for the admittance representation.

5.6 BASIC RELATIONSHIPS FOR NOISY TWO-PORTS

When the two-port is driven by a signal source with the internal admittance $Y_s = G_s + jB_s$, its noise figure is written in the form [25, 27]

$$F = F_{\min} + \frac{R_N}{G_s}\left| Y_s - Y_{\text{opt}} \right|^2 \tag{5.60}$$

The other two equations equivalent to (5.60) are

$$F = F_{\min} + \frac{g_N}{R_s}\left| Z_s - Z_{\text{opt}} \right|^2 \tag{5.61}$$

where $Z_s = R_s + jX_s$ is the internal impedance of the signal source, or

$$F = F_{\min} + N\frac{\left| \Gamma_s - \Gamma_{\text{opt}} \right|^2}{\left(1 - \left| \Gamma_s \right|^2\right)\left(1 - \left| \Gamma_{\text{opt}} \right|^2\right)} \tag{5.62}$$

$$F = F_{\min} + 4\frac{R_N}{Z_0}\frac{\left| \Gamma_s - \Gamma_{\text{opt}} \right|^2}{\left|1 + \Gamma_{\text{opt}} \right|^2\left(1 - \left| \Gamma_s \right|^2\right)} \tag{5.63}$$

where Γ_s is the reflection coefficient of the signal source, and Z_0 the real reference impedance of the input port of the two-port.

According to (5.51) to (5.55), the device noise parameters and the noise figure of a two-port obviously can be expressed as functions of the elements of the correlation matrices, as we can easily predict from a physical point of view.

Admittance Representation

Noise parameters:

$$F_{\min} = 1 + 2\frac{C_{Y22}}{|Y_{21}|^2}\left(G_{\text{opt}} + \text{Re}\left\{Y_{11} - Y_{21}\frac{C_{Y12}}{C_{Y22}}\right\}\right) \tag{5.64}$$

$$G_{\text{opt}} = \sqrt{|Y_{21}|^2\frac{C_{Y11}}{C_{Y22}} - |Y_{21}|^2\frac{|C_{Y12}|^2}{C_{Y22}^2} + \left(\text{Re}\left\{Y_{11} - Y_{21}\frac{C_{12}}{C_{22}}\right\}\right)^2} \tag{5.65}$$

$$B_{\text{opt}} = -\text{Im}\left\{Y_{11} - Y_{21}\frac{C_{Y12}}{C_{Y22}}\right\} \tag{5.66}$$

$$R_N = \frac{C_{Y22}}{|Y_{21}|^2} \tag{5.67}$$

The noise figure:

$$F = 1 + \frac{\mathcal{O}^+\mathbf{H}\mathbf{C}_Y\mathbf{H}\mathcal{O}}{\text{Re } Y_s} \tag{5.68}$$

where

$$\mathcal{O} = \begin{bmatrix} 1 \\ \dfrac{(Y_s + Y_{11})*}{Y_{21}^*} \end{bmatrix} \tag{5.69}$$

$$\mathbf{H} = \begin{bmatrix} 1 & 0 \\ 0 & -1 \end{bmatrix} \tag{5.70}$$

In (5.68) and (5.69) Y_s is the admittance of the signal source.

Chain Matrix Representation

Noise parameters are

$$F_{\min} = 1 + 2\left(C_{A12} + R_N Y_{\text{opt}}\right) \tag{5.71}$$

$$G_{\text{opt}} = \sqrt{\frac{C_{A22}}{C_{A11}} - \left(\text{Im}\left\{\frac{C_{A12}}{C_{A11}}\right\}\right)^2} \tag{5.72}$$

$$B_{\text{opt}} = -\text{Im}\left\{\frac{C_{A_{12}}}{C_{A_{11}}}\right\} \tag{5.73}$$

$$R_N = C_{A_{11}} \tag{5.74}$$

The noise figure is

$$F = 1 + \frac{\mathbf{Z}^+ \mathbf{C}_A \mathbf{Z}}{\text{Re } Z_s} \tag{5.75}$$

where

$$\mathbf{Z} = \begin{bmatrix} 1 \\ Z_s \end{bmatrix} \tag{5.76}$$

In (5.75) and (5.76) Z_s is the impedance of the signal source.

Transfer Scattering Matrix Representation

Noise parameters are

$$F_{\text{min}} = 1 + \frac{1}{2}\left(C_{T_{11}} - C_{T_{22}} + N\right) \tag{5.77}$$

$$\Gamma_{\text{opt}} = \frac{2C_{T_{12}}}{C_{T_{11}} + C_{T_{22}} + N} \tag{5.78}$$

$$N = \sqrt{\left(C_{T_{11}} + C_{T_{22}}\right)^2 - 4\,|\,C_{T_{12}}\,|^2} \tag{5.79}$$

The noise figure is

$$F = 1 + \frac{\Gamma^+ \mathbf{H} \mathbf{C}_T \mathbf{H} \Gamma}{\Gamma^+ \mathbf{H} \Gamma} \tag{5.80}$$

where

$$\Gamma = \begin{bmatrix} 1 \\ \Gamma_s^* \end{bmatrix} \tag{5.81}$$

$$\mathbf{H} = \begin{bmatrix} 1 & 0 \\ 0 & -1 \end{bmatrix} \tag{5.82}$$

In (5.81), Γ_s is the reflection coefficient of the signal source port.

Scattering Matrix Representation

The noise parameters are

$$F_{\min} = 1 + \frac{1}{2}\left[C_{S_{22}}\frac{1 - |S_{11}|^2}{|S_{21}|^2} - C_{S_{11}} + 2\,\mathrm{Re}\left\{C_{S_{12}}\frac{S_{11}^*}{S_{21}^*}\right\} + N\right] \quad (5.83)$$

$$\Gamma_{\mathrm{opt}} = \frac{2\left(C_{S_{12}}^*\dfrac{1}{S_{21}} - C_{22}\dfrac{S_{11}^*}{|S_{21}|^2}\right)}{C_{S_{22}}\dfrac{1 + |S_{11}|^2}{|S_{21}|^2} + C_{S_{11}} - 2\,\mathrm{Re}\left\{C_{S_{12}}\dfrac{S_{11}^*}{S_{21}^*}\right\} + N} \quad (5.84)$$

$$N = \left\{\left[C_{S_{22}}\frac{1 + |S_{11}|^2}{|S_{21}|^2} + C_{S_{11}} - 2\,\mathrm{Re}\left\{C_{S_{12}}\frac{S_{11}^*}{S_{21}^*}\right\}\right]^2\right.$$

$$\left. - 4\left|C_{S_{12}}^*\frac{1}{S_{21}} - C_{S_{22}}\frac{S_{11}^*}{|S_{21}|^2}\right|^2\right\}^{1/2} \quad (5.85)$$

The noise figure is

$$F = 1 + \frac{C_{S_{11}}\left|\dfrac{S_{21}\Gamma_s}{1 - S_{11}\Gamma_s}\right|^2 + C_{S_{22}} + 2\,\mathrm{Re}\left\{C_{S_{12}}\dfrac{S_{21}\Gamma_s}{1 - S_{11}\Gamma_s}\right\}}{\left(1 - |\Gamma_s|^2\right)\left|\dfrac{S_{21}}{1 - S_{11}\Gamma_s}\right|^2} \quad (5.86)$$

5.7 NOISE ANALYSIS OF CASCADED TWO-PORTS

In microwave circuits composed of cascaded two-ports (see Figure 5.10), each element (two-port) has to be specified by its electrical matrix and noise correlation matrix. The appropriate representation for cascaded interconnections is chain matrix **A** or transfer

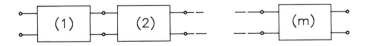

Figure 5.10 Cascade connection of two-port elements.

scattering matrix $\mathbf{T}$ representing individual elements of the cascade. Electrical matrices $\mathbf{A}$ or $\mathbf{T}$ are obtained on a theoretical or experimental basis. The correlation matrices are determined using the methods described in Section 5.5. Noise analysis of a circuit starts with the calculation of the noise correlation and electrical matrices of first two cascaded elements by using the following relations:

$$\mathbf{C}_A = \mathbf{A}^{(1)}\mathbf{C}_A^{(2)}\mathbf{A}^{(1)+} + \mathbf{C}_A^{(1)} \tag{5.87}$$

$$\mathbf{A} = \mathbf{A}^{(1)}\mathbf{A}^{(2)} \tag{5.88}$$

Once matrices $\mathbf{C}_A$ and $\mathbf{A}$ are determined, the matrices of the cascaded first three elements can be calculated by repeating the procedure used for the first two elements. The noise correlation and electrical matrices of the connection of the first two elements are treated as the matrices of an element to which we connect the third element. The resulting matrices are computed by using (5.87) and (5.88). The procedure is repeated until the overall noise correlation matrix $\mathbf{C}_A$ and overall chain matrix $\mathbf{A}$ of the cascade are derived.

The noise analysis of the cascaded two-ports also can be realized by using the transfer scattering matrix noise representation. The principle of the analysis is exactly the same as for the chain matrix representation. The basic equations for the method are

$$\mathbf{C}_T = \mathbf{T}^{(1)}\mathbf{C}_T^{(2)}\mathbf{T}^{(1)+} + \mathbf{C}_T^{(1)} \tag{5.89}$$

$$\mathbf{T} = \mathbf{T}^{(1)}\mathbf{T}^{(2)} \tag{5.90}$$

where, as before, the superscripts (1) and (2) refer to the two-ports to be connected in the cascade in the order indicated. The plus sign means the Hermitian conjugation. In this way the noise correlation and electrical matrices of the overall circuit allow us to calculate any circuit functions.

5.8 NOISE ANALYSIS OF CIRCUITS COMPOSED OF INTERCONNECTED TWO-PORTS

Figure 5.11 shows some examples of circuits whose noise performance can be analyzed using the results of previous sections. When a circuit to be analyzed can be decomposed into basic two-ports connected in cascade, series-series, or parallel-parallel, the noise analysis may be based on the principles discussed in Section 5.4. We now assume that the proper electrical and noise correlation matrices of the basic two-ports can be determined theoretically or experimentally. The basic two-ports are successively interconnected so that the overall two-port finally is obtained. We compute the correlation matrix of the overall network by applying the interconnection rules discussed in Section

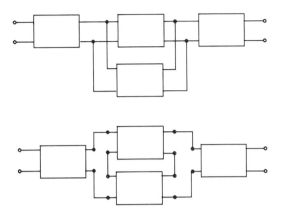

Figure 5.11 Examples of circuits with cascade, parallel-parallel, and series-series connections of two-port elements.

5.4 and given in Tables 5.3 and 5.4 and properly changing the noise representation. Once the correlation matrix of the overall two-port is known, all of its noise parameters can be computed. The noise figure corresponding to the source admittance Y_s, minimum noise figure F_{min}, optimum source admittance Y_{opt}, and the equivalent noise resistance R_N can be determined directly by this noise analysis method.

Example 5.2

The principles of the noise analysis method will be explained by using the circuit presented in Figure 5.12 as an example. It is a single stage MESFET amplifier with parallel and series feedback. First we have to decompose the circuit into basic two-ports that can be described by their electrical and noise correlation matrices. Figure 5.12(b) shows the decomposed version of the example circuit.

As we can easily see, two-ports E1 and E2 are connected in series, two-ports E12 and E3 are connected in parallel. The elements E4, E5, E123, E6, and E7 are connected in cascade. The correlation matrix of the example circuit can be calculated from the matrices of the basic two-ports using the following steps:

1. Compute chain matrix $\mathbf{A}^{(1)}$ and noise correlation matrix $\mathbf{C}_A^{(1)}$ in the chain representation of element E1.
2. Compute chain matrix $\mathbf{A}^{(2)}$ and noise correlation matrix $\mathbf{C}_A^{(2)}$ in the chain representation of element E2.
3. Compute the resulting correlation matrix $\mathbf{C}_A^{(1,2)}$ of cascaded elements E1 and E2:
$$\mathbf{C}_A^{(1,2)} = \mathbf{A}^{(1)}\mathbf{C}_A^{(2)}\mathbf{A}^{(1)} + \mathbf{C}_A^{(1)}.$$

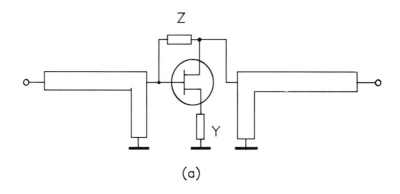

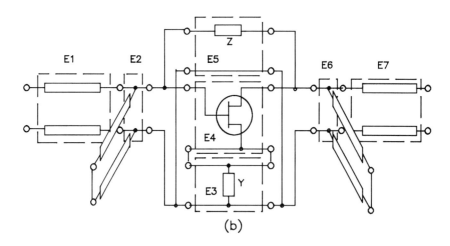

Figure 5.12 Principles of the noise analysis method: (a) example circuit—one stage amplifier with parallel and series feedback; (b) the circuit decomposed into basic two-ports.

4. Multiply the chain matrices of elements E1 and E2 $\mathbf{A}^{(1,2)} = \mathbf{A}^{(1)} \cdot \mathbf{A}^{(2)}$ to obtain chain matrix $\mathbf{A}^{(1,2)}$ of element E12.
5. Compute impedance electrical $\mathbf{Z}^{(3)}$ and the noise $\mathbf{C}_Z^{(3)}$ matrices of element E3.
6. Compute admittance electrical $\mathbf{Z}^{(4)}$ and the noise $\mathbf{C}_Z^{(4)}$ matrices of element E4.
7. Add the electrical $\mathbf{Z}$ matrices of E3 and E4 to obtain $\mathbf{Z}^{(3,4)}$.
8. Add the correlation $\mathbf{C}_Z$ matrices of E3 and E4 to obtain $\mathbf{C}_Z^{(3,4)}$.
9. Transform the resulting impedance electrical $\mathbf{Z}^{(3,4)}$ and noise $\mathbf{C}_Z^{(3,4)}$ matrices into the admittance electrical $\mathbf{Y}^{(3,4)}$ and noise $\mathbf{C}_Y^{(3,4)}$ matrices of element E34.
10. Compute the admittance electrical $\mathbf{Y}^{(5)}$ and noise $\mathbf{C}_Y^{(5)}$ matrices of the element E5.

11. Add the electrical $\mathbf{Y}$ matrices of E34 and E5 to obtain $\mathbf{Y}^{(3,4,5)}$ of element E345.

12. Add the correlation $\mathbf{C}_Y$ matrices of E34 and E5 to obtain $\mathbf{C}_Y^{(3,4,5)}$ of element E345.

13. Transform the resulting admittance electrical $\mathbf{Y}^{(3,4,5)}$ and noise $\mathbf{C}_Y^{(3,4,5)}$ matrices into the chain electrical $\mathbf{A}^{(3,4,5)}$ and noise $\mathbf{C}_A^{(3,4,5)}$ matrices of element E345.

14. Compute correlation matrix $\mathbf{C}_A^{(1,2,3,4,5)}$ of cascaded elements E12 and E345:
$\mathbf{C}_A^{(1,2,3,4,5)} = \mathbf{A}^{(1,2)}\mathbf{C}_A^{(3,4,5)}\mathbf{A}^{(1,2)} + \mathbf{C}_A^{(1,2)}$.

15. Multiply the chain matrices of elements E12 and E345, $\mathbf{A}^{(1,2,3,4,5)} = \mathbf{A}^{(1,2)}\mathbf{A}^{(3,4,5)}$, to obtain chain matrix $\mathbf{A}^{(1,2,3,4,5)}$ of element E12345.

16. Compute chain matrix $\mathbf{A}^{(6)}$ and noise correlation matrix $\mathbf{C}_A^{(6)}$ in the chain representation of element E6.

17. Compute correlation matrix $\mathbf{C}_A^{(1,2,3,4,5,6)}$ of cascaded elements E12345 and E6:
$\mathbf{C}_A^{(1,2)} = \mathbf{A}^{(1,2,3,4,5)}\mathbf{C}_A^{(6)}\mathbf{A}^{(1,2,3,4,5)} + \mathbf{C}_A^{(1,2,3,4,5)}$.

18. Multiply the chain matrices of elements E12345 and E6, $\mathbf{A}^{(1,2,3,4,5,6)} = \mathbf{A}^{(1,2,3,4,5)}\mathbf{A}^{(6)}$, to obtain chain matrix $\mathbf{A}^{(1,2,3,4,5,6)}$ of element E123456.

19. Compute chain matrix $\mathbf{A}^{(7)}$ and noise correlation matrix $\mathbf{C}_A^{(7)}$ in the chain representation of element E7.

20. Compute correlation matrix $\mathbf{C}_A^{(1,2,3,4,5,6,7)}$ of cascaded elements E123456 and E7: $\mathbf{C}_A^{(1,2,3,4,5,6,7)} = \mathbf{A}^{(1,2,3,4,5,6)}\mathbf{C}_A^{(7)}\mathbf{A}^{(1,2,3,4,5,6)} + \mathbf{C}_A^{(1,2,3,4,5,6)}$.

21. Multiply the chain matrices of elements E123456 and E7, $\mathbf{A}^{(1,2,3,4,5,6,7)} = \mathbf{A}^{(1,2,3,4,5,6)}\mathbf{A}^{(7)}$, to obtain chain matrix $\mathbf{A}^{(1,2,3,4,5,6,7)}$ of the whole analyzed circuit.

After the correlation matrix in chain representation has been determined, all noise parameters of the analyzed circuit can be computed using relations discussed in Section 5.6.

5.9 NOISE ANALYSIS OF LINEAR MULTIPORT NETWORKS OF ARBITRARY TOPOLOGY BY USING THE CONNECTION SCATTERING MATRIX

In this method of noise analysis we assume each linear noisy network to be represented as the interconnection of lossy passive multiports that introduce only thermal noise and noisy active two-ports in which the sources of noise are of a more complicated nature. Each linear element in the circuit may be represented by its noiseless equivalent having the same scattering matrix $\mathbf{S}$ of the original network. As it is shown in Figure 5.7, noise generated in an element is represented by mutually correlated noise-wave sources, one source at each port.

In matrix notation, a set of linear equations that relate complex amplitudes of noise waves at ports of a circuit element has the form [9]

$$\mathbf{B}^{(k)} = \mathbf{S}^{(k)}\mathbf{A}^{(k)} + \mathbf{B}_N^{(k)} \qquad (5.91)$$

where $\mathbf{S}^{(k)}$ is the scattering matrix of the kth element, $\mathbf{A}^{(k)}$ and $\mathbf{B}^{(k)}$ vectors of incoming and outgoing noise waves at their ports, and $\mathbf{B}_N^{(k)}$ is a vector of mutually correlated noise wave sources that represents noise generated in the element. The noise waves from these sources radiate out of the ports, and they do not depend on incident noise waves $\mathbf{A}^{(k)}$.

Figure 5.13 presents a general circuit composed of m elements (multiports) connected together by their ports. Considering all m elements of the circuit, we have a set of linear equations whose matrix form is

$$\mathbf{B} = \mathbf{SA} + \mathbf{B}_N \tag{5.92}$$

where

$$
\mathbf{A} = \begin{bmatrix} \mathbf{A}^{(1)} \\ \mathbf{A}^{(2)} \\ \vdots \\ \mathbf{A}^{(k)} \\ \vdots \\ \mathbf{A}^{(m)} \end{bmatrix}
\quad
\mathbf{B} = \begin{bmatrix} \mathbf{B}^{(1)} \\ \mathbf{B}^{(2)} \\ \vdots \\ \mathbf{B}^{(k)} \\ \vdots \\ \mathbf{B}^{(m)} \end{bmatrix}
\quad
\mathbf{B}_N = \begin{bmatrix} \mathbf{B}^{(1)} \\ \mathbf{B}^{(2)} \\ \vdots \\ \mathbf{B}^{(k)} \\ \vdots \\ \mathbf{B}_N^{(m)} \end{bmatrix}
\tag{5.93}
$$

$$
\mathbf{S} = \begin{bmatrix}
\mathbf{S}^{(1)} & \mathbf{0} & \cdots & \mathbf{0} & \cdots & \mathbf{0} \\
\mathbf{0} & \mathbf{S}^{(2)} & & & & \mathbf{0} \\
\vdots & & \ddots & & & \vdots \\
\mathbf{0} & & & \mathbf{S}^{(k)} & & \mathbf{0} \\
\vdots & & & & \ddots & \vdots \\
\mathbf{0} & \cdots & & \mathbf{0} & \mathbf{0} & \mathbf{S}^{(m)}
\end{bmatrix}
\tag{5.94}
$$

The connections between the m elements impose constraints on the vectors $\mathbf{A}$ and $\mathbf{B}$ that can be represented as a matrix equation:

$$\mathbf{B} = \boldsymbol{\Gamma} \mathbf{A} \tag{5.95}$$

where $\boldsymbol{\Gamma}$ is the connection matrix.

In fact, incoming and outgoing noise waves at ports i and j connected together must satisfy the following relation (see Figure 5.14):

$$
\begin{bmatrix} B_i \\ B_j \end{bmatrix} = \frac{1}{Z_i + Z_j} \begin{bmatrix} Z_j - Z_i^* & 2\sqrt{\mathrm{Re}(Z_i)\,\mathrm{Re}(Z_j)} \\ 2\sqrt{\mathrm{Re}(Z_i)\,\mathrm{Re}(Z_j)} & Z_i - Z_j^* \end{bmatrix} \begin{bmatrix} A_i \\ A_j \end{bmatrix}
\tag{5.96}
$$

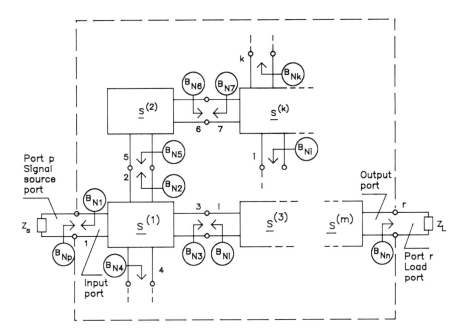

Figure 5.13 Equivalent circuit of a noisy multiport network with noiseless elements and noise wave sources at each port.

where Z_i and Z_j are the reference impedances of the connected ports. This relation defines elements of the connection matrix Γ corresponding to a pair of the connected ports.

In the analysis, we assume for all pairs of the connected ports that

$$Z_i = Z_j^* \tag{5.97}$$

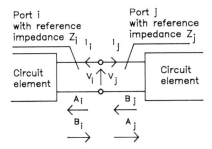

Figure 5.14 Constraints imposed by a connection between adjacent ports.

This means that all port connections in the analyzed circuit are nonreflecting; that is, $A_i = B_j$ and $A_j = B_i$. In such cases, the elements of the connection matrix $\boldsymbol{\Gamma}$ all are zero except the ones in the entries corresponding to pairs of the adjacent ports.

Eliminating vector $\mathbf{B}$ from (5.92) and (5.95), we have

$$\mathbf{WA} = \mathbf{B}_N \tag{5.98}$$

where

$$\mathbf{W} = \boldsymbol{\Gamma} - \mathbf{S} \tag{5.99}$$

is the connection scattering matrix of the analyzed circuit discussed in Chapter 3.

By using (5.98) we are able to get a correlation matrix of the incident noise waves at all circuit ports. Because

$$\mathbf{A} = \mathbf{W}^{-1}\mathbf{B}_N \tag{5.100}$$

it follows that

$$\overline{\mathbf{AA}^+} = \mathbf{W}^{-1}\,\overline{\mathbf{B}_N\mathbf{B}_N^+}\,\left(\mathbf{W}^{-1}\right)^+ = \mathbf{W}^{-1}\mathscr{C}\left(\mathbf{W}^{-1}\right)^+ \tag{5.101}$$

where the superscript pluses indicate the Hermitian complex conjugate of vectors and matrices.

In (5.101)

$$\mathscr{C} = \overline{\mathbf{B}_N\mathbf{B}_N^+} \tag{5.102}$$

is the correlation matrix of the noise wave sources representing noise generated in all circuit elements.

Because the noise-wave sources $\mathbf{B}_N^{(k)}$ of the kth element are not correlated with those of any other circuit element, the correlation matrix $\mathscr{C}$ is a block diagonal matrix of the form:

$$\mathscr{C} = \overline{\mathbf{B}_N\mathbf{B}_N^+} = \begin{bmatrix} \mathscr{C}^{(1)} & \mathbf{0} & \cdots & \mathbf{0} & \cdots & \mathbf{0} \\ \mathbf{0} & \mathscr{C}^{(2)} & & & & \vdots \\ \vdots & & \ddots & & & \vdots \\ \mathbf{0} & & & \mathscr{C}^{(k)} & & \mathbf{0} \\ \vdots & & & & \ddots & \vdots \\ \mathbf{0} & \cdots & & \mathbf{0} & \cdots & \mathscr{C}^{(m)} \end{bmatrix} \tag{5.103}$$

in which $\mathscr{C}^{(1)}$, $\mathscr{C}^{(2)}, \ldots,$ $\mathscr{C}^{(m)}$ are correlation matrices of the noise wave sources of individual circuit elements and the bold zeros represent null matrices.

5.9.1 The Algorithm for Noise Figure Computation of a General Multiport Circuit

To compute the noise figure of a circuit, we assume that the output port is free of noise. Under such conditions, the noise figure of the circuit at frequency f is given by [8]

$$F = 1 + \frac{P_{N_{int}}}{P_{NS}} = 1 + F_e \qquad (5.104)$$

where

$P_{N_{int}}$ = (available) noise power at the output port of the circuit, originating from the noise sources within the circuit;

P_{NS} = (available) noise power at the output port of the circuit, originating from the equivalent thermal noise source of the input port termination.

If r is the number of the load impedance port of the analyzed circuit, the active noise power at the output port is

$$P_N = \left(\overline{\mathbf{AA^+}} \right)_{rr} \left(1 - |S_{rr}|^2 \right) = N_r \left(1 - |S_{rr}|^2 \right) \qquad (5.105)$$

where $N_r = (\overline{\mathbf{AA^+}})_{rr}$ is the rth diagonal element of the correlation matrix $\overline{\mathbf{AA^+}}$,

$$S_{rr} = \frac{Z_L - Z_r^*}{Z_L + Z_r} \qquad (5.106)$$

is the reflection coefficient of the output port load, and

Z_L = the load impedance,
Z_r = the reference impedance of the load impedance port.

The evaluation of N_r can be derived easily from (5.101). In fact, if only the rth diagonal element of the correlation matrix $\overline{\mathbf{AA^+}}$ has to be determined, by letting $\mathcal{B}_r$ be a vector wherein all its elements are zeros except for a one in position r:

$$\mathcal{B}_r^T = [0, \ldots, 0, 1, 0, \ldots, 0] \qquad (5.107)$$

we have

$$N_r = \left(\overline{\mathbf{AA^+}} \right)_{rr} = \mathcal{B}_r^T \overline{\mathbf{AA^+}} \, \mathcal{B}_r$$

$$= \mathcal{B}_r^+ \mathbf{W}^{-1} \mathcal{C} (\mathbf{W^+})^{-1} \mathcal{B}_r$$

$$= \left[(\mathbf{W^+})^{-1} \mathcal{B}_r \right]^+ \mathcal{C} (\mathbf{W^+})^{-1} \mathcal{B}_r \qquad (5.108)$$

Relation (5.107) may be also written in the form:

$$N_r = (\mathbf{AA^+})_{rr} = \mathcal{A}^+ \mathcal{C} \mathcal{A} \qquad (5.109)$$

where a vector:

$$\mathcal{A} = (\mathbf{W}^+)^{-1}\mathcal{B}_r = (\mathbf{W}^{T*})^{-1}\mathcal{B}_r \tag{5.110}$$

is the solution vector of a system of equations:

$$\mathbf{W}^+\mathcal{A} = \mathcal{B}_r \tag{5.111}$$

where the coefficient matrix equals the complex transpose of the connection scattering matrix $\mathbf{W}$ of the analyzed circuit and $\mathcal{B}_r$ is its right-hand vector.

Therefore noise powers $P_{N_{\text{int}}}$ and P_{NS} can be computed using simple matrix multiplications described by (5.109) and (5.110).

The following steps carried out by the numerical algorithm are

1. Connection scattering matrix $\mathbf{W}$ of the analyzed circuit is built and computed.
2. Noise correlation matrices of the individual passive and active circuit elements are found. Noise correlation matrix $\mathscr{C}$ defined by (5.103) is built and computed. Two diagonal elements of $\mathscr{C}$ relative to ports belonging to the signal source impedance and the load impedance are set to zero.

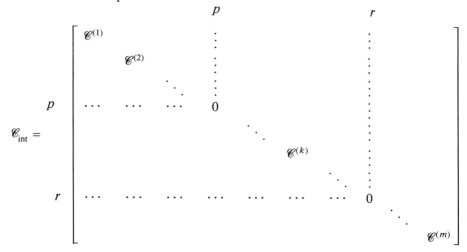

p = signal source port number
r = load impedance port number

Both zeros represent no noise power generated by the signal source and load impedances.

3. Relative to the load impedance port diagonal element $N_{\text{int } r}$ of matrix $\overline{\mathbf{A}\mathbf{A}^+}$ is computed by using (5.109):

$$N_{\text{int } r} = \mathcal{A}^+ \mathscr{C}_{\text{int}} \mathcal{A} \tag{5.112}$$

A value of this element multiplied by $(1 - |S_{rr}|^2)$ equals $P_{N_{\text{int}}}$.

4. The noise correlation matrix $\mathscr{C}$ defined by (5.103) is built and computed once again, this time for the case where noise in the circuit originates only from the equivalent thermal noise source of the input port termination. This means that all elements of matrix $\mathscr{C}$ must equal zero, except an element corresponding to the input port termination. According to (5.50) this element is

$$\mathscr{C}_{pp} = kT\,df\left(1 - |S_{pp}|^2\right) \tag{5.113}$$

where

$$S_{pp} = \frac{Z_s - Z_p^*}{Z_s + Z_p} \tag{5.114}$$

is the reflection coefficient of the input port termination (signal source port), and

Z_s = impedance of the input port termination (signal source impedance),
Z_p = reference impedance of the signal source port.

$$p$$

$$\mathscr{C}_s = p \begin{bmatrix} 0 & & & & & \\ & 0 & & \vdots & & \\ & & \ddots & \vdots & & \\ & \cdots & & \mathscr{C}_{pp} & & \\ & & & & \ddots & \\ & & & & 0 & \ddots \\ & & & & & & 0 \end{bmatrix}$$

p = signal source port number

5. Relative to the load impedance port diagonal element N_{sr} of matrix $\overline{\mathbf{AA}^+}$ is computed by using (5.109)

$$N_{sr} = \mathscr{A}\,\mathscr{C}_s\mathscr{A} \tag{5.115}$$

This time a value of this element multiplied by $(1 - |S_{rr}|^2)$ equals P_{NS}.

6. The noise figure F is found from

$$F = 1 + \frac{N_{\text{int}\,r}}{N_{sr}} \tag{5.116}$$

by using results obtained in steps 3 and 5.

The matrix **W** in (5.98) is the connection scattering matrix of the analyzed circuit, so that a conventional circuit analysis in the frequency domain may be carried out in parallel with the noise analysis and the noise figure computation. This allows a simultaneous optimization of an absolutely general microwave circuit topology with respect to both the noise figure and any of the conventional circuit functions.

The inverse matrices $\mathbf{W}^{-1}$ and $(\mathbf{W}^{-1})^{+}$ of connection scattering matrix **W** can be computed very effectively by using the sparse matrix technique described in Chapter 8.

Example 5.3

A commonly used microwave amplifier integrated circuits for high-power and broadband applications is the balanced amplifier configuration. As shown in Figure 5.15, a balanced amplifier consists of two 3 dB and 90° directional couplers and two conventional two-port transistor amplifiers.

Noise analysis of the example amplifier occurs as follows. The connection scattering **W** of the balanced amplifier circuit is

$$
\mathbf{W} = \begin{array}{c|cccccccccccccccc|}
 & 1 & 2 & 3 & 4 & 5 & 6 & 7 & 8 & 9 & 10 & 11 & 12 & 13 & 14 & 15 & 16 \\
\hline
1 & -S^{(1)} & 0 & 1 & 0 & 0 & 0 & 0 & 0 & 0 & 0 & 0 & 0 & 0 & 0 & 0 & 0 \\
2 & 0 & -S^{(2)} & 0 & 0 & 0 & 1 & 0 & 0 & 0 & 0 & 0 & 0 & 0 & 0 & 0 & 0 \\
3 & 1 & 0 & -\mathbf{S}^{(3)} & & & & 0 & 0 & 0 & 0 & 0 & 0 & 0 & 0 & 0 & 0 \\
4 & 0 & 0 & & & & & 1 & 0 & 0 & 0 & 0 & 0 & 0 & 0 & 0 & 0 \\
5 & 0 & 0 & & & & & 0 & 0 & 1 & 0 & 0 & 0 & 0 & 0 & 0 & 0 \\
6 & 0 & 1 & & & & & 0 & 0 & 0 & 0 & 0 & 0 & 0 & 0 & 0 & 0 \\
7 & 0 & 0 & 0 & 1 & 0 & 0 & -\mathbf{S}^{(4)} & & 0 & 0 & 0 & 0 & 0 & 0 & 0 & 0 \\
8 & 0 & 0 & 0 & 0 & 0 & 0 & & & 0 & 0 & 1 & 0 & 0 & 0 & 0 & 0 \\
9 & 0 & 0 & 0 & 0 & 1 & 0 & 0 & 0 & -\mathbf{S}^{(5)} & & 0 & 0 & 0 & 0 & 0 & 0 \\
10 & 0 & 0 & 0 & 0 & 0 & 0 & 0 & 0 & & & 0 & 0 & 0 & 1 & 0 & 0 \\
11 & 0 & 0 & 0 & 0 & 0 & 0 & 0 & 1 & 0 & 0 & -\mathbf{S}^{(6)} & & & & 0 & 0 \\
12 & 0 & 0 & 0 & 0 & 0 & 0 & 0 & 0 & 0 & 0 & & & & & 1 & 0 \\
13 & 0 & 0 & 0 & 0 & 0 & 0 & 0 & 0 & 0 & 0 & & & & & 0 & 1 \\
14 & 0 & 0 & 0 & 0 & 0 & 0 & 0 & 0 & 0 & 1 & & & & & 0 & 0 \\
15 & 0 & 0 & 0 & 0 & 0 & 0 & 0 & 0 & 0 & 0 & 0 & 1 & 0 & 0 & -S^{(7)} & 0 \\
16 & 0 & 0 & 0 & 0 & 0 & 0 & 0 & 0 & 0 & 0 & 0 & 0 & 1 & 0 & 0 & -\mathbf{S}^{(8)}
\end{array}
$$

In **W**, $\mathbf{S}^{(3)}$ and $\mathbf{S}^{(6)}$ are the scattering matrices of 3 dB/90° directional couplers defined in Table 1.2. Matrices $\mathbf{S}^{(4)}$ and $\mathbf{S}^{(5)}$ are the scattering matrices of two conventional type transistor amplifiers. Matrix $\mathbf{S}^{(1)}$ is the reflection coefficient of the signal generator port. Matrix $\mathbf{S}^{(6)}$ is the reflection coefficient of load impedance Z_{l}. According to the algorithm for noise computation described in Section 5.9.1, elemen

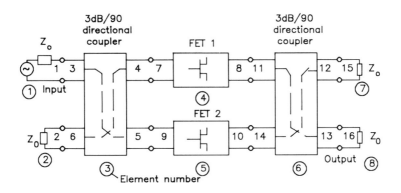

Figure 5.15 Balanced transistor amplifier.

$N_{\text{int }16}$ and N_{s16} (16 is the number of the load port) have to be computed by using (5.111) and (5.115). By using matrix $\mathcal{C}_{\text{int}}$ given as

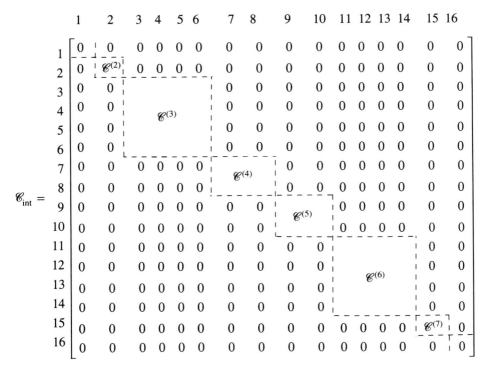

and (5.111), we may compute noise power at the output port of the amplifier originating from the noise sources within the circuit. In $\mathscr{C}_{int}$, $\mathscr{C}^{(3)}$ and $\mathscr{C}^{(6)}$ are noise wave correlation matrices of two directional couplers, whereas $\mathscr{C}^{(4)}$ and $\mathscr{C}^{(5)}$ describe noise properties of two transistor amplifiers. Note that elements $\mathscr{C}_{1,1}$ and $\mathscr{C}_{16,16}$ of the $\mathscr{C}_{int}$ matrix are set to zero (1 is the number of signal generator port and 16 is the number of the load impedance port of the amplifier). The noise wave correlation matrices of passive elements in the analyzed circuit are given in Table 5.5.

Matrix $\mathscr{C}_s$ given as

$$
\mathscr{C}_{int} =
\begin{array}{c}
\begin{array}{cccccccccccccccc}
1 & 2 & 3 & 4 & 5 & 6 & 7 & 8 & 9 & 10 & 11 & 12 & 13 & 14 & 15 & 16
\end{array} \\
\begin{array}{c}
1 \\ 2 \\ 3 \\ 4 \\ 5 \\ 6 \\ 7 \\ 8 \\ 9 \\ 10 \\ 11 \\ 12 \\ 13 \\ 14 \\ 15 \\ 16
\end{array}
\left[
\begin{array}{cccccccccccccccc}
\mathscr{C}^{(1)} & 0 & 0 & 0 & 0 & 0 & 0 & 0 & 0 & 0 & 0 & 0 & 0 & 0 & 0 & 0 \\
0 & 0 & 0 & 0 & 0 & 0 & 0 & 0 & 0 & 0 & 0 & 0 & 0 & 0 & 0 & 0 \\
0 & 0 & 0 & 0 & 0 & 0 & 0 & 0 & 0 & 0 & 0 & 0 & 0 & 0 & 0 & 0 \\
0 & 0 & 0 & 0 & 0 & 0 & 0 & 0 & 0 & 0 & 0 & 0 & 0 & 0 & 0 & 0 \\
0 & 0 & 0 & 0 & 0 & 0 & 0 & 0 & 0 & 0 & 0 & 0 & 0 & 0 & 0 & 0 \\
0 & 0 & 0 & 0 & 0 & 0 & 0 & 0 & 0 & 0 & 0 & 0 & 0 & 0 & 0 & 0 \\
0 & 0 & 0 & 0 & 0 & 0 & 0 & 0 & 0 & 0 & 0 & 0 & 0 & 0 & 0 & 0 \\
0 & 0 & 0 & 0 & 0 & 0 & 0 & 0 & 0 & 0 & 0 & 0 & 0 & 0 & 0 & 0 \\
0 & 0 & 0 & 0 & 0 & 0 & 0 & 0 & 0 & 0 & 0 & 0 & 0 & 0 & 0 & 0 \\
0 & 0 & 0 & 0 & 0 & 0 & 0 & 0 & 0 & 0 & 0 & 0 & 0 & 0 & 0 & 0 \\
0 & 0 & 0 & 0 & 0 & 0 & 0 & 0 & 0 & 0 & 0 & 0 & 0 & 0 & 0 & 0 \\
0 & 0 & 0 & 0 & 0 & 0 & 0 & 0 & 0 & 0 & 0 & 0 & 0 & 0 & 0 & 0 \\
0 & 0 & 0 & 0 & 0 & 0 & 0 & 0 & 0 & 0 & 0 & 0 & 0 & 0 & 0 & 0 \\
0 & 0 & 0 & 0 & 0 & 0 & 0 & 0 & 0 & 0 & 0 & 0 & 0 & 0 & 0 & 0 \\
0 & 0 & 0 & 0 & 0 & 0 & 0 & 0 & 0 & 0 & 0 & 0 & 0 & 0 & 0 & 0 \\
0 & 0 & 0 & 0 & 0 & 0 & 0 & 0 & 0 & 0 & 0 & 0 & 0 & 0 & 0 & 0
\end{array}
\right]
\end{array}
$$

is used to compute noise power at the output port of the amplifier originating from the thermal noise of the signal source impedance Z_s. This time, in matrix $\mathscr{C}_s$, only element $\mathscr{C}_{11} = kT \, df(1 - |S_{11}|^2)$ is the nonzero element (signal source port is numbered as one).

Sparse matrix techniques described in Chapter 8 should be used for computing (5.115).

5.9.2 The Algorithm for Computing the Four Noise Parameters of a General Multiport Circuit

To compute the four noise parameters related to the input and output ports of a general multiport circuit, shown in Figure 5.13, we assume that reflection coefficient S_{pp} of the input port termination (equation (5.106)) and reflection coefficient S_{rr} of the output port load (equation (5.114)) are both zero; both terminations also are assumed to be free of

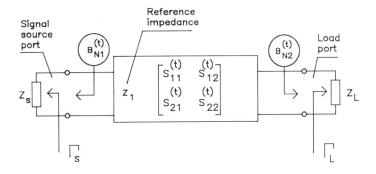

Figure 5.16 Noise waves $B_{n1}^{(t)}$ and $B_{n2}^{(2)}$ at the input and output ports of the overall circuit.

noise. Under such assumptions the noise waves at the input and output ports of the overall circuit can be computed by using (5.101).

The actual steps carried out by the numerical algorithm are as follows:

1. Connection scattering matrix $\mathbf{W}$ of the analyzed circuit is built and computed. Diagonal elements $W_{pp} = -S_{pp}$ and $W_{rr} = -S_{rr}$ of $\mathbf{W}$ must be set to zero (p = signal source port number, r = load impedance port number).

2. The noise correlation matrices of the individual passive and active circuit elements are found. Noise correlation matrix $\mathscr{C}$ defined by (5.103) is built and computed. Two diagonal elements of matrix $\mathscr{C}$ relative to ports belonging to the signal source impedance and the load impedance are set to zero.

3. Three elements $\overline{(\mathbf{AA}^+)}_{pp}$, $\overline{(\mathbf{AA}^+)}_{rr}$, and $\overline{(\mathbf{AA}^+)}_{pr}$ or $\overline{(\mathbf{AA}^+)}_{rp}$, of matrix $\overline{\mathbf{AA}^+}$ are computed by using (5.101).

The correlation matrix of the outgoing noise waves $B_{N1}^{(t)}$ and $B_{N2}^{(t)}$ at the input and output ports of the overall network is given by (see Figure 5.16)

$$\mathscr{C}^{(t)} = \begin{bmatrix} \mathscr{C}_{11}^{(t)} & \mathscr{C}_{12}^{(t)} \\ \mathscr{C}_{21}^{(t)} & \mathscr{C}_{22}^{(t)} \end{bmatrix} = \begin{bmatrix} \overline{B_{N1}^{(t)} B_{N1}^{(t)*}} & \overline{B_{N1}^{(t)} B_{N2}^{(t)*}} \\ \overline{B_{N2}^{(t)} B_{N1}^{(t)*}} & \overline{B_{N2}^{(t)} B_{N2}^{(t)*}} \end{bmatrix}$$

$$= \begin{bmatrix} \overline{\left(\mathbf{AA}^+\right)}_{pp} & \overline{\left(\mathbf{AA}^+\right)}_{pr} \\ \overline{\left(\mathbf{AA}^+\right)}_{pr} & \overline{\left(\mathbf{AA}^+\right)}_{rr} \end{bmatrix} \tag{5.117}$$

After the elements of the noise correlation matrix $\mathscr{C}^{(t)}$ are known, any set of noise parameters of the overall two-port can be computed. We can compute, for example, $F_{\min}$, Γ_{opt}, and N parameters using (5.83) to (5.86). The noise figure of the analyzed circuit now can be computed in two ways. First, the noise figure F of the overall circuit can be computed by using four noise parameters, $F_{\min}$, Γ_{opt}, and N:

$$F = F_{\min} + N \frac{|\Gamma_s - \Gamma_{\text{opt}}|^2}{\left(1 - |\Gamma_s|^2\right)\left(1 - |\Gamma_{\text{opt}}|^2\right)} \tag{5.118}$$

Because the four noise parameters, $F_{\min}$, Γ_{opt}, and N are equivalent to the elements $\mathscr{C}_{11}^{(t)}$, $\mathscr{C}_{12}^{(t)}$, and $\mathscr{C}_{22}^{(t)}$ of the noise correlation matrix, the second way to compute the noise figure F of the circuit is by using the relation:

$$F = 1 + \frac{\mathscr{C}_{11}^{(t)} \left| \dfrac{S_{21}^{(t)} \Gamma_s}{1 - S_{11}^{(t)} \Gamma_s} \right|^2 + \mathscr{C}_{22}^{(t)} + 2\,\mathrm{Re}\left\{ \mathscr{C}_{12}^{(t)} \dfrac{S_{21}^{(t)} \Gamma_s}{1 - S_{11}^{(t)} \Gamma_s} \right\}}{kT_0\, df\left(1 - |\Gamma_s|^2\right) \left| \dfrac{S_{21}^{(t)}}{1 - S_{11}^{(t)} \Gamma_s} \right|^2} \tag{5.119}$$

where $S_{11}^{(t)}$ and $S_{21}^{(t)}$ are the scattering parameters of the overall circuit, and

$$\Gamma_s = \frac{Z_s - Z_1^*}{Z + Z_1} \tag{5.120}$$

is the reflection coefficient of the signal source impedance with respect to the input port reference impedance Z_1.

The two-port scattering parameters $S_{11}^{(t)}$ and $S_{21}^{(t)}$ used in (5.111) and (5.83) to (5.85) are computed from (3.60):

$$\mathbf{W a} = \mathbf{c} \tag{5.121}$$

where $\mathbf{W}$ is the connection scattering matrix given by (5.99), $\mathbf{a}$ is a vector of incident power waves, and $\mathbf{c}$ is a vector of the impressed power waves of the independent sinusoidal signal sources (3.55) and (3.57).

By connecting the input port of the analyzed multiport circuit (see Figure 5.13) to a matched signal source ($S_{pp} = 0$) with the impressed wave $c_p = 1$ and the output port to a matched load ($S_{rr} = 0$), we obtain a case in which the overall scattering parameters $S_{11}^{(t)}$ and $S_{21}^{(t)}$ coincide with the waves a_r and a_p:

$$S_{11}^{(t)} = a_p \big|_{a_1 = 1}; \quad S_{21}^{(t)} = a_r \big|_{a_1 = 1} \tag{5.122}$$

The condition $a_1 = 1$ is imposed by the signal source connected to the input port of the analyzed circuit ($a_1 = c_p = 1$).

5.9.3 Noise Power First-Order Sensitivities

Suppose that parameter p in a given noisy network is to be varied without affecting its topology. Parameter variation will affect the noise power delivered to the load at the rth port of the circuit. Differentiating (5.108) with respect to p and using the relation $\partial\mathbf{W}/\partial p = -\partial\mathbf{S}/\partial p$, we have

$$\frac{\partial N_r}{\partial p} = \frac{\partial\left(\overline{\mathbf{AA^+}}\right)_{rr}}{\partial p} = \mathscr{B}_r^+\mathbf{W}^{-1}\left[\frac{\partial\mathbf{S}}{\partial p}\mathbf{W}^{-1}\mathscr{C} + \left(\frac{\partial\mathbf{S}}{\partial p}\mathbf{W}^{-1}\mathscr{C^+}\right)^+\right](\mathbf{W^+})^{-1}\mathscr{B}_r$$

$$+ \mathscr{B}_r^+\mathbf{W}^{-1}\frac{\partial\mathscr{C}}{\partial p}(\mathbf{W^+})^{-1}\mathscr{B}_r$$

$$= \mathscr{A}^+\left[\frac{\partial\mathbf{S}}{\partial p}\mathbf{W}^{-1}\mathscr{C} + \left(\frac{\partial\mathbf{S}}{\partial p}\mathbf{W}^{-1}\mathscr{C^+}\right)^+\right]\mathscr{A} + \mathscr{A}^+\frac{\partial\mathscr{C}}{\partial p}\mathscr{A} \qquad (5.123)$$

Because noise correlation matrices are Hermitian matrices ($\mathscr{C} = \mathscr{C}^+$), (5.123) simplifies to

$$\frac{\partial N_r}{\partial p} = 2\,\mathrm{Re}\left\{\mathscr{A}^+\frac{\partial\mathbf{S}}{\partial p}\mathbf{W}^{-1}\mathscr{C}\mathscr{A}\right\} + \mathscr{A}^+\frac{\partial\mathscr{C}}{\partial p}\mathscr{A} \qquad (5.124)$$

Further, assuming that a varied parameter is contained in the passive part of the circuit, we have

$$\frac{\partial N_r}{\partial p} = 2\,\mathrm{Re}\left\{\mathscr{A}^+\frac{\partial\mathbf{S}}{\partial p}\mathbf{W}^{-1}\mathscr{C}\mathscr{A}\right\} - 2\mathscr{A}^+\,\mathrm{Re}\left\{\frac{\partial\mathbf{S}}{\partial p}\mathbf{S}^+\right\}\mathscr{A} \qquad (5.125)$$

Using (5.125), we may write

$$\mathbf{G} = \begin{bmatrix} \dfrac{\partial N_r}{\partial p_1} \\[2mm] \dfrac{\partial N_r}{\partial p_2} \\[2mm] \vdots \\[2mm] \dfrac{\partial N_r}{\partial p_n} \end{bmatrix} = 2\,\mathrm{Re}\left\{\mathscr{A}^+ \begin{bmatrix} \dfrac{\partial\mathbf{S}}{\partial p_1} \\[2mm] \dfrac{\partial\mathbf{S}}{\partial p_2} \\[2mm] \vdots \\[2mm] \dfrac{\partial\mathbf{S}}{\partial p_n} \end{bmatrix}\mathbf{W}^{-1}\mathbf{C}\mathscr{A}\right\} - 2\mathscr{A}^+\,\mathrm{Re}\left\{\begin{bmatrix} \dfrac{\partial\mathbf{S}}{\partial p_1} \\[2mm] \dfrac{\partial\mathbf{S}}{\partial p_2} \\[2mm] \vdots \\[2mm] \dfrac{\partial\mathbf{S}}{\partial p_n} \end{bmatrix}\mathbf{S}^+\right\}\mathscr{A}$$

$$(5.126)$$

where $\mathbf{G} = \nabla N_r$ is a vector of sensitivities of noise power dissipated in the load located in the rth port of the analyzed circuit.

Equation (5.125) or (5.126) relates changes in noise power dissipated in the load to changes in parameter values in the passive elements of the circuit. To evaluate the noise power sensitivities, we need to know the partial derivatives of element scattering matrices, the $\mathscr{A}$ vector being the solution of the system of equations (5.111) and the inverse of the connection scattering matrix of the analyzed circuit.

The two first quantities are used as well in sensitivity analysis of circuits excited by the sinusoidal signals discussed in Chapter 4. The sensitivity of sinusoidal wave a_r at rth port is obtained from (4.42):

$$\frac{\partial a_r}{\partial p} = \alpha^T \frac{\partial \mathbf{S}}{\partial p} \mathbf{a} \tag{5.127}$$

where $\mathbf{a}$ is a solution vector of a system of equations (3.60):

$$\mathbf{Wa} = \mathbf{c} \tag{5.128}$$

of the original circuit, and α is a solution vector of a system of equations (4.84):

$$\mathbf{W}^T \alpha = \gamma_r \tag{5.129}$$

of the adjoined circuit, in which γ_r is a vector of the same form as vector $\mathscr{B}_r$ given by (5.107).

By comparing (5.111) and (5.129), we find that the solution vectors of both systems of equations satisfy the relation:

$$\mathscr{A} = \alpha^* \tag{5.130}$$

which means that to have $\mathscr{A}$ we must find α, or *vice versa*. This significantly reduces computational effort when performing noise analysis, noise power sensitivity analysis and sensitivity analysis for sinusoidal signals at the same time.

The partial derivatives of the element scattering matrices required for evaluation of noise power sensitivities can usually be found after some manipulation. Table 5.7 presents scattering matrix partial derivatives for common design components with respect to useful parameters.

The inverse matrices $\mathbf{W}^{-1}$ and $(\mathbf{W}^{-1})^+$ of connection scattering matrix $\mathbf{W}$ can be computed very effectively by using the sparse matrix technique based on the bifactorization method discussed in Chapter 8.

Table 5.7
Differential Scattering Matrices for Some
Elements of Microwave Circuits

Circuit Element	$\partial S/\partial p$	p
Z_N Z	$\dfrac{2Z_N}{Z + Z_N}$	Z
Z Z_N Z_N	$\dfrac{2Z_N}{(Z + 2Z_N)^2}\begin{bmatrix} 1 & -1 \\ -1 & 1 \end{bmatrix}$	Z
Z_N Y	$\dfrac{-2Z_N}{(1 + Z_N Y)^2}$	Y
Y_N Y Y_N	$\dfrac{-2Y_N}{(Y + 2Y_N)^2}\begin{bmatrix} 1 & 1 \\ 1 & 1 \end{bmatrix}$	Y
Z_N Z_0 Z_N Transmission line section $\gamma = \alpha + j\beta,\ \beta = \dfrac{2\pi}{\lambda_g}$	$\dfrac{1}{\left[Z_0^2 + Z_N^2 + 2Z_0 Z_N \coth(\gamma l)\right]^2}$ $\cdot \begin{bmatrix} 4Z_0 Z_N^2 + 2Z_N\left(Z_0^2 + Z_N^2\right)\coth(\gamma l) & 2Z_N\left(Z_N - Z_0^2\right)\operatorname{csch}(\gamma l) \\ 2Z_N\left(Z_N - Z_0^2\right)\operatorname{csch}(\gamma l) & 4Z_0 Z_N^2 + 2Z_N\left(Z_0^2 + Z_N^2\right)\coth(\gamma l) \end{bmatrix}$	Z_0
	$\dfrac{2Z_0 Z_N \gamma \operatorname{csch}^2(\gamma l)}{\left[Z_0^2 + Z_N^2 + 2Z_0 Z_N \coth(\gamma l)\right]^2}$ $\cdot \begin{bmatrix} Z_0^2 - Z_N^2 & -2Z_0 Z_N \sinh(\gamma l) - \left(Z_0^2 + Z_N^2\right)\cosh(\gamma l) \\ -2Z_0 Z_N \sinh(\gamma l) - \left(Z_0^2 + Z_N^2\right)\cosh(\gamma l) & Z_0^2 - Z_N^2 \end{bmatrix}$	l

Table 5.7 (*Continued*)

Circuit Element	$\partial S/\partial p$	p
Short-ended shunt connected stub	$\dfrac{-2Y_N\coth(\gamma l)}{[2Y_N+Y_0\coth(\gamma l)]^2}\cdot\begin{bmatrix}1&1\\1&1\end{bmatrix}$	Y_0
	$\dfrac{2Y_N\gamma}{\sinh^2(\gamma l)[2Y_N+Y_0\coth(\gamma l)]^2}\cdot\begin{bmatrix}1&1\\1&1\end{bmatrix}$	l
Open-ended shunt connected stub	$\dfrac{-2Y_N\tanh(\gamma l)}{[2Y_N+Y_0\tanh(\gamma l)]^2}\cdot\begin{bmatrix}1&1\\1&1\end{bmatrix}$	Y_0
	$\dfrac{2Y_N\gamma}{\cosh^2(\gamma l)[2Y_N+Y_0\tanh(\gamma l)]^2}\cdot\begin{bmatrix}1&1\\1&1\end{bmatrix}$	l
Short-ended series connected stub	$\dfrac{2Z_N\tanh(\gamma l)}{[2Z_N+Z_0\tanh(\gamma l)]^2}\cdot\begin{bmatrix}1&-1\\-1&1\end{bmatrix}$	Z
	$\dfrac{2Z_N\gamma}{\cosh^2(\gamma l)[2Z_N+Z_0\tanh(\gamma l)]^2}\cdot\begin{bmatrix}1&-1\\-1&1\end{bmatrix}$	
Open-ended series connected stub	$\dfrac{2Z_N\coth(\gamma l)}{[2Z_N+Z_0\coth(\gamma l)]^2}\cdot\begin{bmatrix}1&-1\\-1&1\end{bmatrix}$	Z
	$\dfrac{-2Z_N\gamma}{\sinh^2(\gamma l)[2Z_N+Z_0\coth(\gamma l)]^2}\cdot\begin{bmatrix}1&-1\\-1&1\end{bmatrix}$	

Table 5.7 (*Continued*)

Circuit Element	$\partial \mathbf{S}/\partial p$	p
Ideal transformer	$\dfrac{2}{\left[1+p^2\right]^2}\begin{bmatrix} -2p & 1-p^2 \\ 1-p^2 & -2p \end{bmatrix}$	p
Gyrator	$\dfrac{1}{\left[\alpha^2+Z_N^2\right]^2}\begin{bmatrix} 2\alpha Z_N & Z_N^2-\alpha^2 \\ Z_N^2-\alpha^2 & 2\alpha Z_N \end{bmatrix}$	α
Voltage controlled voltage source	$\begin{bmatrix} 0 & 0 \\ 2 & 0 \end{bmatrix}$	k
Voltage controlled current source	$\begin{bmatrix} 0 & 0 \\ -2Z_N & 0 \end{bmatrix}$	g_m
Current controlled voltage source	$\begin{bmatrix} 0 & 0 \\ \dfrac{2}{Z_N} & 0 \end{bmatrix}$	r_m
Current controlled current source	$\begin{bmatrix} 0 & 0 \\ -2 & 0 \end{bmatrix}$	β

5.9.4 Noise Figure Gradient Computation

By differentiating (5.116) with respect to circuit parameter p, by using (5.112), (5.115), and (5.124) describing the noise power sensitivities, we get an expression for noise figure sensitivity relating to parameter p:

$$
\frac{\partial F}{\partial p} = \frac{1}{N_{sr}^2} \left(\frac{\partial N_{\text{int} r}}{\partial p} N_{sr} - \frac{\partial N_{sr}}{\partial p} N_{\text{int} r} \right)
$$

$$
= \frac{1}{\mathscr{A}^+ \mathscr{C}_s \mathscr{A}} \left[2 \operatorname{Re} \left\{ \mathscr{A}^+ \frac{\partial \mathbf{S}}{\partial p} \mathbf{W}^{-1} \mathscr{C}_{\text{int}} \mathscr{A} \right\} + \mathscr{A}^+ \frac{\partial \mathscr{C}_{\text{int}}}{\partial p} \mathscr{A} \right]
$$

$$
- \frac{\mathscr{A}^+ \mathscr{C}_{\text{int}} \mathscr{A}}{\left(\mathscr{A}^+ \mathscr{C}_s \mathscr{A} \right)^2} 2 \operatorname{Re} \left\{ \mathscr{A}^+ \frac{\partial \mathbf{S}}{\partial p} \mathbf{W}^{-1} \mathscr{C}_s \mathscr{A} \right\} \tag{5.131}
$$

In deriving (5.131) we assume that a thermal noise generated by the internal impedance of a signal generator does not depend on varied circuit parameter p.

Consider now a design of a circuit. Suppose we want to optimize (minimize) the noise figure of a circuit. Expanding relation (5.131), we have a gradient vector of the noise figure F considered as an objective function:

$$
\nabla F = \begin{bmatrix} \dfrac{\partial F_r}{\partial p_1} \\[2mm] \dfrac{\partial F_r}{\partial p_2} \\[2mm] \vdots \\[2mm] \dfrac{\partial F_r}{\partial p_n} \end{bmatrix} = \frac{1}{\mathscr{A}^+ \mathscr{C}_s \mathscr{A}} \left[2 \operatorname{Re} \left\{ \mathscr{A}^+ \begin{bmatrix} \dfrac{\partial \mathbf{S}}{\partial p_1} \\[2mm] \dfrac{\partial \mathbf{S}}{\partial p_2} \\[2mm] \vdots \\[2mm] \dfrac{\partial \mathbf{S}}{\partial p_n} \end{bmatrix} \mathbf{W}^{-1} \mathscr{C}_{\text{int}} \mathscr{A} \right\} + \mathscr{A}^+ \begin{bmatrix} \dfrac{\partial \mathscr{C}_{\text{int}}}{\partial p_1} \\[2mm] \dfrac{\partial \mathscr{C}_{\text{int}}}{\partial p_2} \\[2mm] \vdots \\[2mm] \dfrac{\partial \mathscr{C}_{\text{int}}}{\partial p_n} \end{bmatrix} \mathscr{A} \right]
$$

$$
- \frac{\mathscr{A}^+ \mathscr{C}_{\text{int}} \mathscr{A}}{\left(\mathscr{A}^+ \mathscr{C}_s \mathscr{A} \right)^2} 2 \operatorname{Re} \left\{ \mathscr{A}^+ \begin{bmatrix} \dfrac{\partial \mathbf{S}}{\partial p_1} \\[2mm] \dfrac{\partial \mathbf{S}}{\partial p_2} \\[2mm] \vdots \\[2mm] \dfrac{\partial \mathbf{S}}{\partial p_n} \end{bmatrix} \mathbf{W}^{-1} \mathscr{C}_s \mathscr{A} \right\} \tag{5.132}
$$

Equation (5.132) is a general one, applicable to cases when variable parameters are contained in passive as well as in active elements of a circuit.

The preceding discussion of computing the noise figure gradient vector can be extended to obtain a matrix of second-order derivatives (Hessian) of the noise figure F for a two-port with arbitrary internal topology.

5.10 NOISE ANALYSIS OF LINEAR MULTIPORT NETWORKS OF ARBITRARY TOPOLOGY BY USING THE ADMITTANCE MATRIX

We assume that a linear noisy n-port consists of a lossy passive network embedding a number (m) of noisy two-port devices. We also assume that each element of the circuit can be described by its admittance matrix and its noise correlation matrix. As shown in Figure 5.17, the circuit to be analyzed can be treated as the interconnection of a passive noisy multiport and m noisy two-ports. The passive noisy multiport generates only thermal noise and an origin of noise in two-ports is more complicated. Using the results

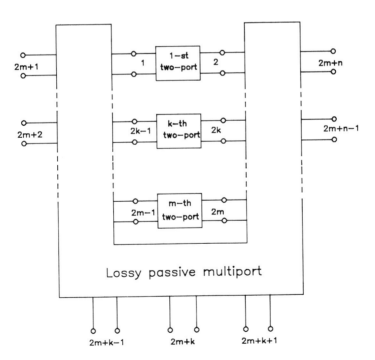

Figure 5.17 Noisy linear circuit decomposed into m noisy active two-ports and a noisy passive multiport with n external ports.

given in Section 5.1 it is obvious that the circuit from Figure 5.17 can be represented by its noiseless equivalent with noise current sources connected across each port. Figure 5.18 presents this equivalent. Here, two different sets of noise current sources are shown. The thermal noise of the passive multiport is represented by the equivalent N noise current sources, and the J sources are the equivalent noise sources of the active two-ports, generally they are not only thermal. The N sources are all correlated. Equation (5.48) describes their correlation matrix $\mathscr{C}_N$. In the case of J sources, only the pairs of noise sources belonging to the same two-port device are correlated, and their normalized correlation matrices $\mathbf{C}_J$ have the form given by (5.33). In accordance with rules of port numbering presented in Figure 5.18, noise sources J_{2k-1} and J_{2k} belong to the kth two-port and are correlated. The correlation matrix of kth two-port device is called $\mathscr{C}_J^{(k)}$.

The network equations can be written in the following way [29]:

$$\begin{bmatrix} I_d \\ I_e \end{bmatrix} = \mathbf{YV} + \mathbf{N} = \begin{bmatrix} \mathbf{Y}_{dd} & \mathbf{Y}_{de} \\ \mathbf{Y}_{ed} & \mathbf{Y}_{ee} \end{bmatrix} \begin{bmatrix} \mathbf{V}_d \\ \mathbf{V}_e \end{bmatrix} + \begin{bmatrix} \mathbf{N}_d \\ \mathbf{N}_e \end{bmatrix} \qquad (5.133)$$

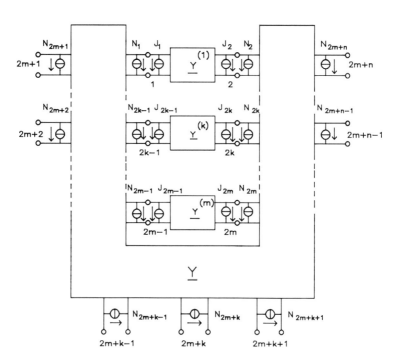

Figure 5.18 Noiseless equivalent of the noisy linear circuit presented in Figure 5.17.

In (5.133) the subscript d refers to the $2m$ device ports, and the subscript e to the n external ports.

The vector $\mathbf{I}_d$ of currents flowing out of the $2m$ device ports equals

$$\mathbf{I}_d = -\mathbf{y}\mathbf{V}_d - \mathbf{J} \tag{5.134}$$

where

$$\mathbf{I}_d = \begin{bmatrix} \mathbf{I}_d^{(1)} \\ \mathbf{I}_d^{(2)} \\ \vdots \\ \mathbf{I}_d^{(k)} \\ \vdots \\ \mathbf{I}_d^{(m)} \end{bmatrix} ; \quad \mathbf{V}_d = \begin{bmatrix} \mathbf{V}_d^{(1)} \\ \mathbf{V}_d^{(2)} \\ \vdots \\ \mathbf{V}_d^{(k)} \\ \vdots \\ \mathbf{V}_d^{(m)} \end{bmatrix} ; \quad \mathbf{J} = \begin{bmatrix} \mathbf{J}^{(1)} \\ \mathbf{J}^{(2)} \\ \vdots \\ \mathbf{J}^{(k)} \\ \vdots \\ \mathbf{J}^{(m)} \end{bmatrix} \tag{5.135}$$

$$\mathbf{y} = \begin{bmatrix} \mathbf{Y}^{(1)} & \mathbf{0} & \cdots & \mathbf{0} & \cdots & \mathbf{0} \\ \mathbf{0} & \mathbf{Y}^{(2)} & & & & \\ \vdots & & \ddots & & & \vdots \\ \mathbf{0} & & & \mathbf{Y}^{(k)} & & \mathbf{0} \\ \vdots & & & & \ddots & \vdots \\ \mathbf{0} & \mathbf{0} & \cdots & \mathbf{0} & \cdots & \mathbf{Y}^{(m)} \end{bmatrix} \tag{5.136}$$

In (5.136), $\mathbf{Y}^{(1)}, \mathbf{Y}^{(2)}, \ldots, \mathbf{Y}^{(m)}$ are the 2×2 device admittance matrices and $\mathbf{J}^{(1)}, \mathbf{J}^{(2)}, \ldots, \mathbf{J}^{(m)}$ are vectors of noise currents related to these two-port devices.

By short-circuiting the external ports ($\mathbf{V}_e = \mathbf{0}$), we can derive the current noise sources of the equivalent circuit of Figure 5.6. Denoting $\mathbf{I}_e$ as $\mathbf{I}_N$, from (5.133) we have

$$-\mathbf{y}\mathbf{V}_d - \mathbf{J} = \mathbf{Y}_{dd}\mathbf{V}_d + \mathbf{N}_d \tag{5.137}$$

$$\mathbf{I}_N = \mathbf{Y}_{ed}\mathbf{V}_d + \mathbf{N}_e \tag{5.138}$$

By eliminating $\mathbf{V}_d$ from these equations, we find the solution vector $\mathbf{I}_N$ in the form:

$$\mathbf{I}_N = \mathbf{H}_N\mathbf{N} + \mathbf{H}_J\mathbf{J} \tag{5.139}$$

where

$$\mathbf{H}_J = -\mathbf{Y}_{ed}(\mathbf{Y}_{dd} + \mathbf{y})^{-1} \tag{5.140}$$

$$\mathbf{H}_N = [\mathbf{H}_J \mathbf{1}_n] \tag{5.141}$$

$$\mathbf{N} = \begin{bmatrix} \mathbf{N}_d \\ \mathbf{N}_e \end{bmatrix} \tag{5.142}$$

In (5.141), $\mathbf{1}_n$ is an $n \times n$ identity matrix, and n is a number of external ports.

By using (5.139) we are able to get a correlation matrix of the equivalent noise source currents $\mathbf{I}_N$. Because the $\mathbf{N}$ and $\mathbf{J}$ noise sources are not correlated to each other, the requested correlation matrix is given by

$$\overline{\mathbf{I}_N \mathbf{I}_N^+} = \mathbf{H}_N \overline{\mathbf{N}\mathbf{N}^+} \mathbf{H}_N^+ + \mathbf{H}_J \overline{\mathbf{J}\mathbf{J}^+} \mathbf{H}_J^+ \tag{5.143}$$

or, by using relations describing correlation matrices of passive multiports (5.48) and active two-ports (5.51) in admittance representation:

$$\overline{\mathbf{I}_N \mathbf{I}_N^+} = 2kT_0\, df\left[\mathbf{H}_N(\mathbf{Y} + \mathbf{Y}^+)\mathbf{H}_N^+ + 2\mathbf{H}_J\mathbf{C}_J\mathbf{H}_J^+\right] \tag{5.144}$$

where $\mathbf{C}_J$ is the correlation matrix of the noise current sources representing noise generated in all two-port devices of the circuit. The term $\mathbf{C}_J$ is the block-diagonal matrix with 2×2 noise current correlation matrices $\mathbf{C}_J^{(k)}$ of active two-ports located on the main diagonal.

As we can see, the admittance matrix of the overall analyzed circuit related to the external ports (n-ports of Figure 5.17) is given by

$$\mathbf{Y}_e = \mathbf{Y}_{ee} + \mathbf{H}_J\mathbf{Y}_{de} \tag{5.145}$$

and can be computed with only slightly more effort. Conventional frequency-domain analysis of the circuit may be carried out simultaneously with the noise analysis of the circuit. This means that a circuit of any general topology may be simultaneously optimized with respect to both the noise properties and any of the conventional circuit functions.

Example 5.4

Figure 5.19 presents the equivalent circuit of a four stage distributed or traveling-wave amplifier. GaAs MESFETs integrated with all the required passive components make this circuit very attractive for ultrabroadband applications. In this example we present

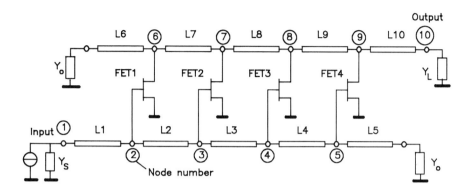

Figure 5.19 Equivalent circuit of four stage distributed amplifier.

the noise analysis method of linear multiport networks of arbitrary topology by using the admittance matrix and the noise current correlation matrix. In Figure 5.20 the amplifier circuit has been partitioned into the lossy passive multiport and four noisy two-ports representing four FETs. The equations of the passive multiport written in the form given by (5.133) are

$$
\begin{array}{c}
\mathbf{I}_d \\
\downarrow
\end{array}
\qquad
\begin{array}{c}
\mathbf{Y}_{dd} \\
\end{array}
\qquad
\begin{array}{c}
\mathbf{Y}_{de} \\
\end{array}
$$

		2	6	3	7 ↓ 4	8	5	9	1 ↓ 10		
I_2	2	Y_{22}	0	Y_{23}	0	0	0	0	0 ⋮ Y_{21}	0	V_2
I_6	6	0	Y_{66}	0	Y_{67}	0	0	0	0 ⋮ 0	0	V_6
I_3	3	Y_{32}	0	Y_{33}	0	Y_{34}	0	0	0 ⋮ 0	0	V_3
I_7	7	0	Y_{76}	0	Y_{77}	0	Y_{78}	0	0 ⋮ 0	0	V_7
I_4	4	0	0	Y_{43}	0	Y_{44}	0	Y_{45}	0 ⋮ 0	0	V_4
I_8	8	0	0	0	Y_{87}	0	Y_{88}	0	Y_{89} ⋮ 0	0	V_8
I_5	5	0	0	0	0	Y_{54}	0	Y_{55}	0 ⋮ 0	0	V_5
I_9	9	0	0	0	0	0	Y_{98}	0	Y_{99} ⋮ 0	Y_{910}	V_9
I_1	1	Y_{12}	0	0	0	0	0	0	0 ⋮ Y_{11}	0	V_1
I_{10}	10	0	0	0	0	0	0	0	Y_{109} ⋮ 0	Y_{1010}	V_{10}

with $\leftarrow \mathbf{V}_d$, $\leftarrow \mathbf{V}_e$ on the right, and

$$
\begin{array}{c}
\uparrow \\
\mathbf{I}_e
\end{array}
\qquad
\begin{array}{c}
\uparrow \\
\mathbf{Y}_{ed}
\end{array}
\qquad
\begin{array}{c}
\uparrow \\
\mathbf{Y}_{ee}
\end{array}
$$

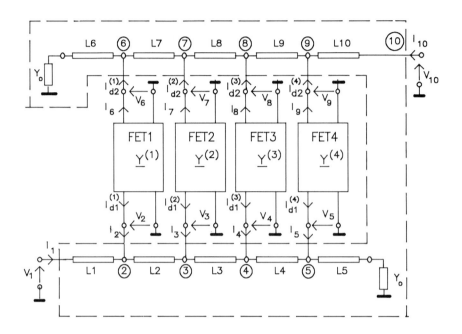

Figure 5.20 Equivalent circuit of the distributed amplifier partitioned into passive multiport and four active two-ports.

where

$$Y_{22} = y_{11}^{L1} + y_{11}^{L2} + y_{12}^{L1} + y_{12}^{L2}$$

$$Y_{13} = Y_{31} = y_{12}^{L1}$$

$$Y_{33} = y_{11}^{L2} + y_{11}^{L3} + y_{12}^{L2} + y_{12}^{L3} \cdots$$

In these relations y_{11}^{Lk} and y_{12}^{Lk}, $k = 1, 2, \ldots, 10$, are terms of the admittance matrices of transmission line sections

$$\mathbf{Y}^{Lk} = \begin{bmatrix} y_{11}^{Lk} & y_{12}^{Lk} \\ y_{12}^{Lk} & y_{11}^{Lk} \end{bmatrix}, \quad k = 1, 2, \ldots, 10$$

Ports 1 and 10 are the external ports of the circuit.

The equations for four FETs in the form given by (5.134) are

$$
\begin{bmatrix} I_2 \\ I_6 \\ \hline I_3 \\ I_7 \\ \hline I_4 \\ I_8 \\ \hline I_5 \\ I_9 \end{bmatrix}
=
\left[
\begin{array}{cc:cc:cc:cc}
\mathbf{Y}^{(1)} & & 0 & 0 & 0 & 0 & 0 & 0 \\
& & 0 & 0 & 0 & 0 & 0 & 0 \\ \hdashline
0 & 0 & \mathbf{Y}^{(2)} & & 0 & 0 & 0 & 0 \\
0 & 0 & & & 0 & 0 & 0 & 0 \\ \hdashline
0 & 0 & 0 & 0 & \mathbf{Y}^{(3)} & & 0 & 0 \\
0 & 0 & 0 & 0 & & & 0 & 0 \\ \hdashline
0 & 0 & 0 & 0 & 0 & 0 & \mathbf{Y}^{(4)} & \\
0 & 0 & 0 & 0 & 0 & 0 & &
\end{array}
\right]
\begin{bmatrix} V_2 \\ V_6 \\ \hline V_3 \\ V_7 \\ \hline V_4 \\ V_8 \\ \hline V_5 \\ V_9 \end{bmatrix}
$$

$$
\mathbf{I}_d =
\begin{bmatrix} \mathbf{I}_d^{(1)} \\ \hline \mathbf{I}_d^{(2)} \\ \hline \mathbf{I}_d^{(3)} \\ \hline \mathbf{I}_d^{(4)} \end{bmatrix}
=
\begin{bmatrix} I_2 \\ I_6 \\ \hline I_3 \\ I_7 \\ \hline I_4 \\ I_8 \\ \hline I_5 \\ I_9 \end{bmatrix}, \quad
\mathbf{V}_d =
\begin{bmatrix} \mathbf{V}_d^{(1)} \\ \hline \mathbf{V}_d^{(2)} \\ \hline \mathbf{V}_d^{(3)} \\ \hline \mathbf{V}_d^{(4)} \end{bmatrix}
=
\begin{bmatrix} V_2 \\ V_6 \\ \hline V_3 \\ V_7 \\ \hline V_4 \\ V_8 \\ \hline V_5 \\ V_9 \end{bmatrix}
$$

and $\mathbf{Y}^{(1)}, \mathbf{Y}^{(2)}, \mathbf{Y}^{(3)}, \mathbf{Y}^{(4)}$ are 2×2 admittance matrices of four FETs.
The correlation matrix $\overline{I_N I_N^+}$ defined by (5.144) is here a 2×2 matrix:

$$
\mathbf{C}_Y = \frac{1}{4kT_0\,\mathrm{d}f}\, \overline{I_N I_N^+} = \frac{1}{4kT_0\,\mathrm{d}f}\, \overline{\begin{bmatrix} I_{N1} \\ I_{N10} \end{bmatrix}\begin{bmatrix} I_{N1} \\ I_{N10} \end{bmatrix}^+}
$$

$$
= \frac{1}{4kT_0\,\mathrm{d}f}\, \overline{\begin{bmatrix} I_{N1}I_{N1}^* & I_{N1}I_{N10}^* \\ I_{N10}I_{N1}^* & I_{N10}I_{N10}^* \end{bmatrix}}
$$

This matrix is related to two external ports numbered 1 and 10. $\mathbf{C}_Y$ may be computed as

$$
\mathbf{C}_Y = \frac{1}{2}\mathbf{H}_N(\mathbf{Y} + \mathbf{Y}^+)\mathbf{H}_N + \mathbf{H}_J\mathbf{C}_J\mathbf{H}
$$

where

$$\mathbf{H}_J = -\mathbf{Y}_{ed}\left[\mathbf{Y}_{dd} + \mathbf{y}\right]^{-1}$$

$$= -\begin{bmatrix} Y_{12} & 0 & 0 & 0 & 0 & 0 & 0 & 0 \\ 0 & 0 & 0 & 0 & 0 & 0 & 0 & Y_{109} \end{bmatrix}$$

$$\begin{bmatrix} \begin{bmatrix} Y_{22} & 0 & Y_{23} & 0 & 0 & 0 & 0 & 0 \\ 0 & Y_{66} & 0 & Y_{67} & 0 & 0 & 0 & 0 \\ Y_{32} & 0 & Y_{33} & 0 & Y_{34} & 0 & 0 & 0 \\ 0 & Y_{76} & 0 & Y_{77} & 0 & Y_{78} & 0 & 0 \\ 0 & 0 & Y_{43} & 0 & Y_{44} & 0 & Y_{45} & 0 \\ 0 & 0 & 0 & Y_{87} & 0 & Y_{88} & 0 & Y_{89} \\ 0 & 0 & 0 & 0 & Y_{54} & 0 & Y_{55} & 0 \\ 0 & 0 & 0 & 0 & 0 & Y_{98} & 0 & Y_{99} \end{bmatrix} \end{bmatrix}$$

$$+ \begin{bmatrix} \begin{array}{cc:cc:cc:cc} \mathbf{Y}^{(1)} & & 0 & 0 & 0 & 0 & 0 & 0 \\ & & 0 & 0 & 0 & 0 & 0 & 0 \\ \hdashline 0 & 0 & \mathbf{Y}^{(2)} & & 0 & 0 & 0 & 0 \\ 0 & 0 & & & 0 & 0 & 0 & 0 \\ \hdashline 0 & 0 & 0 & 0 & \mathbf{Y}^{(3)} & & 0 & 0 \\ 0 & 0 & 0 & 0 & & & 0 & 0 \\ \hdashline 0 & 0 & 0 & 0 & 0 & 0 & \mathbf{Y}^{(4)} & \\ 0 & 0 & 0 & 0 & 0 & 0 & & \end{array} \end{bmatrix}^{-1}$$

$$= \begin{bmatrix} h_{11} & h_{12} & h_{13} & h_{14} & h_{15} & h_{16} & h_{17} & h_{18} \\ h_{21} & h_{22} & h_{23} & h_{24} & h_{25} & h_{26} & h_{27} & h_{28} \end{bmatrix}$$

$$\mathbf{Y} + \mathbf{Y}^+ = 2\,\mathrm{Re}\left\{ \begin{bmatrix} \begin{array}{cccccccc:cc} Y_{22} & 0 & Y_{23} & 0 & 0 & 0 & 0 & 0 & Y_{21} & 0 \\ 0 & Y_{66} & 0 & Y_{67} & 0 & 0 & 0 & 0 & 0 & 0 \\ Y_{32} & 0 & Y_{33} & 0 & Y_{34} & 0 & 0 & 0 & 0 & 0 \\ 0 & Y_{76} & 0 & Y_{77} & 0 & Y_{78} & 0 & 0 & 0 & 0 \\ 0 & 0 & Y_{43} & 0 & Y_{44} & 0 & Y_{45} & 0 & 0 & 0 \\ 0 & 0 & 0 & Y_{87} & 0 & Y_{88} & 0 & Y_{89} & 0 & 0 \\ 0 & 0 & 0 & 0 & Y_{54} & 0 & Y_{55} & 0 & 0 & 0 \\ 0 & 0 & 0 & 0 & 0 & Y_{98} & 0 & Y_{99} & 0 & Y_{910} \\ \hdashline Y_{12} & 0 & 0 & 0 & 0 & 0 & 0 & 0 & Y_{11} & 0 \\ 0 & 0 & 0 & 0 & 0 & 0 & 0 & Y_{109} & 0 & Y_{1010} \end{array} \end{bmatrix} \right\}$$

because, for passive, reciprocal circuit, $\mathbf{Y} = \mathbf{Y}^T$,

$$\mathbf{H}_N = \begin{bmatrix} h_{11} & h_{12} & h_{13} & h_{14} & h_{15} & h_{16} & h_{17} & h_{18} & \vdots & 1 & 0 \\ h_{21} & h_{22} & h_{23} & h_{24} & h_{25} & h_{26} & h_{27} & h_{28} & \vdots & 0 & 1 \end{bmatrix}$$

$$\mathbf{C}_J = \left[\begin{array}{cc:cc:cc:cc} \mathbf{C}_J^{(1)} & & \vdots\, 0 & 0 & \vdots\, 0 & 0 & \vdots\, 0 & 0 \\ & & \vdots\, 0 & 0 & \vdots\, 0 & 0 & \vdots\, 0 & 0 \\ \hdashline 0 & 0 & \mathbf{C}_J^{(2)} & & \vdots\, 0 & 0 & \vdots\, 0 & 0 \\ 0 & 0 & & & \vdots\, 0 & 0 & \vdots\, 0 & 0 \\ \hdashline 0 & 0 & \vdots\, 0 & 0 & \mathbf{C}_J^{(3)} & & \vdots\, 0 & 0 \\ 0 & 0 & \vdots\, 0 & 0 & & & \vdots\, 0 & 0 \\ \hdashline 0 & 0 & \vdots\, 0 & 0 & \vdots\, 0 & 0 & \mathbf{C}_J^{(4)} & \\ 0 & 0 & \vdots\, 0 & 0 & \vdots\, 0 & 0 & & \end{array}\right]$$

$\mathbf{C}_J^{(i)}$, $i = 1, \ldots, 4$ are current noise correlation matrices of four FETs and may be computed from (5.51).

The admittance matrix $\mathbf{Y}_e$ of the overall circuit related to two external ports (port 1 and port 10) is given by (5.145):

$$\mathbf{Y}_e = \begin{bmatrix} Y_{e11} & Y_{e12} \\ Y_{e21} & Y_{e22} \end{bmatrix} = \mathbf{Y}_{ee} + \mathbf{H}_J \mathbf{Y}_{de}$$

where

$$\mathbf{Y}_{ee} = \begin{bmatrix} Y_{1,1} & 0 \\ 0 & Y_{10,10} \end{bmatrix}$$

$$\mathbf{Y}_{de} = \begin{bmatrix} Y_{21} & 0 \\ 0 & 0 \\ 0 & 0 \\ 0 & 0 \\ 0 & 0 \\ 0 & 0 \\ 0 & 0 \\ 0 & Y_{910} \end{bmatrix}$$

The noise figure of the amplifier may be computed by using (5.68):

$$F = 1 + \frac{\mathcal{O}^+ \mathbf{H} \mathbf{C}_Y \mathbf{H} \mathcal{O}}{\operatorname{Re} Y_s}$$

where

$$\mathcal{O} = \begin{bmatrix} 1 \\ \dfrac{(Y_s + Y_{e11})^*}{Y_{e21}^*} \end{bmatrix} \quad \text{and} \quad \mathbf{H} = \begin{bmatrix} 1 & 0 \\ 0 & 1 \end{bmatrix}$$

REFERENCES

[1]   H. Bittel and L. Storm, *Rauschen*, Springer-Verlag, Berlin, 1971.

[2]   A. van der Ziel, *Fluctuation Phenomena in Semiconductors*, Butterworths, London, 1959.

[3]   A. van der Ziel, *Noise, Sources, Characterization, Measurement*, Prentice-Hall, Englewood Cliffs, NJ, 1970.

[4]   W.B. Davenport and W.L. Root, *An Introduction to the Theory of Random Signals and Noise*, McGraw-Hill, New York, 1958.

[5]   A. Papoulis, *Probability, Random Variables and Stochastic Processes*, McGraw-Hill, New York, 1965.

[6]   H. Rothe and W. Dahlke, "Theory of Noisy Fourpoles," *Proc. IRE*, Vol. 44, June 1956, pp. 811–818.

[7]   H. Bauer and H. Rothe, "Der equivalente Rauschvierpol als Wellenvierpol," *Archiv der elektrischen Uebertragung*, Vol. 10, 1956, pp. 241–252.

[8]   H.A. Haus *et al.*, "Representation of Noise in Linear Two-Ports," *Proc. IRE*, Vol. 48, January 1960, pp. 69–74.

[9]   R.P. Hecken, "Analysis of Linear Noisy Two-Ports Using Scattering Waves," *IEEE Trans. Microwave Theory Tech.*, Vol. MTT-29, October 1981, pp. 997–1003.

[10]   R.Q. Lane, "The Determination of the Device Noise Parameters," *Proc. IEEE*, Vol. 116, August 1969, pp. 1461–1462.

[11]   M. Mitama and H. Katoh, "An Improved Computational Method for Noise Parameter Measurement," *IEEE Trans. Microwave Theory Tech.*, Vol. MTT-27, June 1979, pp. 612–615.

[12]   W. Baechtold, W. Kotyczka, and M.J. Strutt, "Computerized Calculation of Small Signal and Noise Properties of Microwave Transistors," *IEEE Trans. Microwave Theory Tech.*, Vol. MTT-17, August 1969, pp. 614–619.

[13]   R.P. Meys, "A Wave Approach to the Noise Properties of Linear Microwave Devices," *IEEE Trans. Microwave Theory Tech.*, Vol. MTT-26, January 1978, pp. 34–37.

[14]   H. Hillbrandt and P.H. Russer, "An Efficient Method for Computer Aided Noise Analysis of Linear Amplifier Networks," *IEEE Trans. Circuits and Systems*, Vol. CAS-23, April 1976, pp. 235–238.

[15]   R. Reiss, "Four-Pole Based Noise Analysis by Digital Computer," *Proc. 1976 European Conf. Circuit Theory and Design*, Genoa, Italy, September 1976, pp. 632–642.

[16]   R. Reiss, "*Schaltungstheorie linearer rauschender Vierpole*," Habilitationsschrift, Fakultaet Fuer Elektrotechnik, Universitaet Fridericiana, Karlsruhe, June 1975.

[17]   H.A. Hause and R.B. Adler, *Circuit Theory of Linear Noisy Networks*, John Wiley and Sons, New York, 1959.

[18]   H. Bosma, "On the Theory of Linear Noisy Systems," *Philips Research Reports Suppl.*, No. 10, 1967.

[19]   R. Rohrer, L. Nagel, R. Meyer, and L. Weber, "Computationally Efficient Electronic-Circuit Calculations," *IEEE Trans. Solid-State Circuits*, Vol. SC-6, August 1971, pp. 204–213.

[20]   H. Fukui, "The Noise Performance of Microwave Transistors," *IEEE Trans. Electron. Devices*, Vol. ED-13, March 1966, pp. 329–341.

[21]   H. Statz, H.A. Haus, and R.A. Pucel, "Noise Characteristics of Gallium Arsenide Field-Effect Transistors," *IEEE Trans. Electron. Devices*, Vol. ED-21, September 1974, pp. 549–562.

[22]   R.A. Pucel, D.J. Masse, and C.K. Krumm, "Noise Performance of Gallium Arsenide Field-Effect Transistors," *IEEE J. Solid-State Circuits*, Vol. SC-11, April 1976, pp. 243–255.

[23]   G. Caruso and M. Sannino, "Computer-Aided Determination of Microwave Two-Port Noise Parameters," *IEEE Trans. Microwave Theory Tech.*, Vol. MTT-26, September 1978, pp. 639–642.

[24]   M.W. Pospieszalski, "On the Measurement of Noise Parameters of Microwave Two-Ports," *IEEE Trans. Microwave Theory Tech.*, Vol. MTT-34, April 1986, pp. 456–458.

[25] J. Lange, ''Noise Characterization of Linear Two-Ports in Terms of Invariant Parameters,'' *IEEE J. Solid-State Circuits*, Vol. SC-2, June 1967, pp. 37–40.

[26] W. Wiatr, ''A Method of Estimating Noise Parameters of Linear Microwave Two-Ports,'' Ph.D. Dissertation, Warsaw University of Technology, Warsaw, Poland, 1980 [in Polish].

[27] J.A. Dobrowolski, ''A CAD Oriented Method for Noise Figure Computation of Two-Ports with Any Internal Topology,'' *IEEE Trans. Microwave Theory Tech.*, Vol. MTT-37, January 1989, pp. 15–20.

[28] Private communication, Z. Nosal, Institute of Electronics Fundamentals, Warsaw University of Technology, Warsaw, Poland.

[29] V. Rizzoli and A. Lipparini, ''Computer-Aided Noise Analysis of Linear Multiport Networks of Arbitrary Topology,'' *IEEE Trans. Microwave Theory Tech.*, Vol. MTT-33, December 1985, pp. 1507–1512.

Chapter 6

Numerical Methods for Solving Systems
of Linear Algebraic Equations

As is clear from the preceding three chapters, many of the problems found in circuit simulation ultimately can be reduced to solving related systems of linear algebraic equations. A set of linear algebraic equations can be written in the matrix form as

$$\mathbf{A}\mathbf{x} = \mathbf{b} \qquad (6.1)$$

where $\mathbf{A} = [a_{ij}]$ is an $n \times n$ matrix of real or complex coefficients, $\mathbf{x}$ is the vector of n unknowns, and $\mathbf{b}$ is the vector of n known right-hand side terms.

Two classes of methods are used to solve the set of equations: direct methods and indirect or iterative methods. Direct methods include those that can solve the system in a fixed and finite number of arithmetic operations. Iterative methods can produce an infinite sequence of approximate solutions that may converge to the correct solution. Application of iterative methods in computer-aided simulation of electronic and microwave circuits is rather limited. Strong conditions on a coefficient matrix are required to get the convergence of the solution [1].

Discussion in this chapter will be of direct methods. Of these approaches, the Gaussian elimination, LU factorization, and bifactorization methods will be discussed in detail. We shall discuss conventional full matrix programming techniques and their error mechanisms [2–7].

Systems of linear equations describing microwave circuits become quite large when we want to analyze sizable circuits, but fortunately they tend also to be very sparse. In the next chapter we will discuss the techniques to efficiently solve these large, sparse systems.

6.1 GAUSSIAN ELIMINATION

No practical scheme has been found that is more efficient than Gaussian elimination in solving the system of linear equations $\mathbf{Ax} = \mathbf{b}$. Gaussian elimination is a recursive algorithm that, at each step, eliminates one unknown and one equation from the original system. This process is repeated $n - 1$ times, until only one equation and one unknown remain. Gaussian elimination is a recursive process of triangularization of $\mathbf{A}$. After the system has been converted into triangular form, it is solved with a process called *backward substitution*, in which the equations are solved successively in reverse order [2-7].

Triangularization

Consider the following system of equations:

$$a_{11}x_1 + a_{12}x_2 + \cdots + a_{1n}x_n = b_1$$

$$a_{21}x_1 + a_{22}x_2 + \cdots + a_{2n}x_n = b_2$$

$$\cdots \tag{6.2}$$

$$a_{n1}x_1 + a_{n2}x_2 + \cdots + a_{nn}x_n = b_n$$

The triangularization starts with the elimination of x_1. The first equation multiplied by $m_{i1} = a_{i1}/a_{11}$ is subtracted from the ith equation (for $i = 2, 3, 4, \ldots, n$). We say that the first equation is eliminated and the remaining $n - 1$ equations are

$$a_{22}^{(2)}x_2 + a_{23}^{(2)}x_3 + \cdots + a_{2n}^{(2)}x_n = b_2^{(n)}$$

$$\cdots \tag{6.3}$$

$$a_{n2}^{(2)}x_2 + a_{n3}^{(n)}x_3 + \cdots + a_{nn}^{(2)}x_n = b_n^{(2)}$$

where

$$a_{ij}^{(2)} = a_{ij} - m_{i1}a_{1j}, \quad i, j = 2, 3, \ldots, n$$

$$b_i^{(2)} = b_i - m_{i1}b_1, \qquad i = 2, 3, \ldots, n. \tag{6.4}$$

In (6.4), the term $m_{i1} = a_{i1}/a_{11}$ is called a *multiplier*. To perform the first step of the triangularization, entry a_{11} of the $\mathbf{A}$ matrix must be different from zero. In the second step of the triangularization we perform identical operations as in the first step.

If $a_{22}^{(2)}$ is different from zero, the unknown x_2 is eliminated from the last $n - 2$ equations. This process leaves $n - 2$ equations with $n - 2$ unknowns. The process is repeated $n - 1$ times, until only one equation and one unknown remain. At each step of elimination the coefficient of the unknown being eliminated in the equation being eliminated must be different from zero. These entries, $a_{kk}^{(k)}$, are called *pivotal elements*, or *pivots* for short.

The resultant triangular system of equations has the form:

$$a_{11}^{(1)} x_1 + a_{12}^{(1)} x_2 + \cdots + a_{1n}^{(1)} x_n = b_1^{(1)}$$

$$a_{22}^{(2)} x_2 + \cdots + a_{2n}^{(2)} x_n = b_2^{(2)}$$

$$\cdots$$

$$a_{nn}^{(n)} x_n = b_n^{(n)}$$

(6.5)

where $a_{1j}^{(1)} = a_{1j}$, for $j = 1, 2, \ldots, n$, and $b_1^{(1)} = b_1$.

The triangularization process can be summarized as follows. The full process requires us to perform $n - 1$ elimination steps. At the kth step, the coefficients of the reduced system of equations:

$$\mathbf{A}^{(k)} \mathbf{x}^{(k)} = \mathbf{b}^{(k)}$$

(6.6)

where

$$\mathbf{A}^{(k)} = \left[a_{ij}^{(k)} \right], \quad i, j = k, \ldots, n,$$

$$\mathbf{x}^{(k)} = \left[x_k, x_{k+1}, \ldots, x_n \right]^T,$$

$$\mathbf{b}^{(k)} = \left[b_k^{(k)}, b_{k+1}^{(k)}, \ldots, b_n^{(k)} \right]^T$$

are modified by simple transformations:

$$a_{ij}^{(k+1)} = a_{ij}^{(k)} - m_{ik} a_{kj}^{(k)}$$

(6.7)

$$b_i^{(k+1)} = b_i^{(k)} - m_{ik} b_k^{(k)}$$

(6.8)

where

$$k = 1, 2, \ldots, n - 1$$

$$i, j = k + 1, k + 2, \ldots, n$$

$$m_{ik} = \frac{a_{ik}^{(k)}}{a_{kk}^{(k)}}$$

$$a_{ij}^{(1)} = a_{ij},$$

$$b_i^{(1)} = b_i$$

Backward Substitution

In backward substitution the triangular system of equations (6.5) is solved in reverse order, from the last to the first equation. The solution for x_n of the last equation is trivial. Once x_n is known, its value is substituted into equation $n - 1$ and the values of x_{n-1} is computed. This process is repeated until the values of all the unknowns are found.

The computation of the unknowns by backward substitution can be summarized in the form:

$$x_n = \frac{b_k^{(k)} - \sum_{j=k+1}^{n} a_{kj}^{(k)} x_j}{a_{kk}^{(k)}}, \quad k = n, n - 1, \ldots, 1 \tag{6.9}$$

where, by definition, the summation is zero when $k = n$ and therefore $j = n + 1$:

$$\sum_{j=n+1}^{n} (\cdot) = 0 \tag{6.10}$$

6.1.1 Operation Count

The computational effort is usually expressed as the total number of multiplications and divisions required. Step k of the triangularization requires $(n - k)$ divisions and $(n - k)(n - k + 1)$ additions and multiplications. The total number of divisions is

$$\sum_{k=1}^{n-1} (n - k) = \left. \frac{n^2 - n}{2} \right|_{n \gg 1} \approx \frac{n^2}{2} \tag{6.11}$$

The number of additions and multiplications is

$$\sum_{k=1}^{n-1} (n - k)(n - k + 1) = \left. \frac{n^3 - n}{3} \right|_{n \gg 1} \approx \frac{n^3}{3} \tag{6.12}$$

This shows that the number of divisions and multiplications required for the triangularization is

$$N = \frac{n^3}{3} + \frac{n^2}{2} - \frac{5n}{6} \tag{6.13}$$

For large systems ($n \gg 1$), N is proportional to the cube of its size n.

From the equation (6.9) we see that backward substitution requires n divisions and

$$\sum_{k=1}^{n} (n - k) = \frac{n^2 - n}{2} \bigg|_{n \gg 1} \approx \frac{n^2}{2} \qquad (6.14)$$

additions and multiplications.

The preceding considerations show that the total number of arithmetic operations required for the solution of a system of linear equations by Gaussian elimination is

$$N_t = \frac{n^3}{3} + n^2 - \frac{n}{3} \qquad (6.15)$$

Note that the operation count as just given is for a system $\mathbf{Ax} = \mathbf{b}$ in which $\mathbf{A}$ is assumed to be full.

6.2 LU DECOMPOSITION

Gaussian elimination is a process of triangularization of the coefficient matrix $\mathbf{A}$ in the system $\mathbf{Ax} = \mathbf{b}$. Although the arithmetic operations involved in Gaussian elimination are the same as in most other matrix solution methods, the Gauss ordering of these operations is too restrictive when we want to solve $\mathbf{Ax} = \mathbf{b}$ with more than one right-hand vector $\mathbf{b}$. In this case, a more flexible solution is LU or triangular decomposition (also called *matrix factorization*). Triangular decomposition enables a simple solution of systems with different right-hand vectors $\mathbf{b}$ as well as transpose systems required in sensitivity computations.

The LU decomposition is equivalent to decomposing $\mathbf{A}$ into a product of a lower triangular matrix $\mathbf{L}$ and an upper triangular matrix $\mathbf{U}$:

$$\mathbf{A} = \mathbf{LU} \qquad (6.16)$$

where $\mathbf{L} = [l_{ij}]$, $l_{ij} = 0$, $i < j$, and $\mathbf{U} = [u_{ij}]$, $u_{ij} = 0$ for $i > j$.

The system $\mathbf{Ax} = \mathbf{b}$ becomes

$$\mathbf{LUx} = \mathbf{b} \qquad (6.17)$$

Letting

$$\mathbf{Ux} = \mathbf{y} \qquad (6.18)$$

so that

$$\mathbf{Ly} = \mathbf{b} \qquad (6.19)$$

We first solve the lower triangular system (6.19) for **y**. This process is called *forward substitution*. Then we solve the upper triangular system (6.18) for **x** with backward substitution. As we see, after LU factorization has been performed, the system of equations can be solved with many different right-hand vectors **b** simply by repeating forward and backward substitutions at a cost of n^2 multiplications and divisions. This is considerably less than the

$$N = \frac{n^3}{3} - \frac{n}{3} \tag{6.20}$$

multiplications and divisions required to decompose the coefficient matrix **A** into **LU** factors. If the number of multiplications and divisions in two substitution steps is added (n^2), the total number of operations required to solve a system of linear equations is

$$N_t = \frac{n^3}{3} + n^2 - \frac{n}{3} \tag{6.15}$$

which is identical to the computational cost of Gaussian elimination.

6.2.1 Gauss's Algorithm

Gaussian elimination is the triangularization of **A**. This process is equivalent to decomposing the coefficient matrix into a product of the lower triangular matrix **L** and an upper triangular matrix **U**. The triangular system of equations (6.5) derived after full matrix triangularization can be written in the form **Ux** = **y** by setting

$$\mathbf{u} = \begin{bmatrix} a_{11}^{(1)} & a_{12}^{(1)} & \cdots & & a_{1n}^{(1)} \\ 0 & a_{22}^{(2)} & \cdots & & a_{2n}^{(2)} \\ & 0 & & & \\ \vdots & \vdots & \ddots & & \vdots \\ 0 & 0 & \cdots & 0 & a_{nn}^{(n)} \end{bmatrix} \quad \text{and} \quad \mathbf{y} = \begin{bmatrix} b_1^{(1)} \\ b_2^{(2)} \\ \vdots \\ b_n^{(n)} \end{bmatrix} \tag{6.21}$$

In the next step, we should show that **y** could be computed by solving an equation of the form **Ly** = **b**. The vector **b** can be written in terms of **y** by recursively using (6.8):

$$b_1 = b_1^{(1)}$$
$$b_2 = m_{21} b_1^{(1)} + b_2^{(2)}$$
$$b_3 = m_{31} b_1^{(1)} + m_{32} b_2^{(2)} + b_3 \tag{6.22}$$
$$\cdots$$
$$b_n = m_{n1} b_n^{(1)} + m_{n2} b_2^{(2)} + \cdots + b_n^{(n)}$$

Relation (6.16) can be rewritten in the matrix form $\mathbf{Ly} = \mathbf{b}$, where

$$\mathbf{L} = \begin{bmatrix} 1 & 0 & & \cdots & 0 \\ m_{21} & 1 & 0 & \cdots & 0 \\ m_{31} & m_{32} & 1 & & 0 \\ \vdots & \vdots & & \ddots & 0 \\ m_{n1} & m_{n2} & & \cdots & 1 \end{bmatrix} \tag{6.23}$$

The factorization of $\mathbf{A}$ into $\mathbf{L}$ and $\mathbf{U}$ is straightforward. According to (6.15) and 6.17), the entries of the $\mathbf{U}$ and $\mathbf{L}$ matrices are

$$u_{ij} = \begin{cases} a_{ij}^{(i)} & \text{for } i \le j \\ 0 & \text{for } i > j \end{cases} \tag{6.24}$$

$$l_{ij} = \begin{cases} 0 & \text{for } i < j \\ 1 & \text{for } i = j \\ m_{ij} & \text{for } i > j \end{cases} \tag{6.25}$$

where $i = 1, 2, 3, \ldots, n$.

The LU decomposition is performed in $n - 1$ steps. The superscripts to the symbols of matrix elements in relations (6.21) to (6.25) indicate the step numbers in which the particular elements of the $\mathbf{L}$ and $\mathbf{U}$ matrices are computed.

This implementation of LU factorization is called Gauss's algorithm. By using 5.7), (6.9), (6.10) and (6.19), we can write schematically the Gauss's algorithm as ollows:

U Decomposition

Step 1 Set $k = 1$.

Step 2 Set the kth row of $\mathbf{U}$ equal to the kth row of the reduced matrix $\mathbf{A}^{(k)}$ ($\mathbf{A}^{(1)} = \mathbf{A}$),

$$u_{kj} = a_{kj}^{(k)}, \quad j = k, k + 1, \ldots, n.$$

Step 3 If $k = n$ then stop.

Step 4 Compute the kth column of $\mathbf{L}$:

$$l_{ik} = \frac{a_{ik}^{(k)}}{u_{kk}}, \quad i = k + 1, \ldots, n.$$

Step 5 Update the reduced matrix from $\mathbf{A}^{(k)}$ to $\mathbf{A}^{(k+1)}$:

$$a_{ij}^{(k+1)} = a_{ij}^{(k)} - l_{ik}u_{kj}, \quad i, j = k + 1, k + 2, \ldots, n.$$

Step 6 $k = k + 1$ and go to step 2.

Forward Substitution

$$y_k = b_k - \sum_{j=1}^{k-1} l_{kj}y_j, \quad k = 1, 2, \ldots, n$$

Backward Substitution

$$x_k = \frac{y_k - \sum_{j=k+1}^{n} u_{kj}x_j}{u_{kk}}, \quad k = n, n - 1, \ldots, 1$$

6.2.2 Doolittle's Algorithm

An alternative form for LU factorization can be found by forming the product of $\mathbf{L}$ and $\mathbf{U}$ and setting each term of the product equal to the corresponding entries in $\mathbf{A}$:

$$\begin{bmatrix} l_{11} & 0 & 0 & \cdots & 0 \\ l_{21} & l_{22} & 0 & \cdots & 0 \\ l_{31} & l_{32} & l_{33} & 0 & \cdots & 0 \\ \vdots & \vdots & & \ddots & & 0 \\ l_{n1} & l_{n2} & \cdots & & & l_{nn} \end{bmatrix} \begin{bmatrix} u_{11} & u_{12} & u_{13} & \cdots & u_{1n} \\ 0 & u_{22} & u_{23} & \cdots & u_{2n} \\ 0 & 0 & u_{33} & \cdots & u_{3n} \\ \vdots & \vdots & & \ddots & \vdots \\ 0 & 0 & \cdots & & 0 & u_{nn} \end{bmatrix}$$

$$\equiv \begin{bmatrix} a_{11} & a_{12} & a_{13} & \cdots & a_{1n} \\ a_{21} & a_{22} & a_{23} & \cdots & a_{2n} \\ a_{31} & a_{32} & a_{33} & \cdots & a_{3n} \\ \vdots & \vdots & & \ddots & \vdots \\ a_{n1} & a_{n2} & a_{n3} & \cdots & a_{nn} \end{bmatrix} \qquad (6.2\text{€}$$

This relation generates the system of n^2 equations, whose number equals the number of elements in $\mathbf{A}$. The number of unknowns equals the number of entries of and $\mathbf{U}$: $n^2 + n$. Because the generated system of equations is underdetermined, elements of $\mathbf{L}$ and $\mathbf{U}$ matrices may be chosen freely. This is accomplished in two way by setting the diagonal elements of either **L** or **U** equal to one. Setting the diagonal elements of **U** to one leads to Crout's algorithm for LU factorization. When the diagonal elements of **L** are set to one it leads to Dootlittle's algorithm for LU factorization.

The computation of the elements of **L** and **U** matrices proceeds in a straightforward manner. To show this, we use (6.20) and write

$$l_{11}u_{11} = a_{11}$$

$$l_{11}u_{12} = a_{12}$$

$$\ldots \tag{6.27}$$

$$l_{11}u_{1n} = a_{1n}$$

Because $l_{11} = 1$, we can immediately determine the first row of **U**. It is simply set equal to the first row of **A**.

$$u_{1j} = a_{1j}, \quad j = 1, 2, \ldots, n \tag{6.28}$$

Now, we can calculate the first column of **L**. From

$$l_{21}u_{11} = a_{21}$$

$$l_{31}u_{11} = a_{31}$$

$$\ldots \tag{6.29}$$

$$l_{n1}u_{11} = a_{n1}$$

we have

$$l_{21} = a_{21}/u_{11}$$

$$l_{31} = a_{31}/u_{11}$$

$$\ldots \tag{6.30}$$

$$l_{n1} = a_{n1}/u_{11}$$

The second row of **U** can be computed next from

$$l_{21}u_{12} + l_{22}u_{22} = a_{22}$$

$$l_{21}u_{13} + l_{22}u_{23} = a_{23}$$

$$\ldots \tag{6.31}$$

$$l_{21}u_{1n} + l_{22}u_{2n} = a_{2n}$$

Because $l_{22} = 1$, we have

$$u_{22} = a_{22} - l_{21}u_{12}$$

$$u_{23} = a_{23} - l_{21}u_{13}$$

$$\ldots$$

$$u_{2n} = a_{2n} - l_{21}u_{1n}$$

(6.32)

Next, we calculate the second column of **L**. The procedure of computation, alternating a row of **U** and a column of **L**, is repeated until the all elements of **L** and **U** are determined. In general, we can write the procedure in the form:

$$k = 1, 2, \ldots, n$$

$$u_{kj} = a_{kj} - \sum_{p=1}^{k-1} l_{kp}u_{pj}, \quad j = k, k+1, \ldots, n \qquad (6.33)$$

$$l_{ik} = \frac{a_{ik} - \sum_{p=1}^{k-1} l_{ip}u_{pk}}{u_{kk}}, \quad i = k+1, \ldots, n \qquad (6.34)$$

where

$$\sum_{p=1}^{0} (\,\cdot\,) \triangleq 0$$

The computer implementation of the algorithm proceeds directly from (6.27) and (6.28). Note that the procedure is halted by the zero-valued pivot u_{kk}, $k = 1, 2, \ldots, n$. Doolittle's algorithm can be written schematically as follows.

LU Decomposition

Step 1 Set $k = 1$.
Step 2 Compute the kth row of **U**:

$$u_{kj} = a_{kj} - \sum_{p=1}^{k-1} l_{kp}u_{pj}, \quad j = k, k+1, \ldots, n$$

Step 3 If $k = n$ then stop.

Step 4 Compute the kth column of $\mathbf{L}$:

$$l_{ik} = \frac{a_{ik} - \sum_{p=1}^{k-1} l_{ip} u_{pk}}{u_{kk}}, \quad i = k, k+1, \ldots, n$$

Step 5 $k = k + 1$ and return to step 2.

Forward Substitution

Compute auxiliary unknowns:

$$y_k = b_k - \sum_{j=1}^{k-1} l_{kj} y_j, \quad k = 1, 2, \ldots, n$$

Backward Substitution

Compute the unknowns:

$$x_n = \frac{y_k - \sum_{j=k+1}^{n} u_{kj} x_j}{u_{kk}}, \quad k = n, n-1, n-2, \ldots, 1$$

6.2.3 Crout's Algorithm

In Crout's algorithm, the diagonal elements of $\mathbf{U}$ are chosen to be one. Of course, $\mathbf{L}$ and $\mathbf{U}$ are not the same in Crout's algorithm as in Doolittle's algorithm. By using (6.20), we can show that the computation of $\mathbf{L}$ and $\mathbf{U}$ proceeds in the following way. The first column of $\mathbf{L}$ can be calculated from

$$l_{11} u_{11} = a_{11}$$
$$l_{21} u_{11} = a_{21}$$
$$\ldots \tag{6.35}$$
$$l_{n1} u_{11} = a_{n1}$$

Because $u_{11} = 1$, the first column of $\mathbf{L}$ is simply set equal to the first column of $\mathbf{A}$:

$$l_{i1} = a_{i1}, \quad i = 1, 2, \ldots, n \tag{6.36}$$

Now, we compute the first row of $\mathbf{U}$. From

$$l_{11}u_{12} = a_{12}$$
$$l_{11}u_{13} = a_{13}$$
$$\cdots$$
$$l_{11}u_{1n} = a_{1n}$$

(6.37)

we have

$$u_{12} = a_{12}/l_{11}$$
$$u_{13} = a_{13}/l_{11}$$
$$\cdots$$
$$u_{1n} = a_{1n}/l_{11}$$

(6.38)

The second column of $\mathbf{L}$ can be computed next from

$$l_{21}u_{12} + l_{22}u_{22} = a_{22}$$
$$l_{31}u_{12} + l_{32}u_{22} = a_{32}$$
$$\cdots$$
$$l_{n1}u_{12} + l_{n2}u_{22} = a_{n2}$$

(6.39)

Because $u_{22} = 1$, we have

$$l_{22} = a_{22} - l_{21}u_{12}$$
$$l_{32} = a_{32} - l_{31}u_{12}$$
$$\cdots$$
$$l_{n2} = a_{n2} - l_{n1}u_{12}$$

(6.40)

Next, we compute the second row of $\mathbf{U}$, then the third column of $\mathbf{L}$, and so on The computation, alternately, of a column of $\mathbf{L}$ and then of a row of $\mathbf{U}$ is repeated unti all the elements of $\mathbf{L}$ and $\mathbf{U}$ are determined. The procedure can be written in the form

$$k = 1, 2, \ldots, n$$

$$l_{ik} = a_{ik} - \sum_{p=1}^{k-1} l_{ip}u_{pk}, \quad i = k, k+1, \ldots, n$$

(6.41)

$$u_{kj} = \frac{a_{kj} - \sum_{p=1}^{k-1} l_{kp}u_{pj}}{l_{kk}}, \quad j = k+1, k+2, \ldots, n$$

(6.42)

where

$$\sum_{p=1}^{0} (\cdot) \triangleq 0$$

The computer implementation of the algorithm can be written directly from (6.41) and (6.42). Note that the pivots l_{kk}, $k = 1, 2, \ldots, n$ must be different from zero.

By using (6.41) and (6.42) we can schematically write Crout's algorithm as follows.

LU Decomposition

Step 1 Set $k = 1$.
Step 2 Compute the kth column of **L**:

$$l_{ik} = a_{ik} - \sum_{p=1}^{k-1} l_{ip} u_{pk}, \quad i = k, k + 1, \ldots, n$$

Step 3 If $k = n$ then stop.
Step 4 Compute the kth row of **U**:

$$u_{kj} = \frac{a_{kj} - \sum_{p=1}^{k-1} l_{kp} u_{pj}}{l_{kk}}, \quad j = k + 1, k + 2, \ldots, n$$

Step 5 $k = k + 1$ and return to step 2.

Forward Substitution

$$y_k = \frac{b_k - \sum_{j=1}^{k-1} l_{kj} y_j}{l_{kk}}, \quad k = 1, 2, \ldots, n$$

Backward Substitution

$$x_k = y_k - \sum_{j=k+1}^{n} u_{kj} x_j, \quad k = n, n - 1, \ldots, 1$$

Comparing Gauss's algorithm with Doolittle's algorithm we see that the operations are the same in both, but performed in a different order. The elements of **L** and **U** are identical for both algorithms and the number of arithmetic operations required to decompose and to solve the system $\mathbf{Ax} = \mathbf{b}$ is also the same, but there are important differences among these algorithms. The required entries of **L** and **U** in Doolittle's algorithm are computed at some previous stage of the algorithm. Further, in Doolittle's algorithm the only entry a_{ij} required is that which belongs to the row or to the column that is being eliminated. In Gauss's algorithm, every entry a_{ij} that has not been eliminated is updated at each stage of the algorithm. This is illustrated in Figure 6.1. It makes Gauss's algorithm superior when considering pivoting. For this reason, among others, Gauss's algorithm works better for sparse matrices. However, Doolittle's algorithm requires less memory for reference values than Gauss's algorithm.

Crout's algorithm is basically the same as Doolittle's. Because in Crout's algorithm the diagonal elements of **U** rather than **L** are chosen to be one, the entries of **L** and **U** are not the same for the two algorithms.

To summarize our discussion, the important features of LU decomposition are as follows:

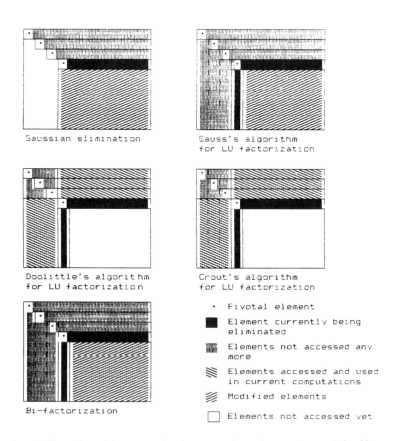

Gaussian elimination

Gauss's algorithm for LU factorization

Doolittle's algorithm for LU factorization

Crout's algorithm for LU factorization

Bi-factorization

- Fivotal element
- Element currently being eliminated
- Elements not accessed any more
- Elements accessed and used in current computations
- Modified elements
- Elements not accessed yet

Figure 6.1 Graphic illustration of the computational processes for various methods and algorithms.

1. No memory beyond that used to store the original coefficient matrix **A** is needed because the entries of the decomposed matrices **L** and **U** can overwrite the previous values a_{ij} and be stored in the same place (there is no need to store the ones on the main diagonal of **L** [Doolittle and Gauss] or **U** [Crout]).
2. If only the right-hand vector **b** is changed, there is no need to recalculate the decomposition. Only the forward and back substitutions have to be performed from the beginning.
3. In the sensitivity analysis, discussed in Chapter 4, we had to solve the transposed system of the form $\mathbf{A}^T\mathbf{x} = \mathbf{b}$, which can be solved using the same triangular factors **U** and **L**.
4. The determinant of **A** can be calculated from the relation:

$$\det \mathbf{A} = \prod_{i=1}^{n} l_{ii} \quad \text{(Gauss's and Doolittle's algorithms)} \tag{6.43}$$

or

$$\det \mathbf{A} = \prod_{i=1}^{n} u_{ii} \quad \text{(Crout's algorithm)} \tag{6.44}$$

6.3 BIFACTORIZATION

In the bifactorization method, the inverse of the coefficient matrix **A** in the system of linear equations (6.1) is expressed as a multiple product of $2n$ factor matrices [7]:

$$\mathbf{A}^{-1} = \mathbf{R}^{(1)}\mathbf{R}^{(2)} \cdots \mathbf{R}^{(n-1)}\mathbf{R}^{(n)}\mathbf{L}^{(n)}\mathbf{L}^{(n-1)} \cdots \mathbf{L}^{(2)}\mathbf{L}^{(1)} \tag{6.45}$$

where

$\mathbf{L}^{(i)}$, $i = 1, 2, \ldots, n$ are left-hand factor matrices,
$\mathbf{R}^{(i)}$, $i = 1, 2, \ldots, n$ are right-hand factor matrices.

The left-hand factor matrices $\mathbf{L}^{(k)}$ differ from the unity matrix only in the kth column:

$$\mathbf{L}^{(k)} = \begin{bmatrix} 1 & 0 & \cdots & 0 & \cdots & 0 \\ 0 & 1 & & \vdots & & \vdots \\ & & \ddots & & & \\ & & & l_{kk}^{(k)} & & \\ & & & l_{k+1,k}^{(k)} & 1 & \\ \vdots & & & \vdots & & \ddots \\ 0 & \cdots & & l_{n,k}^{(k)} & \cdots & 1 \end{bmatrix} \tag{6.46}$$

Note that the diagonal term $l_{kk}^{(k)}$ of $\mathbf{L}^{(k)}$ does not equal one.

The right-hand factor matrices $\mathbf{R}^{(k)}$ differ from the unity matrix only in the kth row as well:

$$\mathbf{R}^{(k)} = \begin{bmatrix} 1 & 0 & & & & & & \cdots & & 0 \\ 0 & \ddots & & & & & & & & \\ & & 1 & & & & & & & \\ 0 & 0 & 0 & 1 & r^{(k)}_{k,k+1} & r^{(k)}_{k,k+2} & \cdots & & & r^{(k)}_{k,n} \\ & & & & 1 & & & & & 0 \\ & & & & & \ddots & & & & \vdots \\ & & & & & & 1 & & & 0 \\ \vdots & & & & & & & & & \\ 0 & 0 & \cdots & & & & & & 0 & 1 \end{bmatrix} \quad (6.47)$$

In $\mathbf{R}^{(k)}$ all diagonal elements are equal to one and thus $\mathbf{R}^{(n)} = \mathbf{I}$.

The factor matrices $\mathbf{L}^{(i)}$ and $\mathbf{R}^{(i)}$, $i = 1, 2, \ldots, n$ can be found from the equation:

$$\mathbf{L}^{(n)}\mathbf{L}^{(n-1)} \cdots \mathbf{L}^{(2)}\mathbf{L}^{(1)}\mathbf{A}\mathbf{R}^{(1)}\mathbf{R}^{(2)} \cdots \mathbf{R}^{(n-1)}\mathbf{R}^{(n)} = \mathbf{I} \quad (6.48)$$

written as a sequence of the following intermediate matrices:

$$\mathbf{A}^{(0)} = \mathbf{A}$$
$$\mathbf{A}^{(1)} = \mathbf{L}^{(1)}\mathbf{A}^{(0)}\mathbf{L}^{(1)}$$
$$\mathbf{A}^{(2)} = \mathbf{L}^{(2)}\mathbf{A}^{(1)}\mathbf{L}^{(2)}$$
$$\cdots \qquad (6.49)$$
$$\mathbf{A}^{(k)} = \mathbf{L}^{(k)}\mathbf{A}^{(k-1)}\mathbf{L}^{(k)}$$
$$\cdots$$
$$\mathbf{A}^{(n)} = \mathbf{L}^{(n)}\mathbf{A}^{(n-1)}\mathbf{L}^{(n)}$$

These equations represent the step-by-step transformation of the initial matrix $\mathbf{A} = \mathbf{A}^{(0)}$ to the identity matrix $\mathbf{I}$. This is done by forming the successive inner product $\mathbf{L}^{(k)}\mathbf{A}^{(k-1)}\mathbf{R}^{(k)}$ for $k = 1, 2, \ldots, n$.

Using (5.46) to (6.49) we can find the elements of the reduced matrix $\mathbf{A}^{(k)}$ computed as

$$a^{(k)}_{kk} = 1, \quad a^{(k)}_{kj} = 0, \quad a^{(k)}_{ik} = 0 \qquad (6.50)$$

$$a^{(k)}_{ij} = a^{(k-1)}_{ij} - \frac{a^{(k-1)}_{ik} a^{(k-1)}_{kj}}{a^{(k-1)}_{kk}} \qquad (6.51)$$

where k is the pivotal index, and $i, j = (k+1), \ldots, n$.

The elements of the left-hand factor matrices are

$$l_{kk}^{(k)} = \frac{1}{a_{kk}^{(k-1)}} \tag{6.52}$$

$$l_{ik}^{(k)} = -\frac{a_{ik}^{(k-1)}}{a_{kk}^{(k-1)}}, \quad i = (k+1), \ldots, n \tag{6.53}$$

In a similar way, we compute the elements of the right-hand factor matrices

$$r_{kk}^{(k)} = 1 \tag{6.54}$$

$$r_{kj}^{(k)} = -\frac{a_{kj}^{(k-1)}}{a_{kk}^{(k-1)}}, \quad j = (k+1), \ldots, n \tag{6.55}$$

As in other methods discussed earlier, the pivots $a_{kk}^{(k-1)}$, $k = 1, 2, \ldots, n$ must be different from zero.

The vector of unknowns, the solution vector, may be computed from the relation:

$$\begin{aligned}
\mathbf{x} &= \mathbf{A}^{-1}\mathbf{b} \\
&= \mathbf{R}^{(1)}\mathbf{R}^{(2)} \cdots \mathbf{R}^{(n-1)}\mathbf{R}^{(n)}\mathbf{L}^{(n)}\mathbf{L}^{(n-1)} \cdots \mathbf{L}^{(2)}\mathbf{L}^{(1)}\mathbf{b}
\end{aligned} \tag{6.56}$$

According to this relation, the right-hand vector $\mathbf{b}$ is transformed stepwise to the solution vector by successive multiplications of the factor matrix by the vector.

The computer implementation of the algorithm can be written directly from (6.50) to (6.56) as follows.

Bifactorization

Step 1 Set $k = 1$.
Step 2 Compute the kth column of $\mathbf{L}^{(k)}$ by setting

$$l_{kk}^{(k)} = \frac{1}{a_{kk}^{(k-1)}}$$

$$l_{ik}^{(k)} = -a_{ik}^{(k-1)}l_{kk}^{(k)}, \quad i = (k+1), \ldots, n$$

where $a_{ij}^{(k-1)}$ are entries of the reduced matrix $\mathbf{A}^{(k-1)}$ and $\mathbf{A}^{(0)} = \mathbf{A}$.
Step 3 If $k = n$ then stop.
Step 4 Update the reduced matrix from $\mathbf{A}^{(k-1)}$ to $\mathbf{A}^{(k)}$ by setting $a_{ij}^{(k)} = a_{ij}^{(k-1)} + a_{kj}^{(k-1)}l_{ik}^{(k)}$ where $i, j = k+1, k+2, \ldots, n$.
Step 5 Obtain the kth row of $\mathbf{R}^{(k)}$ by setting $r_{kj}^{(k)} = -a_{kj}^{(k-1)}l_{kk}^{(k)}$.
Step 6 $k = k + 1$ and return to step 2.

Computation of the Solution Vector

Set the solution vector to the right-hand vector:

$$x_i = b_i, \quad i = 1, 2, \ldots, n$$

multiply the left-hand factor matrix by the vector in the following steps:

Step 1 Set $k = 1$.
Step 2 Update the solution vector by multiplying it by the left-hand factor matrix:

$$x_k = l_{kk}^{(k)} x_k$$

$$x_i = x_i + l_{ik}^{(k)} x_k, \quad i = k + 1, \ldots, n$$

Step 3 If $k = n$ then go to step 5.
Step 4 $k = k + 1$ and return to step 2.

Multiply the right-hand factor matrix by the vector in the following steps:

Step 5 Set $k = n - 1$.
Step 6 Update the solution vector by multiplying it by the right-hand factor matrix:

$$x_k = x_k + \sum_{j=k+1}^{n} r_{kj}^{(k)} x_j$$

Step 7 If $k = 1$ then stop.
Step 8 $k = k - 1$ and return to step 6.

By comparing the LU factorization with bifactorization, we see that the operations in both methods are the same, but performed in a different order. The number of arithmetic operations required in both methods also are the same. As in the Gauss's algorithm for LU factorization, at every stage of the bifactorization, every entry a_{ij} that has not been eliminated is updated in accordance with (6.51). This is illustrated in Figure 6.1. It is an important feature of the bifactorization when considering pivoting.

6.4 PIVOTING

As already mentioned, the Gaussian elimination or the LU decomposition process cannot be performed if an element in the pivot location is valued zero. The pivot is the element of the reduced matrix by which we divide, and division by zero is not possible. This problem can be overcome by permuting either rows or columns or both. We should note here that interchanging two rows requires simultaneous interchanging of the corresponding two elements in right-hand vector **b**. Similarly, interchanging two columns requires simultaneous interchanging of the corresponding elements in the unknown vector. There is yet another reason for pivoting: the accuracy of the solution. Because of finite precision arithmetic, it is possible—and not at all that rare—to get a significant error in the solution as a result from rounding off.

Consider as an example the following system of linear equations:

$$\begin{bmatrix} 0.01 & 10 & -0.01 \\ 10 & 10 & -1 \\ 10 & 1 & 1 \end{bmatrix} \begin{bmatrix} x_1 \\ x_2 \\ x_3 \end{bmatrix} = \begin{bmatrix} 10 \\ 19 \\ 12 \end{bmatrix}$$

and assume that the computer has three-digit accuracy. This system of equations will be stored in the computer memory as

$$\begin{bmatrix} .100 \cdot 10^{-1} & .100 \cdot 10^2 & -.100 \cdot 10^{-1} \\ .100 \cdot 10^2 & .100 \cdot 10^2 & -.100 \cdot 10^1 \\ .100 \cdot 10^2 & .100 \cdot 10^1 & .100 \cdot 10^1 \end{bmatrix} \begin{bmatrix} x_1 \\ x_2 \\ x_3 \end{bmatrix} = \begin{bmatrix} .100 \cdot 10^2 \\ .190 \cdot 10^2 \\ .120 \cdot 10^2 \end{bmatrix}$$

The Gaussian reduction leads to the resultant triangular system of equations in the form:

$$\begin{bmatrix} .100 \cdot 10^{-1} & .100 \cdot 10^2 & -.100 \cdot 10^{-1} \\ 0 & .100 \cdot 10^5 & -.900 \cdot 10^1 \\ 0 & 0 & .200 \cdot 10^2 \end{bmatrix} \begin{bmatrix} x_1 \\ x_2 \\ x_3 \end{bmatrix} = \begin{bmatrix} .100 \cdot 10^2 \\ .100 \cdot 10^5 \\ .000 \cdot 10^0 \end{bmatrix}$$

The computed solution is $x_1 = 0$, $x_2 = .1 \cdot 10^1$, $x_3 = 0$, whereas the true solution is $x_1 = x_2 = x_3 = .1 \cdot 10^1$.

If rows 1 and 2 are interchanged, we have

$$\begin{bmatrix} .100 \cdot 10^{-1} & .100 \cdot 10^2 & -.100 \cdot 10^{-1} \\ .100 \cdot 10^2 & .100 \cdot 10^2 & -.100 \cdot 10^1 \\ .100 \cdot 10^2 & .100 \cdot 10^1 & .100 \cdot 10^1 \end{bmatrix} \begin{bmatrix} x_1 \\ x_2 \\ x_3 \end{bmatrix} = \begin{bmatrix} .100 \cdot 10^2 \\ .190 \cdot 10^2 \\ .120 \cdot 10^2 \end{bmatrix}$$

After two steps of Gaussian elimination, we have

$$\begin{bmatrix} .100 \cdot 10^2 & .100 \cdot 10^2 & -.100 \cdot 10^1 \\ 0 & .100 \cdot 10^2 & -.900 \cdot 10^{-2} \\ 0 & 0 & .199 \cdot 10^1 \end{bmatrix} \begin{bmatrix} x_1 \\ x_2 \\ x_3 \end{bmatrix} = \begin{bmatrix} .190 \cdot 10^2 \\ .100 \cdot 10^2 \\ .198 \cdot 10^1 \end{bmatrix}$$

and the solution is $x_1 = x_2 = x_3 = .1 \cdot 10^1$, which is correct.

As we have shown in this example, the rounding-off errors are minimized when the elements with the largest absolute values are chosen as pivots [8]. Two pivoting strategies are most often used, partial pivoting and complete pivoting, and these allow us to achieve that goal. If we select such an element from among the coefficients in the first column in the reduced matrix, this is partial pivoting. By interchanging two rows we move the desired element into the pivot position. Partial pivoting can also be performed by interchanging columns. In complete pivoting, we scan for the largest absolute value element in the whole reduced matrix. To move the desired element into the pivot position we interchange two rows and two columns. Because complete pivoting requires much more work, it generally is not preferred. Improvement in the upper bound on the solution error by using complete pivoting has been shown not to be significant for most practical problems and the added computational effort of renumbering the unknowns as a result of column permutation is not justified [9].

Crout's and Doolittle's algorithms do not update the entire reduced matrix. The matrix element that has not been updated cannot be chosen as a pivot because its value can change very drastically when it eventually is updated. Therefore, in these algorithms the only candidates for pivots are elements of the row and column actually being eliminated after they are updated. We should realize that only partial pivoting may be used when using the Crout's or Doolittle's algorithms. In Crout's algorithm the pivot must be chosen from the first column; and in Doolittle's algorithm, from the first row in the reduced matrix after it is updated. Gauss's algorithm for LU factorization and the bifactorization method should be used if complete pivoting is required.

Note that, if the coefficient matrix is diagonally dominant, that is,

$$|a_{ii}| \geq \sum_{\substack{j=1 \\ j \neq i}}^{n} |a_{ij}|, \quad i = 1, 2, \ldots, n \qquad (6.57)$$

then pivoting becomes unnecessary. The node admittance matrices and connection scattering matrices with rows (columns) ordered to place ones on the main diagonal are diagonally dominant.

In the case of large sparse matrices, the pivoting for accuracy is usually replaced by pivoting to preserve sparsity. These problems will be discussed in the following sections of this chapter.

6.5 NUMERICAL PROBLEMS AND ERROR MECHANISMS

6.5.1 Numerical Conditioning of a System of Linear Equations

Consider a system of linear equations (6.1). Because computer word length is limited, the solution $\mathbf{x} = \mathbf{A}^{-1}\mathbf{b}$, even with exact arithmetic, will be in error due to round-off errors in the computer representation of $\mathbf{A}$ and $\mathbf{b}$. If the system of equations is ill-conditioned, its solution will be excessively sensitive to the disturbances in the input data, a result which occurs when the system is nearly singular. We will consider now the sensitivity of the solution with respect to small uncertainty in $\mathbf{A}$ and $\mathbf{b}$, due to approximately linearly dependent rows in $\mathbf{A}$.

Let us assume first that $\mathbf{A}$ is known exactly and only $\mathbf{b}$ is changed to $\mathbf{b} + \Delta\mathbf{b}$. Let $\mathbf{x} + \Delta\mathbf{x}$ be the resulting solution of the system (6.1). Then,

$$\mathbf{A}(\mathbf{x} + \Delta\mathbf{x}) = \mathbf{b} + \Delta\mathbf{b} \tag{6.58}$$

or

$$\Delta\mathbf{x} = \mathbf{A}^{-1}\,\Delta\mathbf{b} \tag{6.59}$$

By taking norms on both sides, we have

$$\|\Delta\mathbf{x}\| \le \|\mathbf{A}^{-1}\| \cdot \|\Delta\mathbf{b}\| \tag{6.60}$$

By using (6.59) and (6.60) we obtain a bound on the norm of $\Delta\mathbf{x}$:

$$\frac{\|\Delta\mathbf{x}\|}{\|\mathbf{x}\|} \le \|\mathbf{A}\| \cdot \|\mathbf{A}^{-1}\| \frac{\|\Delta\mathbf{b}\|}{\|\mathbf{b}\|} \tag{6.61}$$

The standard measure of the ill-conditioning in a system is the condition number, which is defined as [1, 3, 10]

$$\text{cond}(\mathbf{A}) = \|\mathbf{A}\| \cdot \|\mathbf{A}^{-1}\| \tag{6.62}$$

By using (6.59) and (6.60), we have

$$\frac{\|\Delta\mathbf{x}\|}{\|\mathbf{x}\|} \le \text{cond}(\mathbf{A}) \frac{\|\Delta\mathbf{b}\|}{\|\mathbf{b}\|} \tag{6.63}$$

Because we know that the condition number can be expressed as (see Appendix 1)

$$\text{cond}(\mathbf{A}) = \left|\frac{\sigma_1}{\sigma_2}\right| = \frac{(\text{maximum singular value of } \mathbf{A})}{(\text{minimum singular value of } \mathbf{A})} \ge 1 \tag{6.64}$$

From (6.63), we see that the relative error of the vector $\mathbf{x}$ is the relative error of the right-hand vector magnified by the condition number of the coefficient matrix $\mathbf{A}$.

As an example, consider the system:

$$x_1 + x_2 = 3$$

$$0.999\,x_1 + x_2 = 2.998$$

The exact solution is $x_1 = 2$, $x_1 = 1$. If right-hand vector **b** is changed by $\Delta\mathbf{b} = [0.0, 0.001]^T$, we have

$$x_1 + x_2 = 3$$

$$0.999x_1 + x_2 = 2.999$$

which has the exact solution $x_1 = 1$, $x_2 = 2$. The change in the solution is $\Delta\mathbf{x} = [-1, 1]^T$, and $\|\Delta\mathbf{x}\|/\|\mathbf{x}\| = \sqrt{2/5} = 0.6325$, whereas $\|\Delta\mathbf{b}\|/\|\mathbf{b}\| = 0.0002357$. The largest singular value $\sigma_1 \cong 1.9995$ and the smallest $\sigma_2 = 0.0005$ so that the condition number is about 3999. If we calculate $\text{cond}(\mathbf{A}) \cdot \|\Delta\mathbf{b}\|/\|\mathbf{b}\| = 0.9426$. We now see that the relative error of the solution vector **x**, $\|\Delta\mathbf{x}\|/\|\mathbf{x}\|$ is very close to the bound. The determinant of **A** is equal 10^{-3}, and its reciprocal is of the same order as the condition number.

We now consider the sensitivity of the solution vector **x** to a small change in **A**. We assume that **b** is known exactly, that **A** is changed by $\Delta\mathbf{A}$, and the resulting change in **x** is $\Delta\mathbf{x}$. Then,

$$(\mathbf{A} + \Delta\mathbf{A})(\mathbf{x} + \Delta\mathbf{x}) = \mathbf{b} \tag{6.65}$$

$$\Delta\mathbf{x} = (\mathbf{A} + \Delta\mathbf{A})^{-1}\mathbf{b} - \mathbf{A}^{-1}\mathbf{b} \tag{6.66}$$

By using the identity $\mathbf{B}^{-1} - \mathbf{A}^{-1} = \mathbf{A}^{-1}(\mathbf{A} - \mathbf{B})\mathbf{B}^{-1}$, we have

$$(\mathbf{A} + \Delta\mathbf{A})^{-1} - \mathbf{A}^{-1} = -\mathbf{A}^{-1}\Delta\mathbf{A}(\mathbf{A} + \Delta\mathbf{A})^{-1} \tag{6.67}$$

$$\Delta\mathbf{x} = -\mathbf{A}^{-1}\Delta\mathbf{A}(\mathbf{x} + \Delta\mathbf{x}) \tag{6.68}$$

By using (6.62) and (6.68), we find

$$\frac{\|\Delta\mathbf{x}\|}{\|\mathbf{x} + \Delta\mathbf{x}\|} \le \|\mathbf{A}^{-1}\| \cdot \|\mathbf{A}\| \frac{\|\Delta\mathbf{A}\|}{\|\mathbf{A}\|} = \text{cond}(\mathbf{A})\frac{\|\Delta\mathbf{A}\|}{\|\mathbf{A}\|} \tag{6.69}$$

The relation (6.69) is similar to (6.63), which indicates that the condition number of coefficient matrix **A** is the magnifying coefficient between the relative error of **x** and the relative error in **A**.

Returning to the example discussed earlier, if the system is changed to

$$x_1 + x_2 = 3$$

$$0.998x_1 + x_2 = 2.998$$

the exact solution becomes $x_1 = 1$, $x_2 = 2$, which is completely wrong and intolerable. The condition number of **A** is now $\simeq 999$.

Numerical conditioning of systems of equations has an interesting geometric interpretation. Each equation in the system can be interpreted as a hyperplane in the hyperspace of the dimension equal to the number of unknowns (of equations in the system). The solution of the system of linear equations is the point at which all the hyperplanes intersect. Without a point of intersection, there is no unique solution to the system. This means that some of the hyperplanes are parallel or coincident. The system is singular.

In ill-conditioned systems, two or more hyperplanes must be nearly parallel. Figure 6.2 shows a geometric interpretation of the conditioning of a system of two equations. The two hyperplanes and their intersection shown in Figure 6.2(a) correspond to an ill-conditioned system:

$$x_1 + x_2 = 3$$

$$0.999 x_1 + x_2 = 2.998$$

The well-conditioned system:

$$x_1 + x_2 = 3$$

$$x_1 - x_2 = 1$$

has hyperplanes shown in Figure 6.2(b). In the two-dimensional problem, the hyperspace is a two-dimensional space and the hyperplanes are lines.

As is well understood by now, the solution of an ill-conditioned system is so

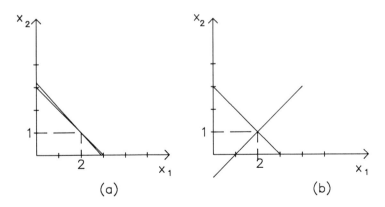

Figure 6.2 Hyperplanes (lines) and their intersection: (a) an ill-conditioned system of two equations; (b) a well-conditioned system of two equations.

sensitive to small changes in **A** or **b**. According to Figure 6.2(a), slight change in the position of a line in the ill-conditioned system can result in a very large change of the location of the intersection point. In the case of the well-conditioned system slight moves of a line do not change much the location of the intersection point.

If the system is ill-conditioned, uncertainty in its solution is unavoidable. The solution uncertainty can be reduced, however, if the problem is solved by using double precision arithmetic.

6.5.2 Round-off Error Growth and Proper Choice of Pivots

The growth of round-off errors due to numeric instability in the solution algorithm is the second cause of the solution uncertainty. The main reason why round-off errors can become excessive is the growth in size of the elements as the matrix reduction continues. Growth in the magnitude of matrix elements occurs during the process of subtracting the product of the multiplier $m_{ik} = a_{ik}^{(k)}/a_{kk}^{(k)}$ and the elements of the row being eliminated from the elements of some other row in the reduced matrix. The modification of the reduced matrix is performed according to (6.7). An element in the reduced matrix is updated as

$$a_{ij}^{(k+1)} = a_{ij}^{(k)} - m_{ik}a_{kj}^{(k)} = a_{ij}^{(k)} - \frac{a_{ik}^{(k)}}{a_{kk}^{(k)}}a_{kj}^{(k)} \tag{6.70}$$

If the subtracted constituent $m_{ik}a_{kj}^{(k)}$ is much larger in magnitude than the minuend $a_{ij}^{(k)}$, the growth of $a_{ij}^{(k+1)}$ occurs. The effect of the round-off error aggravation can be seen by expressing the entries of the **L** and **U** matrices as

$$\bar{l}_{ik} = l_{ik} + \epsilon_{ik} \tag{6.71}$$

$$\bar{u}_{kj} = u_{kj} + \epsilon_{kj} \tag{6.72}$$

where the overbars denote values of l_{ik} or u_{kj} that have been distorted by round-off errors ϵ_{ik} or ϵ_{kj}.

As the calculated factor matrices **L** and **U** are not exact, we must write (6.11) as

$$\mathbf{A} + \mathbf{E} = \bar{\mathbf{L}}\bar{\mathbf{U}} \tag{6.73}$$

where $\mathbf{E} = [e_{ij}]$ is the matrix of error terms. From (6.71) to (6.73), we have

$$a_{ij} + e_{ij} = \sum_{k=1}^{n} \bar{l}_{ik}\bar{u}_{kj}$$

$$= \sum_{k=1}^{n} \left(l_{ik}u_{kj} + l_{ik}\epsilon_{kj} + u_{kj}\epsilon_{ik} + \epsilon_{ik}\epsilon_{kj} \right) \qquad (6.74)$$

$$e_{ij} = \sum_{k=1}^{n} \left(l_{ik}\epsilon_{kj} + u_{kj}\epsilon_{ik} + \epsilon_{ik}\epsilon_{kj} \right) \qquad (6.75)$$

From (6.75), we see that error e_{ij} is directly proportional to the absolute size of the round-off errors ϵ_{kj} and ϵ_{ik} of the elements of the L and U matrices. Because the round-off errors ϵ_{kj} and ϵ_{ik} are proportional to the magnitude of the L and U matrix entries, then to minimize their effect, we must minimize the growth of these entries. Note from (6.75) that even one large term in L or U can contribute errors to many elements of the A matrix.

The minimization of the growth may be achieved by careful pivoting. Choosing large pivotal elements $a_{kk}^{(k)}$ makes the multipliers small, which, in turn, makes the subtrahend small in magnitude. Partial or complete pivoting is the best way to keep elements of the reduced matrix small, because they guarantee that the magnitude of the multipliers $m^{(k)}$ in each reduction step will be less than one.

6.6 COMPLEX MATRIX EQUATIONS

The analysis of linear microwave circuits in frequency domain generates systems of linear equations with complex coefficients and complex unknowns. Two different approaches may be used for solving these complex matrix equations. The first one is evident: the systems with complex coefficient matrices may be solved in a manner identical to real matrices using complex instead of real number arithmetic. In the second approach, we rewrite each complex equation as two equivalent real equations and apply a real coefficient matrix method to solve this new system of equations.

In the first approach, one aspect of complex arithmetic may cause problems when performed by a computer. The standard way to divide two complex numbers is

$$\frac{a + jb}{c + jd} = \frac{(a + jb)(c - jd)}{(c + jd)(c - jd)} = \frac{ac + bd + j(bc - ad)}{c^2 + d^2} \qquad (6.76)$$

The squaring in the denominator of (6.76) can result in unnecessary arithmetic overflow and underflow. This disadvantage may be removed by dividing both the numerator and the denominator by the larger of either c or d:

$$\frac{a + jb}{c + jd}\bigg|_{|c| > |d|} = \frac{a + b\dfrac{d}{c}}{c + d\dfrac{d}{c}} + j\frac{b - a\dfrac{d}{c}}{c + d\dfrac{d}{c}} \tag{6.77}$$

$$\frac{a + jb}{c + jd}\bigg|_{|c| < |d|} = \frac{a\dfrac{c}{d} + b}{c\dfrac{c}{d} + d} + j\frac{b\dfrac{c}{d} - a}{c\dfrac{c}{d} + d} \tag{6.78}$$

The second approach is based on the transformation of each complex equation into two real equations.

Without loss of generality, consider the following two equations:

$$\begin{bmatrix} a_{11} + jb_{11} & a_{12} + jb_{12} \\ a_{21} + jb_{21} & a_{22} + jb_{22} \end{bmatrix} \begin{bmatrix} x_1 + jy_1 \\ x_2 + jy_2 \end{bmatrix} = \begin{bmatrix} c_1 + jd_1 \\ c_2 + jd_2 \end{bmatrix} \tag{6.79}$$

These equations can be reorganized by writing $(a + jb)(x + jy) = (c + jd)$ as

$$\begin{matrix} ax - by = c \\ bx + ay = d \end{matrix} \quad \text{or} \quad \begin{bmatrix} a & -b \\ b & a \end{bmatrix} \begin{bmatrix} x \\ y \end{bmatrix} = \begin{bmatrix} c \\ d \end{bmatrix} \tag{6.80}$$

Now, two complex equations (6.79) can be written as four equations with real coefficients and real unknowns:

$$\begin{bmatrix} a_{11} & -b_{11} & a_{12} & -b_{12} \\ b_{11} & a_{11} & b_{12} & a_{12} \\ a_{21} & -b_{21} & a_{22} & -b_{22} \\ b_{21} & a_{21} & b_{22} & a_{22} \end{bmatrix} \begin{bmatrix} x_1 \\ y_1 \\ x_2 \\ y_2 \end{bmatrix} = \begin{bmatrix} c_1 \\ d_1 \\ c_2 \\ d_2 \end{bmatrix} \tag{6.81}$$

This matrix equation can be solved by using only real arithmetic.

REFERENCES

[1] G.E. Forsythe and C.M. Moler, *Computer Solution of Linear Algebraic Equations*, Prentice-Hall, Englewood Cliffs, NJ, 1967.

[2] G.H. Golub and C.F. Van Loan, *Matrix Computation*, Johns Hopkins University Press, Baltimore, 1983.

[3] G. Strong, *Linear Algebra and Its Applications*, Academic Press, New York, 1980.

[4]   D.A. Calahan, *Computer-Aided Network Design*, McGraw-Hill, New York, 1972.

[5]   K.C. Gupta, R. Garg, and R. Chadha, *Computer-Aided Design of Microwave Circuits*, Artech House, Norwood, MA, 1981.

[6]   J. Vlach and K. Singhal, *Computer Methods for Circuit Analysis and Design*, Van Nostrand Reinhold, New York, 1983.

[7]   K. Zollenkopf, "Bi-Factorisation—Basic Computational Algorithm and Programming Techniques," in *Large Sparse Sets of Linear Equations*, J.K. Reid, ed., Proc. Oxford Conf. Inst. Mathematics and Its Applications, April 1970, Academic Press, New York, 1971.

[8]   J.H. Wilkinson, *Rounding Errors in Algebraic Processes*, Prentice-Hall, Englewood Cliffs, NJ, 1963.

[9]   J.H. Wilkinson, *The Algebraic Eigenvalue Problem*, Oxford University Press, London, 1965.

[10] G. Dahlquist and A. Bjoerck, *Numerical Methods*, Prentice-Hall, New York, 1974.

[11] A.R. Newton and A.L. Sangiovanni-Vincentelli, "Relaxation-Based Electrical Simulation," *IEEE Trans. Electron Devices*, Vol. ED-30, September 1983, pp. 1184–1207.

Chapter 7
Sparse Matrix Techniques

The solution of the system of linear equations $\mathbf{Ax} = \mathbf{b}$ by any of the methods discussed in Chapter 6 requires $n^3/3 + n^2 - n/3$ multiplications-divisions and storage for $n^2 + n$ complex numbers. Because the systems of circuit equations used in microwave circuit simulation procedures (frequency domain analysis, sensitivity analysis, tolerance analysis) have to be solved tens or hundreds of times, computational effort becomes large. In the optimization methods used in nominal design procedures, tolerance design, centering and tuning the analysis of the designed circuit, the circuit equations may have to be solved even thousands of times.

As shown in Chapter 3, where the nodal admittance matrix or the connection scattering matrix is used as the coefficient matrix of systems of linear equations of microwave circuits, many entries are zero. The matrices of microwave circuits are sparsely filled with elements of nonzero value.

Sparse matrix techniques were devised to take advantage of the sparsity in circuit equations. In effect they reduce both the number of arithmetic operations and the storage requirements. The reduction often is quite dramatic, depending on the sparsity of the matrix and the ingenuity of the programmer.

Nodal admittance matrices and connection scattering matrices are very sparse, and it is advantageous to exploit their sparsity even when the number of equations is small. Exploiting the sparsity of the circuit matrices allows us to drastically reduce the required computation time. Typically for sparse matrix techniques, the number of multiplications-divisions required to solve the system grows approximately linearly with the size n, as compared to $n^3/3 + n^2 - n/3$ for full matrices. By storing only nonzero elements of the matrix, the computer memory required is also drastically reduced.

Three basic ideas are used in sparse matrix techniques. The first is not to store matrix entries that have zero values. The primary and most obvious reason for this is the reduction of required computer memory for storing the matrix data. Moreover, because the zeros are not stored, there is no need for access to them, which reduces execution time somewhat.

The second goal in the sparse matrix techniques is to perform only nontrivial arithmetic operations with nonzero elements. Because zero elements are not stored, multiplications by zero and additions with zero-valued elements may be eliminated by using appropriate programming techniques.

To exploit the matrix's sparsity, it must not be destroyed during the decomposition procedure. The number of nonzero elements in the matrix may grow during its reduction. The new nonzero elements coming into existence in empty places of a matrix are called *fill-ins*. The growth of the number of fill-ins can be minimized by proper pivoting strategy. Maintaining sparsity during the reduction of a matrix is essential.

7.1 STORAGE SCHEMES FOR SPARSE MATRICES

There is no optimal scheme for storing the nonzero entries of a sparse matrix. The best scheme in any particular case depends on the nonzero structure of a matrix and the use to which the matrix will be put. A good storage scheme of a sparse matrix should be characterized by two basic features: the nonzero elements should be readily accessible and the amount of the memory used should be kept low.

Many data structures for efficient storing of nonzero elements of sparse matrices have been developed over the years [1–12]. Out of all of these, two storing schemes, called *ordered lists* and *linked lists*, are used most often.

7.1.1 Static Storage Schemes with Ordered Lists

In static storage schemes, the nonzero element values along with their coordinates are stored in either an arbitrary order or are ordered, for example, by increasing the value of the column index i and then the row index j [2, 3, 7, 13, 14]. A common method requires three one-dimensional arrays: (1) a real (or a complex) array holding the values of the nonzero entries of the matrix. The nonzero entries may be stored as encountered when scanning down the columns (or along the rows) in order; (2) an integer array of the same length holding the corresponding row indices (or column indices); and (3) an integer array of the dimension $n + 1$ (n – the order of the matrix) holding in the position i (or position j) the address in the two earlier mentioned arrays of the first nonzero element in column i (or the first nonzero entry in row j). The $(n + 1)$th entry is an integer number greater by one than the number of nonzero elements in the matrix. Figure 7.1, by means of a small example ($n = 4$), presents a static storage scheme with a row-oriented ordered list.

This kind of static storage with a row pointer–column index (or column pointer–row index) scheme characterizes the ease of access and economy of memory space. These schemes might be used when storing sparse matrix information in a file or when doing operations that do not change the nonzero structure of the matrix (like matrix by vector multiplications).

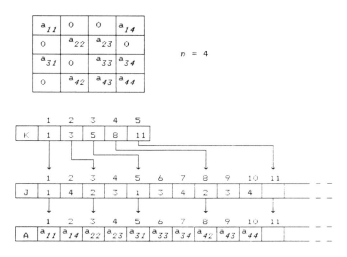

Figure 7.1 Static storage scheme with row oriented ordered list: K is an array of address pointers, which in the Ith location holds the address of the first nonzero term in the Ith row of the matrix, stored in arrays J and A; $K(5)$ is the address pointer of the first empty location in arrays J and A; J is the array of column indices of nonzero terms stored rowwise in table A.

7.1.2 Dynamic Storage Schemes with Linked Lists

The second important class of sparse matrix storage schemes are dynamic schemes with linked lists [1, 6, 12, 14]. The nonzero elements of the coefficient matrix are not stored in any particular order. With each nonzero entry of the matrix is associated a set of certain attributes stored in parallel arrays.

These attributes can be some from the following:

1. row index,
2. column index,
3. pointer to the next element in the row,
4. pointer to the next element in the column,
5. pointer to the previous element in the row,
6. pointer to the previous element in the column.

Two or more additional arrays of length n are usually used to indicate:

1. the first element in the row,
2. the first element in the column,
3. number of nonzero elements in the row,
4. number of nonzero elements in the column.

A possible variant of the dynamic storage schemes with linked lists is presented in

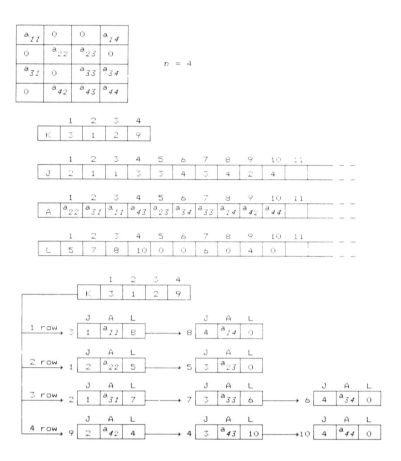

Figure 7.2 Dynamic storage scheme with linked list: K is the array of address pointers, which in the Ith location holds the address of the first nonzero term of the Ith row, stored in array A; J is the array of column indices of nonzero terms stored in Table A; L is the array of address pointers linking elements of the same row.

Figure 7.2. The array A contains nonzero elements stored in any order. An integer array J holds the column indices of corresponding elements in A. Integer array K holds the addresses in array A of the first nonzero elements in succeeding rows. Array L links the elements belonging to the same row and contains *pointers* (in FORTRAN; they are called *records* in Pascal) to the next element in the row. A zero entry in that array indicates the last element in a row.

The linked storage scheme presented in Figure 7.2 is only a simple example. Usually, the linked list requires two arrays: one containing pointers to the first nonzero element in every row, and the other containing pointers to the first element in each column. Thus the nonzero element values are stored with two orientations, rowwise and columnwise.

This data structure allows easy access to any element by starting at the beginning of a row or column and going along it until the desired element is found. Insertion of new nonzero elements that are generated during the reduction of the matrix is also straightforward. Numerical values of fill-ins may be stored at the end of the list in empty locations of array A. The other benefit of this storage scheme is that it is very easy to delete elements and reutilize any free space caused by such deletion. All that is required to insert or delete an element is a proper adjustment of the links.

7.2 PIVOT SELECTION STRATEGIES FOR SPARSE MATRICES

In classical algorithms working with full matrices, pivots are selected to maintain numerical stability and usually chosen to be the element with the largest absolute value in the first column of the reduced matrix (partial pivoting) or the largest (in magnitude) of all elements of the reduced matrix (complete pivoting). In the case of sparse matrices, such techniques for selection of pivots cannot be adopted because they conflict with the need to maintain matrix sparsity. The objective in pivot selection for sparse matrices is to choose a sequence of pivot elements such that the number of nontrivial numerical operations or the number of fill-ins is minimized. To understand fully the problem of how the pivot sequence affects sparsity, let us consider the following example, often quoted in the literature, shown in Figure 7.3 [15]. If the element $(1, 1)$ is chosen as the first pivot, then after the first step of the reduction the reduced matrix is completely full. The sparsity of the matrix is destroyed in the first step. However, if the pivots are chosen from the main diagonal so that the element $(1, 1)$ is chosen last, no fill-ins occur. Figure 7.4 presents sparseness structure of the resultant matrix, when the elements $(1, 1)$ and $(5, 5)$ have been interchanged and the matrix reduction performed.

To find globally optimal ordering that produced optimal sequence of pivots in some well-defined sense would be desirable. However, finding the optimal pivoting order is practically impossible (there are $(n!)^2$ orderings in the case of a full matrix), so

$$
\begin{bmatrix}
X & X & X & X & X \\
X & X & O & O & O \\
X & O & X & O & O \\
X & O & O & X & O \\
X & O & O & O & X
\end{bmatrix}
\longrightarrow
\begin{bmatrix}
X & X & X & X & X \\
X & X & F & F & F \\
X & F & X & F & F \\
X & F & F & X & F \\
X & F & F & F & X
\end{bmatrix}
$$

(a)   (b)

X – original nonzero elements
O – original zero-valued elements
F – fill-ins

Figure 7.3 (a) The original sparse matrix; and (b) the structure of the matrix after the first step of the reduction, where the element $(1, 1)$ has been taken as the pivot.

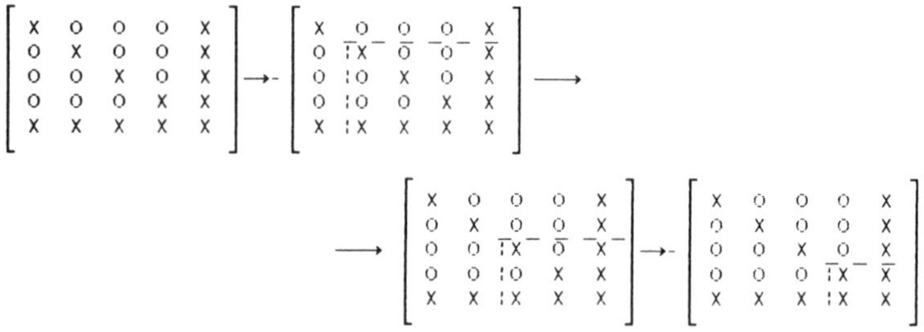

$$
\begin{bmatrix}
X & O & O & O & X \\
O & X & O & O & X \\
O & O & X & O & X \\
O & O & O & X & X \\
X & X & X & X & X
\end{bmatrix}
\rightarrow
\begin{bmatrix}
X & O & O & O & X \\
O & X & O & O & X \\
O & O & X & O & X \\
O & O & O & X & X \\
X & X & X & X & X
\end{bmatrix}
\rightarrow
$$

$$
\rightarrow
\begin{bmatrix}
X & O & O & O & X \\
O & X & O & O & X \\
O & O & X & O & X \\
O & O & O & X & X \\
X & X & X & X & X
\end{bmatrix}
\rightarrow
\begin{bmatrix}
X & O & O & O & X \\
O & X & O & O & X \\
O & O & X & O & X \\
O & O & O & X & X \\
X & X & X & X & X
\end{bmatrix}
$$

Figure 7.4 Sparseness structure of the matrix after the elements $(1, 1)$ and $(5, 5)$ have been interchanged; no fill-ins occur after each reduction step.

heuristics are applied to find an "optimal" ordering. Importantly, we must realize that the word *optimal* used in the literature does not mean optimal pivoting strategy in a global sense for general sparse matrices. Most pivoting selection strategies may be divided into two categories: static orderings and dynamic orderings.

7.2.1 Static (*a priori*) Ordering

In static ordering, determination of the pivoting order for all steps of the matrix reduction takes place before the reduction procedure is started. Static ordering strategies do not take into account changes in the nonzero structure that originates in each reduction step of the coefficient matrix. Static orderings give results that are far from optimal; however, they provide a great improvement over not ordering for sparsity at all [16]. They are very easily programmed, but not as effective as dynamic ordering strategies.

Common selection criteria to determine pivotal sequence include ordering the columns in a sequence corresponding to the growing number of nonzero elements in each column or ordering columns by increasing total number of nonzero elements in rows having a nonzero entry in the given column. The data needed for this is information on the number of nonzero elements in each row and each column of the matrix. Other static ordering strategies are discussed by I.S. Duff and J.K. Reid [16] and R.P. Tewarson [17].

7.2.2 Dynamic Ordering

In dynamic ordering, the choice of the pivot for each step of the reduction of the coefficient matrix **A** takes place immediately before performing the actual reduction step. In the choice of the pivot, the changes in the sparseness structure that occurred in earlier reduction steps are taken into consideration. Dynamic ordering requires simulation of the matrix reduction (decomposition) in which only positions of nonzero elements and not their values are considered, because the reduced matrix may have been modified by fill-ins.

Most dynamic ordering strategies are based on the minimization of the fill-ins in each reduction step, so the pivot selected at each reduction step is the one that produces the fewest fill-ins. This is called the *minimum local fill-in criterion.* Of course, this pivot ordering strategy is difficult in programming, and the execution of the symbolic matrix reduction is rather long. Much easier in programming and faster in execution are pivot ordering strategies based on the minimization of a number of arithmetic operations required to perform in each reduction (decomposition) step.

L.W. Nagel [18] thoroughly tested and compared the four pivot-ordering methods of Markowitz [19], Berry [20], Nakhla, Singhal, and Vlach [21], and Hsieh and Ghausi [22]. He found that Markowitz's method, which is based on the principle of minimization of the number of arithmetic operations, produces only about 5% more fill-ins than the best of the other methods, but it needs much less time for performing optimal pivot ordering. At each step of matrix reduction, the Markowitz ordering algorithm chooses as a pivot that nonzero element of the reduced matrix for which the product of the number of other nonzero elements in the candidate's row and column is minimum.

For the candidate $a_{ij}^{(k)}$ as a pivot of the kth reduction step, consider a product $(r_i - 1)(c_j - 1)$, where r_i is the number of nonzero elements in row i of the reduced matrix, and c_j is the number of nonzero elements in column j of the reduced matrix. This product equals the number of multiplications-additions. It also represents the maximum number of fill-ins that might be created if that element were selected as a pivot. Choosing the pivot candidate with the smallest product $(r_i - 1)(c_j - 1)$ minimizes the number of multiplications-additions to perform the actual reduction step. For the matrix:

	1	2	3	4
1	a_{11}	0	0	a_{14}
2	0	a_{22}	a_{23}	0
3	a_{32}	0	a_{33}	a_{34}
4	0	a_{42}	a_{43}	a_{44}

we can create the following matrix of the Markowitz's products:

	1	2	3	4
1	1			2
2		1	2	
3	2		4	4
4		2	4	4

According to Markowitz's ordering method two elements a_{11} and a_{22} are candidates for the pivot in the first reduction step. To break this tie, additional criteria may be used to select the pivot. For example, the largest of the contesting elements may be chosen to be the pivot [2]; this will prevent the growth of excessive round-off errors in the solution. Another idea is to choose the element with the smallest column count, to avoid the divisions required in multiplier's computations [18]. We assume here that the multipliers are in the **L** matrix. If the multipliers are in the **U** matrix the pivot with the smallest row count is preferred.

As discussed in the previous chapter LU decomposition or bifactorization requires that none of the pivots may equal zero. Moreover, not only should the pivots be different from zero, they should be sufficiently large not to result in excess round-off errors. As suggested by Markowitz, a threshold for the element magnitude should be considered in parallel with the sparsity criterion in choosing the pivot element. The candidate of a pivotal element below the threshold cannot be chosen as a pivot element. An element $a_{rs}^{(k)}$ of reduced matrix $\mathbf{A}^{(k)}$ may be considered as a pivot if it satisfies one of the following conditions:

$$| a_{rs}^{(k)} | \geq u \cdot \max_{i=k,\ldots,n} | a_{is}^{(k)} | \tag{7.1}$$

$$| a_{rs}^{(k)} | \geq u \cdot \max_{j=k,\ldots,n} | a_{rj}^{(k)} | \tag{7.2}$$

$$| a_{rs}^{(k)} | \geq u \cdot \max_{i,j=k,\ldots,n} | a_{ij}^{(k)} | \tag{7.3}$$

where $0 < u \leq 1$ is the relative pivot threshold parameter. For $u = 1$, threshold pivoting becomes partial pivoting (constraints (7.1) or (7.2)) or complete pivoting (constraint (7.3)).

In any strategy for choosing a pivoting order for solution accuracy, we should drift to reduce round-off error by avoiding growth in magnitude of nonzero element $a_{ij}^{(k)}$ of reduced matrix $\mathbf{A}^{(k)}$. To date, we do not know of a reasonably accurate relation between the growth of the magnitude of the elements and the resulting round-off error of the solution and the size of the threshold parameter u. A.R. Curtis and J.K. Reed have tried to find this relation experimentally [23]. They suggest using u from the range $0.05 \leq u \leq 0.25$ to keep the round-off error from contaminating more than about three to four digits. Because the amount of error growth is a function of the matrix, to determine a good threshold we recommend monitoring the growth on a series of matrices of given application while changing the threshold value within a certain range.

7.3 IMPLEMENTATION OF SPARSE MATRIX TECHNIQUES

To this point, we have shown that the use of sparse matrix techniques can substantially reduce the execution time and memory requirements even for relatively small problems. In this section we will discuss several distinct methods for the practical implementation of sparse matrix techniques [31].

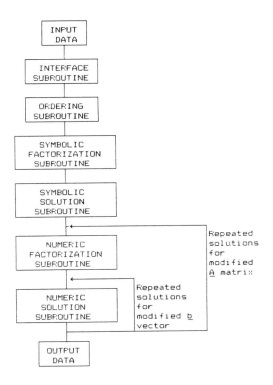

Figure 7.5 Distinct segments of a typical sparse matrix solver.

Figure 7.5 presents a structure of a typical sparse matrix solver. It consists of a number of separate subroutines that are executed sequentially and then repeated in a manner dependent on the problem solved. The separate routines that create a solver are as follows:

1. A subroutine or a set of subroutines that provide an interface between the natural and simple input data and the complicated data structures used in sparse matrix techniques.
2. A static ordering routine that renumbers the equations and variables. The execution of this routine is usually performed only once for a given and constant zero-nonzero structure of coefficient matrix **A**. The order often is independent of actual values of the matrix entries.
3. A symbolic factorization routine that step by step simulates matrix reduction and determines the zero-nonzero structures of **L** and **U** (LU factorization) or $\mathbf{L}^{(k)}$ and $\mathbf{R}^{(k)}$, $k = 1, 2, \ldots, n$ (bifactorization). At this stage, compiled code, looping index, or interpretable code are generated. Such a code will perform the LU factorization or bifactorization of matrices with specified zero-nonzero structure. For a given zero-nonzero structure of the coefficient matrix, this subroutine is

executed only once. In dynamic ordering, symbolic factorization and ordering are implemented and performed in one routine.

4. A symbolic solution routine that simulates the computation of the whole solution vector or only some elements of the solution vector. The routine analyzes the zero-nonzero structure of the right-hand vector and determines operands and operations to be performed to generate the desired "outputs." The routine may generate a compiled, looping indexed, or interpretable code for the solution phase. This subroutine is performed only once for a given zero-nonzero structure of the right-hand vector and desired outputs. This routine must be executed each time the structure of the right-hand vector b changes or new outputs are specified.

5. A numerical LU factorization or bifactorization routine that computes nonzero elements of the L and U matrices or $L^{(k)}$ and $R^{(k)}$ matrices, $k = 1, 2, \ldots, n$, using the code generated in phase 3. This subroutine is invoked each time the values of the A matrix elements are modified.

6. A numerical solution routine that computes the desired elements of the solution vector by using the solution code generated in phase 4.

Although a detailed discussion of different code approaches is clearly out of place, we will provide some information on these problems as follows.

7.3.1 Compiled Code Techniques

In the compiled code approach, a symbolic factorization routine and symbolic solution routine analyze the nonzero structure of the initial matrix, the decomposed matrix, or the right-hand vector and generate a loop-free code that can be used for repeated solution of systems with matrices of this structure. The compiled code is extremely fast in execution because no testing or branching is needed and there is direct access to every operand. The main disadvantage of the approach is that the compiled code can be very long [24, 25]. The code may be generated in a higher-level language, but to increase computational efficiency the loop-free code often is generated in machine code, which makes the program much less portable.

As an example illustrating principles of the sparse matrix techniques based on the compiled code approach, let us consider the sparse matrix A presented in Figure 7.6 and assume that the system Ax = b is solved by using LU factorization.

$$
\underline{A} = \begin{bmatrix} a_{11} & a_{12} & a_{13} \\ 0 & a_{22} & 0 \\ a_{31} & 0 & a_{33} \end{bmatrix} = \begin{bmatrix} l_{11} & 0 & 0 \\ 0 & l_{22} & 0 \\ l_{31} & l_{32} & l_{33} \end{bmatrix} \begin{bmatrix} 1 & u_{12} & u_{13} \\ 0 & 1 & 0 \\ 0 & 0 & 1 \end{bmatrix}
$$

Figure 7.6 LU factorization of the sparse matrix A.

A sequence of arithmetic operations performed on nonzero elements of **A**, leading to the computation of the nonzero elements of the **L** and **U** (Crout's algorithm) is

$$l_{11} = a_{11}$$

$$l_{31} = a_{31}$$

$$u_{12} = a_{12} / l_{11}$$

$$u_{13} = a_{13} / l_{11}$$

$$l_{22} = a_{22}$$

$$l_{32} = -l_{31} u_{12}$$

$$l_{33} = a_{33} - l_{31} u_{13}$$

Corresponding to these operations, a compiled code generated as FORTRAN statements has the form presented in Figure 7.7. After compilation, the binary form of this code is very fast in execution and may be executed many times in response to changing numerical values of the nonzero elements of **A**.

7.3.2 Looping Indexed Code Techniques

In the looping indexed code approach, an appropriate routine simulates the matrix factorization, analyzes the structure of the matrix, and generates a sequence of ordering arrays and indexing vectors that can be used to further process matrices of the same structure [23, 26–28]. This integer information is then used to perform the numeric factorization and finally the numeric solution of the equations with the coefficient matrix

```
C(1) = A(1)
C(2) = A(2)
C(3) = A(3)/C(1)
C(4) = A(5)/C(1)
C(5) = A(4)
C(6) = -C(2)/C(3)
C(7) = A(6)-C(2)*C(4)
```

	A
1	a_{11}
2	a_{31}
3	a_{12}
4	a_{22}
5	a_{13}
6	a_{33}

(a)

(b)

Figure 7.7 (a) Compiled code performing LU factorization of matrix A presented in Figure 7.6, and (b) a one-dimensional array with nonzero elements of A stored columnwise.

of the same zero-nonzero structure. The basic advantage of this approach is that the technique may be implemented in any higher-level language. In comparison to the compiled code approach, the looping indexed approach requires less work in the simulation of the factorization and solution processes and, moreover, the generated ordering arrays and indexing vectors require much less memory space. It is slower in execution because the variables are addressed indirectly.

A detailed presentation of the sparse matrix solver based on the looping indexed approach applicable to the solution of sparse systems of equations for microwave circuits will be presented in the next chapter.

7.3.3 Interpretable Code Techniques

Expressing a sequence of nontrivial arithmetic operations leading to the computation of the solution of a sparse system in a form of interpretable code is midway between the two previous approaches. Typically, the interpretable code is a sequence of sets of integer numbers defining the operation type and addresses of the associated operands. The data to the factorization program thus consist of a sequence of operation codes and addresses to be executed sequentially [29, 30]. The code is generated by a routine simulating the factorization and solution procedures for a sparse system of equations. For example, in the LU factorization procedure we can distinguish seven different operations performed with two or three operands. Each of these operations may be coded by an integer number in the range 1–7 and two or three integer numbers form the addresses of operands involved in these operations. Table 7.1 presents an example of

Table 7.1

Example of an Interpretable Code

		Interpretable Code		
	Operations Coded as FORTRAN Statements	Operation Type	Addresses	
LU Factorization	$A(I) = A(I) - A(J)*A(K)$	1	I J	K
	$A(I) = -A(J)*A(K)$	2	I J	K
	$A(I) = A(I)/A(J)$	3	I J	—
Forward Substitution $Ly = b$	$Y(I) = Y(I) - A(J)*Y(K)$	4	I J	K
	$Y(I) = Y(I)/A(J)$	5	I J	—
Backward Substitution $Ux = Y$	$Y(I) = Y(I) - U(J)*X(K)$	6	I J	K
	$X(I) = Y(J)$	7	I J	—

the interpretable code for the LU factorization method. Of course, the interpretable code presented in this table is one of many possible implementations.

The execution speed of the interpretable codes lies between the other approaches discussed. The same may be said about the amount of time for the initial analysis of the zero-nonzero structure of coefficient matrix **A**. The storage required is usually much less than for the compiled code approach, but often greatly in excess of that needed for the looping indexed approach. It is easily coded in higher-lever languages and can be adopted to deal with microwave circuit analysis problems [11].

REFERENCES

[1]   D.E. Knuth, *The Art of Computer Programing*, Volume 1. *Fundamental Algorithms*, Addison-Wesley, Reading, MA, 1973.

[2]   O. Osterby and Z. Zlatev, *Direct Methods for Sparse Matrices*, Springer-Verlag, Berlin, 1983.

[3]   A.R. Curtis and J.K. Reid, "The Solution of Large Sparse Unsymmetrical Systems of Linear Equations," *J. Inst. Mathematics and Its Application*, Vol. 8, No. 3, December 1971, pp. 344–353.

[4]   D.J. Rose and R.A. Willoughby, eds., *Sparse Matrices and Their Applications*, Plenum Press, New York, 1972.

[5]   I.S. Duff and J.K. Reid, "Some Design Features of a Sparse Matrix Code," *ACM Trans. Mathematical Software*, Vol. MS-5, No. 1, March 1979, pp. 18–35.

[6]   W.M. Gentleman and A. George, "Sparse Matrix Software," in *Sparse Matrix Computations*, J.R. Bunch, D.J. Rose, eds., Academic Press, New York, 1976.

[7]   F.G. Gustavson, "Some Basic Techniques for Solving Sparse Matrices," in [4], pp. 41–52.

[8]   L.O. Chua and Pen-Min Lin, *Computer-Aided Analysis of Electronic Circuits: Algorithms and Computational Techniques*, Prentice-Hall, Englewood Cliffs, NJ, 1975.

[9]   K.C. Gupta, R. Garg, and R. Chadha, *Computer-Aided Design of Microwave Circuits*, Artech House, Norwood, MA, 1981.

10]   J. Vlach and K. Singhal, *Computer Methods for Circuit Analysis and Design*, Van Nostrand Reinhold, New York, 1983.

11]   F. Bonfatti, V.A. Monaco, and P. Tiberio, "Microwave Circuit Analysis by Sparse-Matrix Techniques," *IEEE Trans. Microwave Theory Tech.*, Vol. MTT-22, No. 3, March 1974, pp. 264–269.

12]   J.A. Dobrowolski, "Algorithms and Storage Schemes in the Sparse Matrix Approach to Computer-Aided Analysis of Microwave Circuits," *Proc. Conf. on Computer-Aided Design of Electronic and Microwave Circuits and Systems*, Hull, London, 1977, pp. 122–127.

13]   D.M. Brandon, Jr., "The Implementation and Use of Sparse Matrix Techniques in General Simulation Programs," *Comput. J.*, Vol. 17, 1974, pp. 165–170.

14]   F.G. Gustavson, W.M. Liniger, and R.A. Willoughby, "Symbolic Generation of an Optimal Crout Algorithm for Sparse Systems of Linear Equations," *J. Assoc. Comput. Mach.*, Vol. 17, 1970, pp. 87–109.

15]   I.S. Duff, "A Survey of Sparse Matrix Research," *Proc. IEEE*, Vol. 65, No. 4, April 1977, pp. 500–535.

16]   I.S. Duff and J.K. Reid, "A Comparison of Sparsity Orderings for Obtaining a Pivotal Sequence in Gaussian Elimination," *J. Inst. Math. Appl.*, Vol. 14, 1974, pp. 281–291.

17]   R.P. Tewarson, *Sparse Matrices*, Academic Press, New York, 1973.

18]   L.W. Nagel, "SPICE2: A Computer Program to Simulate Semiconductor Circuits," Ph.D. Dissertation, University of California, Berkeley, May 1975.

[19]   H.M. Markowitz, "The Elimination Form of Inverse and Its Application to Linear Programming," *Management Science*, Vol. 3, April 1957, pp. 255-269.

[20]   R.D. Berry, "An Optimal Ordering of Electronic Circuit Equations for a Sparse Matrix Solution," *IEEE Trans. Circuit Theory*, Vol. CT-19, January 1971, pp. 40-50.

[21]   M. Nakhla, K. Singhal, and J. Vlach, "An Optimal Pivoting Order for the Solution of Sparse Systems of Equations," *IEEE Trans. Circuits and Systems*, Vol. CAS-21, March 1974, pp. 222-225.

[22]   H.Y. Hsieh and M.S. Ghausi, "On Optimal Pivoting Algorithms in Sparse Matrices," *IEEE Trans Circuit Theory*, Vol. CT-19, January 1972, pp. 93-96.

[23]   A.R. Curtis and J.K. Reid, "The Solution of Large Sparse Unsymmetrical Systems of Linear Equations," *J. Inst. Mathematics and Its Application*, Vol. 8, No. 3, December 1971, pp 344-353.

[24]   J.K. Reid, ed., *Large Sparse Sets of Linear Equations*, Proc. Oxford Conf. Inst. Mathematics and Its Applications, April 1970, Academic Press, New York, 1971.

[25]   R.A. Willoughby, "Sparse Matrix Algorithms and Their Relation to Problem Classes and Compute Architecture," in *ibid*.

[26]   D.J. Rose and R.A. Willoughby, "Sparse matrices and their applications," *Proc. Conf. at IBM Research Center, New York, September 9-10, 1971*, Plenum Press, New York, 1972.

[27]   K. Zollenkopf, "Bi-Factorization—Basic Computational Algorithm and Programming Techniques," pp. 75-96, in Reid, [24].

[28]   F.G. Gustavson, "Some Basic Techniques for Solving Sparse Systems of Linear Equations," pp 41-52, in [26].

[29]   B. Dembart and A.M. Erisman, "Hybrid Sparse Matrix Methods," *IEEE Trans. Circuit Theory* Vol. CT-20, 1973, pp. 641-649.

[30]   R.A. Willoughby, ed., *Proceedings of the Symposium on Sparse Matrices and Their Applications*, Yorktown Heights, NY, IBM Report RAI (#11707), 1969.

[31]   K.S. Kundert, "Sparse Matrix Techniques," in A.E. Ruehli, ed., *Circuit Analysis, Simulation, and Design*, Elsevier Science Publishers B.V., North-Holland, Amsterdam, 1986.

Chapter 8
Sparse Matrix Techniques for Analysis of Microwave Circuits Described by the Connection Scattering Matrix

In Chapters 3, 4, and 5, we developed matrix equations for the analysis of microwave circuits:

$$\mathbf{Wa} = \mathbf{c} \tag{8.1}$$

$$\mathbf{W}^T\alpha = \gamma \tag{8.2}$$

$$\mathbf{W}\frac{\partial\mathbf{a}}{\partial p} = \delta \tag{8.3}$$

$$\mathbf{WA} = \mathbf{B}_n \tag{8.4}$$

Equation (8.1) corresponds to the frequency domain analysis, whereas (8.2) and (8.3) are related to the sensitivity analysis of the microwave circuits. The last matrix (8.4) is used in the computer-aided noise analysis of microwave circuits.

In the computer-aided procedures of analysis and design, these equations have to be solved many times. The zero-nonzero structure of the connection scattering matrix $\mathbf{W}$ of a circuit with constant topology does not change, whereas the numerical values of the nonzero elements of the matrix or of the right-hand vectors are modified as the analysis progresses.

Computation time and needed memory space may be reduced significantly with proper use of the sparse matrix techniques. The techniques described in this chapter are from a sparse matrix technique based on the bifactorization method with the application of a looping indexed code approach. The equation solver takes advantage of the characteristics of the connection scattering matrix.

8.1 CHARACTERISTICS OF CIRCUIT EQUATIONS WITH THE CONNECTION SCATTERING MATRIX

Connection scattering matrix formulation has been described in detail in Chapter 3. The circuit equations (8.1) to (8.4) with the connection scattering matrix as the coefficient matrix possess the following characteristics:

1. $\mathbf{W}$ is a sparse complex square matrix whose size n equals the sum of the number of ports of all circuit elements.

2. In the general case, $\mathbf{W}$ is an asymmetric matrix; that is, the nonzero elements are located symmetrically with respect to the main diagonal, but the values of the symmetrical elements are not equal to each other ($w_{ij} \neq w_{ji}$, $i \neq j$). Only when all circuit elements are reciprocal is the connection scattering matrix a symmetrical matrix.

3. The diagonal elements of $\mathbf{W}$ represent the reflection coefficients of the circuit element ports ($w_{ii} = -s_{ii}$, $i = 1, 2, \ldots, n$). Therefore, the diagonal elements of $\mathbf{W}$ have their values near or equal to zero. They may not be chosen as pivots.

4. Most of the off-diagonal terms of $\mathbf{W}$ are valued zero. The nonzero elements w_{ij} correspond to ports belonging to a particular element or connected ports of different circuit elements. In the first case they are equal to the transmission coefficients between ports of a certain circuit element ($w_{ij} = -s_{ij}$). In the second case they are entries of the connection matrix $\boldsymbol{\Gamma}$ and equal to ones.

5. Nonzero structures of $\mathbf{W}$ and $\mathbf{W}^T$ are identical, and if all elements of a circuit are reciprocal then $\mathbf{W} = \mathbf{W}^T$.

6. The nonzero structure of $\mathbf{W}$ and $\mathbf{W}^T$ is constant. It does not depend on frequency and circuit parameters.

7. Each row and each column of $\mathbf{W}$ contains one constant element equal to one. It derives from the $\boldsymbol{\Gamma}$ matrix.

8. The right-hand vectors $\mathbf{c}$, $\boldsymbol{\gamma}$, $\boldsymbol{\delta}$, and $\mathbf{B}_n$ have different zero-nonzero structures. The structures of these vectors may change.

To exploit an important characteristic that systems of equations with identical zero-nonzero structure of the coefficient matrix tend to be solved repeatedly, we need to create a sparse matrix solver that consists of four distinct segments:

1. Generation of the indexing and addressing data and their ordering,
2. Simulation and further ordering,
3. Reduction,
4. Solution.

The first segment of the sparse matrix solver interprets the topological data of the analyzed circuit and generates integer-type arrays containing indexing and addressing information describing the zero-nonzero structure of the connection scattering matrix of the microwave circuit to be analyzed.

The second routine simulates the reduction of the coefficient matrix of the system of equations. The optimal ordering of the matrix is performed before each step of a symbolic reduction of the matrix. The indexing and addressing data contained in the arrays are modified to the form determining the zero-nonzero structure of the left-hand and right-hand factor matrices.

The third subprogram uses the indexing and addressing information received from the simulation and ordering subprogram and performs the numerical reduction of the actual coefficient matrix. The subprogram performs only the nonzero operations leading to computation of nonzero elements of the left-hand and right-hand factor matrices.

The fourth segment of the solver computes the solution vector $\mathbf{x}$. The actual right-hand vector of the system of circuit equations is transformed stepwise to the solution vector by successive multiplications of factor matrix by vector. In the numerical solution routine, only the arithmetic operations with nonzero elements of the $\mathbf{R}$ and $\mathbf{L}$ factor matrices are performed. The zero-nonzero structures of the vectors $\mathbf{c}$, $\boldsymbol{\gamma}$, $\boldsymbol{\delta}$, and $\mathbf{B}_n$ are not taken into account in the solution vector computation. The applied programming technique and storage scheme are like the techniques used for the symmetrical matrices [1–3].

For a given microwave circuit described by a particular connection scattering matrix $\mathbf{W}$ two first segments of the described sparse matrix solver are executed only once. Repeated analysis of the same circuit is performed by executing only the reduction and solution subprograms (frequency domain analysis, change of circuit parameters, sensitivity analysis, noise analysis) or only the solution subprogram (sensitivity analysis of microwave circuits composed of reciprocal elements only, change of the location of the generator ports).

8.2 CONNECTION SCATTERING MATRIX ORDERING STRATEGY

The total number of nonzero elements of the connection scattering matrix is

$$n_t = \sum_{i=1}^{m} \left(n_{e_i}^2 + n_{e_i} \right) \tag{8.5}$$

where n_{e_i} is the number of ports of the ith element of the circuit and m is the number of elements (multiports) of the circuit. The second term in (8.5) corresponds to the nonzero elements of the connection matrix $\boldsymbol{\Gamma}$.

Matrix $\mathbf{M}$ is defined as

$$\mathbf{M} = \sum_{j=1}^{n} \mathbf{L}^{(j)} + \sum_{j=1}^{n} \mathbf{R}^{(j)} - \mathbf{I} \tag{8.6}$$

where n is the order of the connection scattering matrix, which contains all nonzero elements of the left-hand factor matrices $\mathbf{L}^{(j)}$, $j = 1, 2, \ldots, n$ and the right-hand factor matrices $\mathbf{R}^{(j)}$, $j = 1, 2, \ldots, n$.

The number of nonzero elements in $\mathbf{M}$ is usually bigger than in the $\mathbf{W}$ matrix, because at each reduction step of $\mathbf{W}$ new nonzero elements (fill-ins) may be created at places that were empty in the original $\mathbf{W}$ matrix. Analyzing (6.44) to (6.49) we see that the element $r_{ji}^{(j)}$, $i = (j + 1), \ldots, n$ of the jth right-hand factor matrix $\mathbf{R}^{(j)}$ (or the element $l_{kj}^{(j)}$, $k = (j + 1), \ldots, n$ of the left-hand factor matrix $\mathbf{L}^{(j)}$) may be nonzero, even if the corresponding to it element $w_{ji}^{(0)}$ (or $w_{kj}^{(0)}$) of the original matrix $\mathbf{W} = \mathbf{W}^{(0)}$ equals zero, if for at least one earlier reduction step the product $w_{ji}^{(l-1)} w_{li}^{(l-1)}$, $l = 1, 2, \ldots, (j - 1)$ (or $w_{lj}^{(l-1)} w_{il}^{(l-1)}$, $l = 1, \ldots, (j - 1)$) is nonzero.

The objective of optimal ordering is to minimize the total number of fill-ins. Sparsity must be maintained as far as possible. For connection scattering matrix $\mathbf{W}$, as described in Section 8.2, the diagonal elements representing the reflection coefficients of the circuit component ports may have their numerical values very close to zero or even be zero. They cannot be chosen as pivots. However, every row of $\mathbf{W}$ contains the constant one, deriving from $\boldsymbol{\Gamma}$, that can be an ideal pivot. A simple row interchange may place all ones of $\boldsymbol{\Gamma}$ on the main diagonal. These row interchanges can be performed and completed before the reduction of the matrix $\mathbf{W}$ begins. The ones of $\boldsymbol{\Gamma}$ located on the main diagonal make the connection scattering matrix diagonally dominant. The ones of $\boldsymbol{\Gamma}$ located on the main diagonal are ideal candidates for pivots because they allow great precision. As has been practically proven, less than 50% of the ones on the main diagonal are modified in the course of the reduction of the connection scattering matrices of practical microwave circuits. With such a preordered $\mathbf{W}$ matrix, the pivots may be chosen from the elements that lie along the main diagonal.

The Markowitz criterion is a greedy strategy for maintaining and exploiting the sparsity of the $\mathbf{W}$ matrix. The principle of the strategy is at each step of the reduction process to select that diagonal element w_{ii} for which the Markowitz product $(c_i - 1)(r_i - 1)$, where c_i is a number of nonzero terms in ith column and r_i is a number of nonzero terms in ith row, is the smallest. Any diagonal element that meets this criterion may be selected. Such a pivot will result in the fewest number of arithmetic operations in that reduction step. This pivot also has the least capacity to create fill-ins during that step. This strategy requires current bookkeeping of the number of nonzero terms in each column and each row and little additional computation.

8.3 STORAGE SCHEME OF THE CONNECTION SCATTERING MATRIX

The numerical values of nonzero terms and structural data of the connection scattering matrix are stored using a packed matrix storage scheme with linked list [4].

The nonzero terms are stored in three arrays DE, CE, and RE. Diagonal elements, initially all equal to one, are stored in the table DE.

The off-diagonal terms, the scattering parameters of circuit elements, are stored in both directions (i.e., in table CE, they are stored ordered columnwise, and stored CE rowwise in the parallel table). A set of integer-type tables determines the zero-nonzero structure of the **W** matrix.

A table ITAGC, parallel to CE, contains row indices of the terms stored in CE. The accompanying table LNXTC contains the location of the next nonzero term in CE, linking these terms in ascending columnwise order. The entry 0 in LNXTC indicates the last term of a column. The addresses of the first terms of each column in CE are stored in a table LCOL.

Similarly, the column indices of the elements in RE are stored in table ITAGR. Parallel to RE and ITAGR, table LNXTR contains the location of the next nonzero term in RE, linking these terms in ascending rowwise order. The entry 0 in LNXTR indicates the last term of a row. The addresses of the first elements of each row in RE are stored in table LROW. When a column or row does not have off-diagonal terms, the respective position in table LCOL or LROW must contain zero.

Three next tables NOZEC, NOZER, and NSEQ are used for storing information required for optimal ordering of the matrix according to the algorithm described earlier. Table NOZEC contains the numbers of nonzero off-diagonal terms in each column, and table NOZER has the numbers of nonzero off-diagonal elements in each row. The third table, NSEQ, initially must contain the integer numbers 1 to n in ascending sequence. At the end of the simulation procedure table NSEQ contains the pivotal sequence as it results from the used-ordering strategy.

As we can see from the example, locations of the arrays LNXTC and LNXTR, corresponding to unused storage positions of the arrays CE and RE must contain proper initial values. The last positions of the tables LNXTC and LNXTR must contain zeros, whereas the other vacant positions of LNXTC and LNXTR must be numbered consecutively.

We should mention here that the address of the first vacant location in tables CE, RE, ITAGC, ITAGR, LNXTC, and LNXTR has to be stored in LF. Figure 8.1 presents the described storage scheme for the connection scattering matrix.

The dual storage of the off-diagonal nonzero terms of **W** in the tables CE and RE allows very fast, easy access to any element in a column or row. The other benefits of this scheme are the extreme ease of inserting or deleting elements and the reutilization of any free space caused by such a deletion.

The dual storage of the off-diagonal terms of **W** does not waste memory space either. After completing the execution of the simulation and ordering subroutine, each nonzero element of factor matrices **L** and **R** is stored only once. The nonzero terms of **L** are stored in CE, and nonzero terms of **R** are stored in RE. The storage positions in CE and RE that become vacant in the course of matrix reduction are utilized for storing fill-in terms.

276

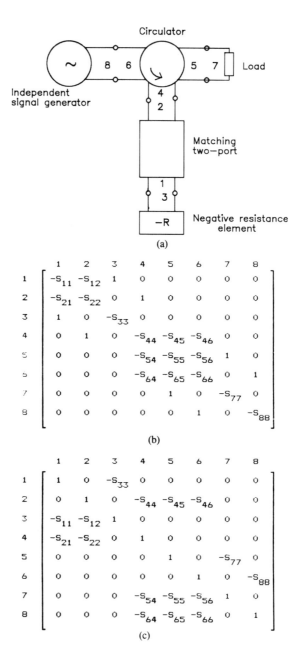

Figure 8.1 Packed storage scheme of the connection scattering matrix: (a) example of a microwave circuit; (b) connection scattering matrix W of the circuit; (c) preordered W matrix with ones on the main diagonal; (d) storage of nonzero terms of the connection scattering matrix.

	NSEQ	LROW	NOZER	LCOL	NOZEC
1	1	14	1	1	2
2	2	5	3	2	2
3	3	1	2	14	1
4	4	3	2	5	3
5	5	15	1	6	3
6	6	16	1	7	3
7	7	8	3	15	1
8	8	11	3	16	1

LF = 17

	RE	ITAGR	LNXTR	CE	ITAGC	LNXTC
1	S(1,1)	1	2	S(1,1)	3	3
2	S(1,2)	2	0	S(1,2)	3	4
3	S(2,1)	1	4	S(2,1)	4	0
4	S(2,2)	2	0	S(2,2)	4	0
5	S(4,4)	4	6	S(4,4)	2	8
6	S(4,5)	5	7	S(4,5)	2	9
7	S(4,6)	6	0	S(4,6)	2	10
8	S(5,4)	4	9	S(5,4)	7	11
9	S(5,5)	5	10	S(5,5)	7	12
10	S(5,6)	6	0	S(5,6)	7	13
12	S(6,4)	4	12	S(6,4)	8	0
13	S(6,5)	5	13	S(6,5)	8	0
14	S(6,6)	6	0	S(6,6)	8	0
15	S(3,3)	3	0	S(3,3)	1	0
16	S(7,7)	7	0	S(7,7)	5	0
17	–	0	18	–	0	18
18	–	0	19	–	0	19
19	–	0	20	–	0	20
20	–	0	0	–	0	0

(d)

Figure 8.1 *Continued*

8.4 PROCEDURE FOR GENERATION OF THE INDEXING, ADDRESSING, AND ORDERING ARRAYS

The procedure for generation of the indexing, addressing, and ordering arrays is a subroutine that forms an interface between the simple input data describing the topology of a circuit to be analyzed and the complicated data structure used by the next sparse matrix routines. The procedure generates sequences of indices and pointers describing the zero-nonzero structure of connection scattering matrix **W** of the given circuit in accordance with the storage scheme defined in Section 8.3. The input data for the procedure are the topological data of the circuit: *NS* indicates the number of nonzero terms of scattering matrix **S** of the whole circuit, *IS(NS)* indicates column indices, and *JS(NS)* indicates row indices of the nonzero terms of scattering matrix **S** of the entire circuit, *NP* indicates the number of all ports of the circuit, and *ICON(I)* and *JCON(I)*, *I* = 1, 2, . . . , *N*/2, are numbers of pairs of interconnected ports.

Let us return to the microwave circuit example presented in Figure 8.1(a), its scattering matrix **S**, and connection matrix *Γ* given in Figure 8.1(b) and (c), respectively. The topological data of this circuit is presented in Figure 8.2. These data are the input data for the procedure for generation of the indexing, addressing, and ordering data arrays.

Figure 8.3 presents detailed schemes of the procedure. It generates sequences of indices and pointers stored in the tables NSEQ, LROW, NOZER, LCOL, NOZEC, ITAGR, LNXTR, ITAGC, and LNXTC. The output data of the procedure for the circuit given in Figure 8.1(a) are given in Figure 8.1(c).

8.5 SIMULATION AND ORDERING PROCEDURE

The simulation and ordering procedure simulates, step by step, the reduction of connection scattering matrix **W**. Detailed schemes of the subroutine are presented in

NS = 16   NP = 8

	IS	JS
1	1	1
2	1	2
3	2	1
4	2	2
5	3	3
6	4	4
7	4	5
8	4	6
9	5	4
10	5	5
11	5	6
12	6	4
13	6	5
14	6	6
15	7	7
16	8	8

	ICON	JCON
1	6	8
2	2	4
3	1	3
4	5	7

Figure 8.2 Input data for the generation of the indexing, addressing, and ordering arrays for the example circuit in Figure 8.1(a).

Figure 8.4(a), (b), and (c). Simulation of each reduction step is divided into two parts: pivotal search and indexing and addressing modifications.

8.5.1 Pivotal Search—Matrix Ordering

The pivotal element for the actual step of the matrix reduction is selected from among all diagonal elements that were not pivotal elements earlier. The ordering strategy of connection scattering matrix **W** has been described in Section 8.2. If more than one

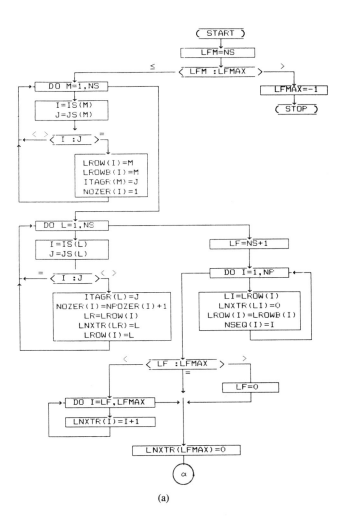

(a)

Figure 8.3 Schemes of the procedures for the generation of the indexing, addressing, and ordering arrays describing the sparseness structure of the W matrix.

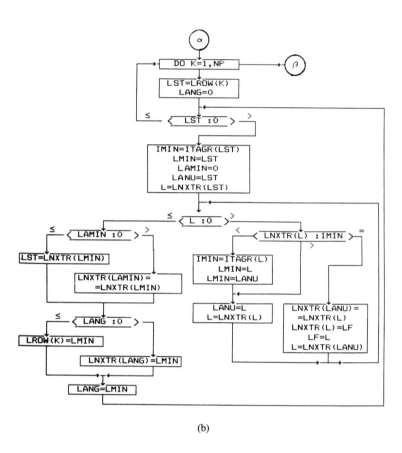

(b)

Figure 8.3 *Continued*

diagonal element meets the pivot criterion the element number in the first location of table NSEQ is selected as the pivotal index. Having determined the pivotal index *KP*, no interchange of rows and columns is actually done. Only two respective indices in table NSEQ are interchanged. In this way, step by step, the optimal sequence of pivot indices is created in NSEQ. The pivotal search algorithm scheme is presented in Figure 8.4(a).

8.5.2 Indexing and Addressing Modifications

In each step of the simulation of the **W** matrix reduction, first, the pivotal row *KP* is compared term by term with each row whose row index *K* is contained in the pivota

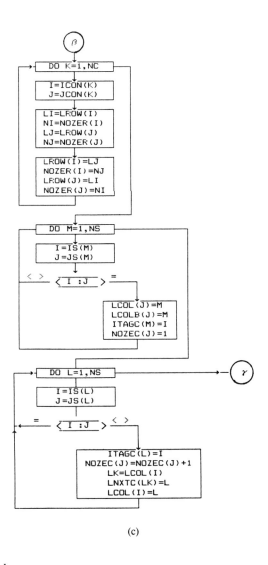

(c)

Figure 8.3 *Continued*

column *KP*. During this comparison the following indexing and addressing information is altered.

1. In the *K*th row compared with the pivotal row, the nonzero term of **W** (stored in RE), with the column index equal to the pivotal index $(I + KP)$ is cancelled. This location is used to store fill-in terms of reduced matrix **W**.

2. If any column index *IP* in the pivotal row is not contained in a compared actual

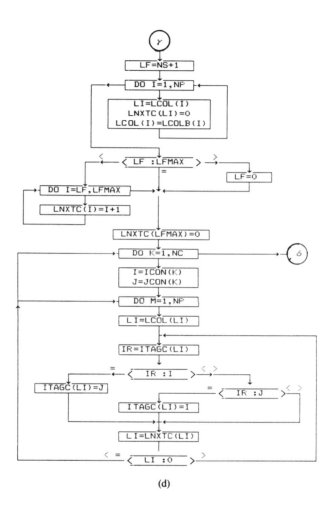

(d)

Figure 8.3 *Continued*

row $(I > IP$, and $IP > K)$, this index is added to the column indices in the table ITAGR (fill-in). The fill-ins are stored, first, in the locations of table RE vacated during the simulation process, and then, in the vacant locations at the ends of these tables. The first vacant location in RE, ITAGR, and LNXTR is always indicated by LFR.

3. Each time a term is cancelled or added, the addressing information contained in LNXTR and LROW is modified respectively. The entries of NOZER equal numbers of nonzero elements in each row of the reduced matrix are also updated.

After finishing the comparison of the pivotal row with appropriate row, the pivotal column KP is compared, term by term, with each column whose column index

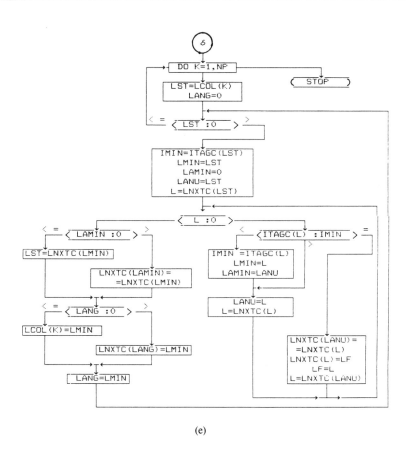

(e)

Figure 8.3 *Continued*

K is contained in the pivotal row KP. The following indexing and addressing modifications are made during this column comparison:

1. In the kth column compared with the pivotal column, the nonzero element of the coefficient matrix **W**, stored in CE, with the row index equal to the pivotal index $(I + KP)$ is cancelled. As this location becomes free, it is used to store fill-ins of reduced matrix **W**.

2. Any row index IP in the pivotal column not contained in a compared actual column ($IP > I$ and $IP > K$) is added to the row indices in table ITAGC (fill-in term). Similarly as for rows, the new fill-in terms are stored not only in the vacant locations at the end of tables CE, ITAGC, and LNXTC, but also in the locations of these tables vacated during the simulation process. The first vacant location in these tables is always indicated by LFC.

3. Similarly as for rows, each time a term is added or cancelled in table CE, the addressing information contained in the tables LNXTC and LCOL are modified.

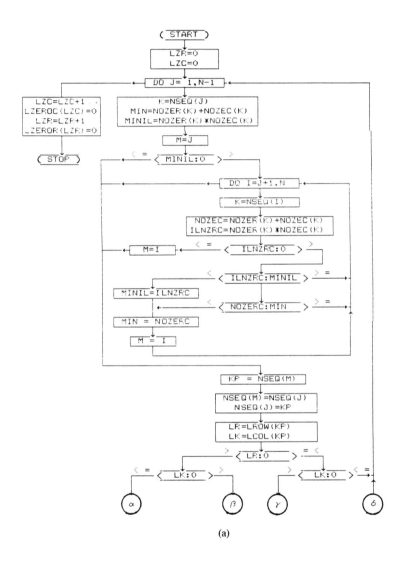

(a)

Figure 8.4 Schemes of the simulation and ordering procedure: (a) the pivotal element search; (b) comparison of the pivotal row with rows containing nonzero term in pivotal column; (c) comparison of the pivotal column with columns containing nonzero term in pivotal row.

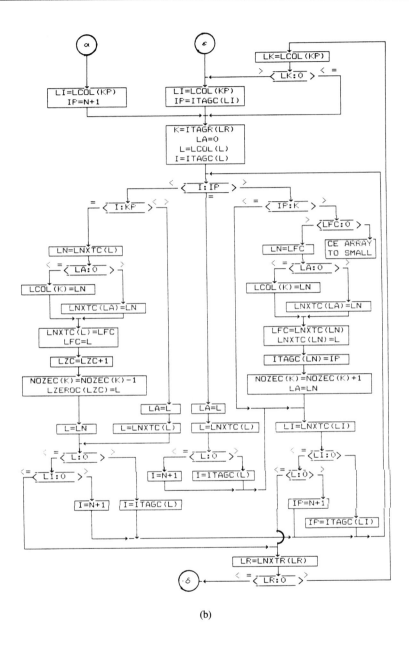

(b)

Figure 8.4 *Continued*

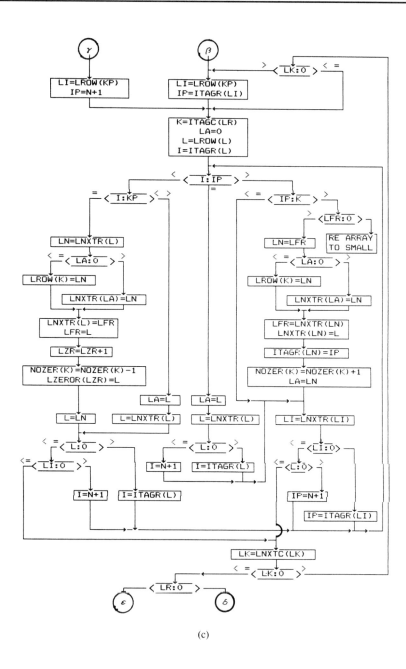

(c)

Figure 8.4 *Continued*

The entries in NOZEC equal to numbers of nonzero elements in each column in reduced matrix **W** are also updated.

Tables LZEROR and LZEROC contain addresses of these entries, respectively, in RE and CE, which have to be set to zero each time the reduction procedure is repeated with new values of the **W** matrix entries. The fill-in terms are stored in these locations. The schemes of the simulation procedure performing symbolic reduction of the sparse coefficient matrix **W** are presented in Fig. 8.4(b, c).

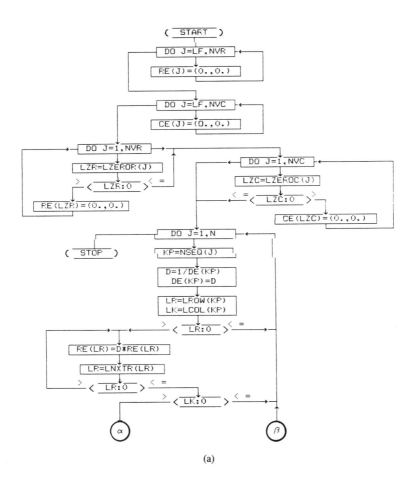

(a)

Figure 8.5 Schemes of the reduction subroutine: (a) computation of the diagonal terms and nonzero terms of the right-hand factor matrix; (b) computation of the nonzero terms of the reduced matrix located above the main diagonal; (c) computation of the nonzero terms of the reduced matrix located at or below the main diagonal.

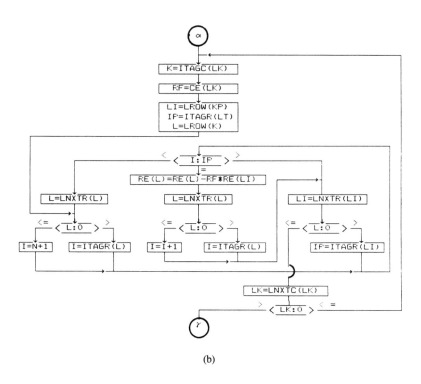

(b)

Figure 8.5 *Continued*

After performing the simulation and ordering procedure, tables LCOL, ITAGC, and LNXTC contain information on the nonzero structure of the left-hand factor matrices $\mathbf{L}^{(k)}$, $k = 1, 2, \ldots, n$, and tables LROW, ITAGR, and LNXTR contain the same information about the right-hand factor matrices $\mathbf{R}^{(k)}$, $k = 1, 2, \ldots, n$. Table NSEQ contains the near-optimal sequence of pivot indices.

8.6 REDUCTION PROCEDURE

The reduction procedure performs the numeric reduction of the actual $\mathbf{W}$ matrix. The reduction of the coefficient matrix is guided by the pivotal sequence contained in table NSEQ. At each step of the reduction, the pivotal element $DE(KP)$ is computed first and then the nonzero terms of the jth right-hand factor matrix $\mathbf{R}^{(j)}$. These are stored in array RE. Next, the procedure computes the nonzero terms of the reduced matrix $\mathbf{W}^{(j)}$ located above the main diagonal of $\mathbf{W}^{(j)}$. This part of the reduction procedure is illustrated in Figure 8.5(a) and (b). Then the nonzero terms of reduced matrix $\mathbf{W}^{(j)}$ located below the main diagonal are computed. Figure 8.5(c) shows this part of the

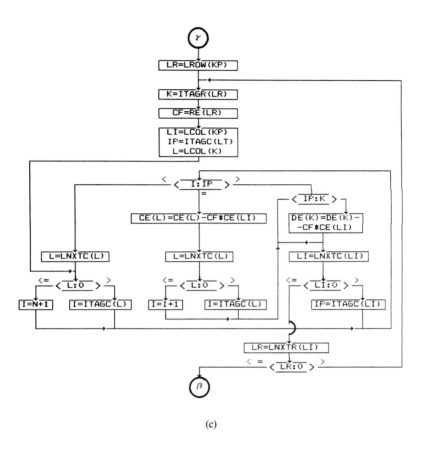

(c)

Figure 8.5 *Continued*

reduction procedure. The nonzero terms of the jth left-hand factor matrix $\mathbf{L}^{(j)}$ (multiplication of the nonzero terms of the pivotal column by the reciprocal of its diagonal term) are not computed during the reduction procedure, but rather in the solution procedure. The numeric reduction process of the actual coefficient matrix is performed in accordance with the indexing and addressing information received from the simulation and ordering subprogram.

8.7 SOLUTION PROCEDURE

Initially, the right-hand vector $\mathbf{c}$, $\boldsymbol{\gamma}$, $\boldsymbol{\delta}$, or $\mathbf{B}_n$ of the system or circuit equation is stored in array V. Then, in the solution procedure, the right-hand vector is transformed stepwise to the solution vector by successive multiplication of factor matrix by vector. The scheme of the solution procedure is presented in Figure 8.6. After the solution procedure, array V contains the solution vector.

A package of the sparse matrix solver discussed in this chapter is given in Appendix 2. The package is suitable for the solution of microwave circuit equations in which the coefficient matrix is nonsymmetric and diagonally dominant. The sparse matrix solver significantly reduces the amount of memory required to store the matrix **W** and the execution time of the repeated analysis of microwave circuits.

Large systems of equations **Wa** = **c** solved by using conventional full matrix algorithms require storage for $n^2 + n$ complex numbers and execution of the order of n^3 long arithmetic operations. For connection scattering matrices of the order of over 10 the amount of computer memory required by the sparse matrix data structure becomes smaller than that used to store the full matrix.

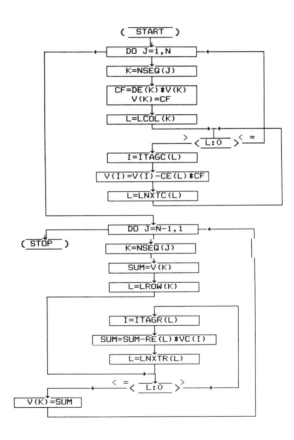

Figure 8.6 A scheme of the solution subroutine.

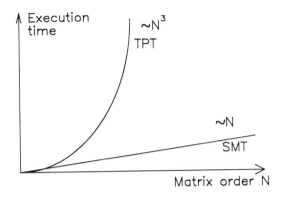

Figure 8.7 Execution time to solve a system **Wa** = **c** by conventional programming technique, *CPT*, and by sparse matrix technique, *SMT*, as a function of the **W** matrix order.

Figure 8.7 shows the relation between execution time of the repeated solutions of microwave circuit equations and the problem size. In the presented sparse matrix solver, execution time varies approximately directly with problem size.

REFERENCES

[1]   K. Zollenkopf, "Bi-Factorization—Basic Computational Algorithm and Programming Techniques," in *Large Sparse Sets of Linear Equations*, J.R. Reid, ed., *Proc. Oxford Conf. Inst. Mathematics and Its Application*, April 1970, Academic Press, New York, 1971.

[2]   R.D. Berry, "An Optimal Ordering of Electronic Circuit Equations for a Sparse Matrix Solution," *IEEE Trans. Circuit Theory*, Vol. CT-18, 1971, pp. 40–50.

[3]   W.T. Weeks, A.J. Jimenez, G.W. Mahoney, H. Qassemzadah, and T.R. Scott, "Algorithms for ASTAP—A Network Analysis Program," *IEEE Trans. Circuit Theory*, Vol. CT-20, 1973, pp. 628–634.

[4]   J.A. Dobrowolski, "Algorithms and Storage Scheme in the Sparse Matrix Approach to Computer-Aided Analysis of Microwave Circuits," *Proc. Conf. Computer-Aided Design of Electronic and Microwave Circuits and Systems*, Hull, UK, 1977, pp. 122–127.

Chapter 9
Tolerance Analysis of Microwave Circuits

The approach of nominal circuit design corresponds to a certain kind of idealized situation. Unfortunately, in reality, many uncertainties must be considered. In the realistic design of a microwave circuit, we must consider that the circuit parameters are varying, during manufacture (tolerances), during aging, or due to environmental factors (for example, temperature). Tolerance analysis is a process that allows us to study the effects of parameter variations on the circuit behavior.

Tolerance analysis evolves from the need to transform the idealized circuit model into a manufacturing product. The goal of the tolerance analysis is to predict the quality of manufactured networks. This chapter presents some techniques for computer-aided tolerance analysis based on deterministic and statistical approaches. The worst-case viewpoint is to find the largest variations of the circuit caused by the assumed variations of designable parameters of the circuit. The worst case is usually attributed to parameter tolerance limits.

In the statistical analysis, circuit parameters are assumed random variables. The goal of the statistical analysis is to estimate the *probability density function* (pdf) (or statistical parameters of pdf) of circuit functions on the ground of knowledge of pdf (or its statistical parameters) of the circuit parameters.

9.1 FUNDAMENTAL CONCEPTS

We consider a set of m circuit functions F_i (circuit responses) of n real variables (circuit parameters) p_j that have nominal values p_j^0:

$$\mathbf{F}(\mathbf{p}) = \left[F_1(\mathbf{p}), F_2(\mathbf{p}), \ldots, F_m(\mathbf{p}) \right]^T \tag{9.1}$$

$$\mathbf{p} = \left[p_1, p_2, \ldots, p_n \right]^T \tag{9.2}$$

Tolerances can be defined in terms of a set of independent random variables, some of which are the tolerances on the individual elements and others are functions of independent random variables representing temperature or manufacturing spread. Thus, a physical design of a circuit can be described by a vector of nominal parameters $\mathbf{p}^0$ and a vector of tolerances ϵ, where

$$
\mathbf{p}^0 \triangleq \begin{bmatrix} p_1^0 \\ p_2^0 \\ \vdots \\ p_n^0 \end{bmatrix}, \quad \epsilon = \begin{bmatrix} \epsilon_1 \\ \epsilon_2 \\ \vdots \\ \epsilon_n \end{bmatrix} \tag{9.3}
$$

Tolerance vector ϵ defines the extremes of the tolerance region or the standard deviations. The parameters values, $\mathbf{p}$, of the actual circuit design represents a point in the parameter space given by [1–3]

$$
\mathbf{p} = \mathbf{p}^0 + \mathbf{E}\mu \tag{9.4}
$$

where $\mathbf{p}^0$ is a the vector of nominal parameters given by (9.3):

$$
\mathbf{E} = \begin{bmatrix} \epsilon_1 & 0 & \cdots & 0 \\ 0 & \epsilon_2 & & 0 \\ \vdots & & \ddots & \vdots \\ 0 & \cdots & & \epsilon_n \end{bmatrix} \tag{9.5}
$$

$$
\mu = [\mu_1, \mu_2, \ldots, \mu_n]^T \tag{9.6}
$$

μ is a random vector distributed according to the joint pdf, $h(\mathbf{p})$, of the designable circuit parameters. In general, the pdf is defined in the infinite range of variable parameters, but for practical cases a finite tolerance region R_T such that

$$
\int_{R_T} h(\mathbf{p}) \, dp_1 \, dp_2 \cdots dp_k \approx 1 \tag{9.7}
$$

is considered.

In the design we have no control over μ. Considering the tolerance region R_T as a finite region, it may be defined in the following, general way:

$$
R_T \triangleq \{ \mathbf{p} \mid \mathbf{p} = \mathbf{p}^0 + \mathbf{E}\mu, \ \mu \in R_\mu \} \tag{9.8}
$$

where

$$
R_\mu \triangleq \{ \mu \mid -1 \leq \mu_i \leq 1, \ i = 1, 2, \ldots, n \} \tag{9.9}
$$

The tolerance region R_T is an orthotope of n dimensions, with sides of length $2\epsilon_i$, $i = 1, 2, \ldots, n$ and centered at $\mathbf{p}^0$. Tolerance regions for two- and three-dimensional problems are shown in Figure 9.1. The extreme points of R_T, called *vertices*, are defined by setting $\mu_i = \mp 1$. The set of vertices is defined by

$$R_v \triangleq \{ p \mid p_i = p_i + \epsilon_i \mu_i, \; \mu_i \in \{-1, 1\}, \; i = 1, 2, \ldots, n \} \qquad (9.10)$$

The number of these points is 2^n. The two-dimensional tolerance region of Figure

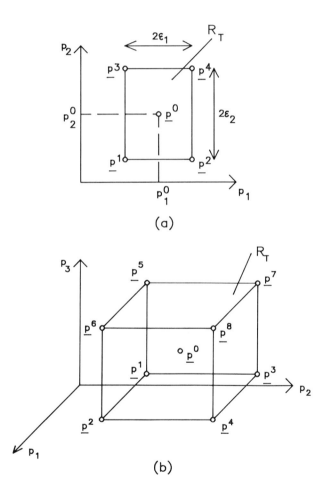

(a)

(b)

Figure 9.1 Tolerance regions R_T: (a) two-dimensional case; (b) three-dimensional case.

9.1(a) has four vertices, and the three-dimensional tolerance region of Figure 9.1(b) has eight vertices. A vertex

$$\mathbf{p}_r = \mathbf{p}^0 + \mathbf{E}\boldsymbol{\mu}_r, \quad \mu_r \in \{-1, 1\} \tag{9.11}$$

may be given a number r according to the following formula [1]:

$$r = 1 + \sum_{i=1}^{n} \left(\frac{\mu_i^r + 1}{2} \right) 2^{i-1} \tag{9.12}$$

The vertices of R_T in Figure 9.1 are numbered according to this formula.

In a typical design problem of a microwave circuit, a given response function $F(\mathbf{p}, \omega)$ must satisfy design specification $S(\omega)$ at a number of discrete frequency points ω_i. This means that a number of inequality constraints given by

$$g_i(\mathbf{p}) \triangleq w_i\big(F(\mathbf{p}, \omega_i) - S(\omega_i)\big) \leq 0, \quad i = 1, 2, \ldots, m \tag{9.13}$$

should be satisfied. The constraints on circuit performance partition the parameter space into regions of acceptable and unacceptable performance. The region of acceptable performance of the circuit defined as

$$R_A \triangleq \{\mathbf{p} \mid g_i(\mathbf{p}) \leq 0, i = 1, 2, \ldots, m\} \tag{9.14}$$

is usually referred to as the acceptable (feasible, constrained) region. Figure 9.2

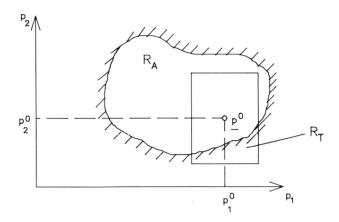

Figure 9.2 Illustration of feasible region R_A and tolerance region R_T.

presents an example of the feasible region and typical tolerance region about nominal design $\mathbf{p}^0$.

If the tolerance region R_T is entirely within the feasible region R_A, then the yield of the design is 100%. Otherwise the yield is less than 100%. Figure 9.2 presents a general situation in which the yield $Y < 100\%$. For a design with the given nominal point $\mathbf{p}^0$, a given tolerance region R_T, acceptable region R_A, and the joint pdf, $h(\mathbf{p})$, of the design parameters $\mathbf{p}$, the yield is given by

$$Y(\mathbf{p}^0) = \int_{R_T \cap R_A} h(\mathbf{p}, \mathbf{p}^0) \, d\mathbf{p} = pr\{\mathbf{F}(\mathbf{p}) \in R_T \cap R_A\} \qquad (9.15)$$

The yield is the probability that the designed circuit will work (i.e., it will meet the desired specification given by (9.13)).

9.2 DETERMINISTIC TOLERANCE ANALYSIS

9.2.1 Worst-Case Tolerance Analysis by the Sensitivity Approach

In a typical worst-case tolerance analysis problem, we want to find the set of extremes of the output response of a circuit $F(\mathbf{p}, \omega)$ at a number of discrete frequency points ω_i for circuit parameters $\mathbf{p}$ belonging to the tolerance region R_T defined by (9.9) and (9.10).

The changes in $F_j(\mathbf{p}) = F(\mathbf{p}, \omega_j)$ due to changes in $\mathbf{p}$ around $\mathbf{p}^0$ may be written as

$$\Delta F_j = \sum_{i=1}^{n} \frac{\partial F_j(\mathbf{p})}{\partial p_i} \bigg|_{\mathbf{p}=\mathbf{p}^0} \cdot \Delta p_i \qquad (9.16)$$

where $\Delta p_i = p_i - p_i^0$.

Now, consider the problem of finding the largest change in F_j when the circuit parameters may receive any values from the tolerance region defined by (9.9) and (9.10). We may expect that the extremes of $F_j(\mathbf{p})$ correspond to vertices of R_T. Thus, we can write

$$\Delta F_j^{\max} = \sum_{i=1}^{n} \frac{\partial F_j(\mathbf{p})}{\partial p_i} \bigg|_{\mathbf{p}=\mathbf{p}^0} \cdot (\epsilon_i \mu_i) \qquad (9.17)$$

where the value of μ_i is chosen either $+1$ or -1 depending on the sign of the derivative (sensitivity), so that each term of the summation is a positive number.

Of course, we have no absolute assurance that a vertex of the tolerance region pointed by the sensitivity signs corresponds to the extreme value of the circuit function $F_j(\mathbf{p})$. Indeed, the true worst-case behavior of $F_j(\mathbf{p})$ may not occur at a vertex of R_T. The sensitivity describes a local behavior of $F_j(\mathbf{p})$ due to one parameter and cannot predict large change effects or the coupling effects of the large variations of many circuit parameters.

The confidence of a correct choice of parameter extreme (a correct choice of a sign of μ_i in (9.17)) may be significantly improved by recalculating the parameter sensitivities at the extreme parameter values. If these sensitivities have the same signs as calculated for nominal parameter values $\mathbf{p}^0$, the extreme values of F_j calculated from (9.17) may be accepted. If any sensitivity sign has changed, the alternative parameter extreme value is assigned.

In practice, relation (8.17) should not be used to compute $\Delta F_j^{\max}$ if the tolerances were assumed large. Once the values of μ_i ($+1$ or -1) have been chosen, the network should be analyzed again with the extreme parameter values.

9.2.2 Worst-Case Tolerance Analysis by the Large Change Sensitivity Approach

In Section 3.2.1, we considered the connection scattering matrix of a microwave circuit. The incoming and outgoing wave variables at all n ports of a circuit composed of m elements (multiports) describe the relations:

$$\mathbf{b} = \mathbf{Sa} + \mathbf{c} \qquad (9.18)$$

$$\mathbf{b} = \boldsymbol{\Gamma}\mathbf{a} \qquad (9.19)$$

which lead to the system of linear equations:

$$(\boldsymbol{\Gamma} - \mathbf{S})\mathbf{a} = \mathbf{c} \quad \text{or} \quad \mathbf{Wa} = \mathbf{c} \qquad (9.20)$$

where $\mathbf{W}$ is the connection scattering matrix of the circuit.

Let us now assume a large change in the parameter value of the kth element of the circuit. This parameter perturbation will cause changes in scattering matrix terms of the kth element:

$$\mathbf{S}^{(k)\Delta} = \mathbf{S}^{(k)} + \Delta\mathbf{S}^{(k)} \qquad (9.21)$$

In (9.21), Δ indicates a quantity altered by the parameter perturbation.

Provided that the perturbed circuit parameter does not change the impressed wave variables $\mathbf{c}$ of the circuit, (9.18) for the perturbed circuit element now will be written as

$$\mathbf{b}^{(k)\Delta} = \mathbf{S}^{(k)\Delta}\mathbf{a}^{(k)\Delta} + \mathbf{c}^{(k)} \qquad (9.22)$$

or, by using (9.21),

$$\mathbf{b}^{(k)\Delta} = \mathbf{S}^{(k)}\mathbf{a}^{(k)\Delta} + \Delta\mathbf{S}^{(k)}\mathbf{a}^{(k)\Delta} + \mathbf{c}^{(k)}$$

$$= \mathbf{S}^{(k)}\mathbf{a}^{(k)\Delta} + \left(\mathbf{c}^{(k)} + \Delta\mathbf{c}^{(k)}\right) \tag{9.23}$$

where

$$\Delta\mathbf{c}^{(k)} = \Delta\mathbf{S}^{(k)}\mathbf{a}^{(k)\Delta} \tag{9.24}$$

By comparing (9.18) and (9.23), we recognize that the perturbed circuit element with the scattering matrix $\mathbf{S}^{(k)\Delta}$ is formally replaced through the original element with matrix $\mathbf{S}^{(k)}$ by a set of supplementary impressed waves at its ports, defined by (9.24).

For the whole circuit with a perturbed kth element, we have

$$\mathbf{W}\mathbf{a}^{\Delta} = \mathbf{c} + \Delta\mathbf{c} \tag{9.25}$$

or

$$\mathbf{a}^{\Delta} = \mathbf{W}^{-1}\mathbf{c} + \mathbf{W}^{-1}\Delta\mathbf{c} \tag{9.26}$$

which, by using (9.20), reduces to

$$\mathbf{a}^{\Delta} = \mathbf{a} + \mathbf{W}^{-1}\Delta\mathbf{c} \tag{9.27}$$

Because the vector $\Delta\mathbf{c}$ has nonzero terms only at rows corresponding to the ports of the perturbed kth element of the circuit, (9.27) may be written as

$$\mathbf{a}^{\Delta} = \mathbf{a} + \mathbf{W}^{-1}\Big|_{\left(\begin{smallmatrix} 1,\ldots,N \\ k_1,\ldots,k_l \end{smallmatrix}\right)} \cdot \Delta\mathbf{c}^{(k)} \tag{9.28}$$

where the notation:

$$\mathbf{M}\Big|^{\left(\begin{smallmatrix} m,\ldots,n \\ p,\ldots,q \end{smallmatrix}\right)}$$

means a matrix $\mathbf{M}$ reduced to the rows $m, \ldots, n$ and columns $p, \ldots, q$.

In (9.28), $k_1, \ldots, k_l$ denote numbers of l ports of the perturbed kth element of the circuit, and $\mathbf{W}^{-1}$ contains only columns corresponding to numbers of ports of the kth element.

Because incoming wave variables at l ports of the kth element are contained in rows $k_1, \ldots, k_l$ of the vector $\mathbf{a}^{\Delta}$, we may write

$$\mathbf{a}^{(k)\Delta} = \mathbf{a}^{(k)} + \mathbf{W}^{-1}\Big|_{\left(\begin{smallmatrix} k_1,\ldots,k_l \\ k_1,\ldots,k_l \end{smallmatrix}\right)} \cdot \Delta\mathbf{c}^{(k)} \tag{9.29}$$

This equation multiplied by $\Delta S^{(k)}$ reduces to the relation:

$$\Delta c^{(k)} = \Delta S^{(k)} a^{(k)} + \Delta S^{(k)} W^{-1} \Big|_{\left(\begin{smallmatrix} k_1, \ldots, k_l \\ k_1, \ldots, k_l \end{smallmatrix}\right)} \cdot \Delta c^{(k)} \qquad (9.30)$$

from which, we can obtain $\Delta c^{(k)}$ as

$$\Delta c^{(k)} = \left[I - \Delta S^{(k)} W^{-1} \Big|_{\left(\begin{smallmatrix} k_1, \ldots, k_l \\ k_1, \ldots, k_l \end{smallmatrix}\right)} \right]^{-1} \Delta S^{(k)} a^{(k)} \qquad (9.31)$$

In (9.31) I denotes the identity matrix of the order l. Derived from (9.31) vector $\Delta c^{(k)}$ can be substituted in (9.28) to compute incoming wave variables at all ports of the circuit. Computation of the vector $\Delta c^{(k)}$ requires inversion of a matrix whose order equals the number of ports of the perturbed kth element of the circuit. Because standard elements of microwave circuits are multiports with a maximum of four ports, the additional computation needed for large change sensitivity evaluation, in comparison to full analysis, is rather small.

If the perturbation of a parameter influences more than one element of the circuit or if perturbations of parameters are introduced in more than one circuit element, the nonzero entries of Δc will occur in rows corresponding to ports of all perturbed elements of the circuit. This means that the size of a matrix to be inverted will be larger and equal to the sum of numbers of ports of all perturbed circuit elements.

9.3 STATISTICAL TOLERANCE ANALYSIS

In statistical tolerance analysis a set of m circuit functions $F(p)$ and a set of n circuit parameters p defined by (9.1) and (9.2) are assumed to be multidimensional random variables. Equation (9.1) defines transformation of the random variable p into random variable F. A joint pdf $h_p(p)$ of circuit parameters is assumed known.

The goals of statistical tolerance analysis are

1. Determination of a joint pdf $h_F(F)$ of the random variable $F(p)$ for a given joint pdf $h_p(p)$ of the random variable p. This complicated problem is usually avoided by focusing on the following simpler problems.
2. Determination of some statistical parameters of pdf $h_F(F)$ of network function such as the first moment (the mean of the distribution of F), the second moment (the variance), and the third moment (the skewness measure about the mean) for given pdf $h_p(p)$ or for given statistical moments of the pdf of circuit parameters.
3. Determination of the yield of the design (equation (9.15)) for a given pdf or for given statistical moments of the circuit parameters p.

The most difficult problem is the first. Sometimes the solution of the first problem may be found by solving the second problem; that is, by first finding statistical moments of **F** random variable. A pdf may be determined from knowledge of the statistical moments of a distribution [4].

9.3.1 The Method of Statistical Moments—Computation of Statistical Parameters of Circuit Functions

By using this method we can determine statistical moments of circuit function distribution $h_F(\mathbf{F})$, knowing the statistical moments of the circuit parameter distribution $h_p(\mathbf{p})$. Most often we are interested in the expected value (M1) and standard deviation (M2). If the *central limit theorem* is satisfied, $h_F(\mathbf{p})$ is a multidimensional normal distribution [4, 5].

The interpretation of the standard deviation is well known. In intervals $\pm\sigma$, $\pm 2\sigma$, and $\pm 3\sigma$, it is contained to, respectively, 68.26%, 94.45%, and 99.70% of the whole distribution (for a normal distribution). If the type of the distribution is not known, the Chebyschev inequality [4, 5]

$$\mathrm{pr}\{x - E[x] < (k\sigma)\} \geq 1 - \frac{1}{k^2}, \quad k > 0 \qquad (9.32)$$

is valid for any (one-dimensional) distribution of a random variable x (in the preceding relation $E[\cdot]$ is the expected value (mean) operator).

Because of its simplicity the moment method is used very often in practice when circuit parameter deviations are small. The method may be used also as a part of computer-aided procedures for optimal tolerance assignment [6, 7].

Computation of Statistical Parameters of Circuit Functions

A Taylor series expansion may be used to approximate the circuit function in the vicinity of a point $\mathbf{p} = \mathbf{p}^0$:

$$F(\mathbf{p}) \approx F(\mathbf{p}^0) + \nabla^T F(\mathbf{p}^0) \cdot \Delta\mathbf{p} + \frac{1}{2}\Delta\mathbf{p}^T \mathbf{H}(\mathbf{p}^0)\,\Delta\mathbf{p} \qquad (9.33)$$

where $\mathbf{p}^0 = [p_1^0, p_2^0, \ldots, p_n^0]$ is a vector of nominal values of circuit parameters, $\Delta\mathbf{p} = \mathbf{p} - \mathbf{p}^0$, $\nabla F(\mathbf{p})$ is a gradient vector at $\mathbf{p} = \mathbf{p}^0$:

$$\nabla F(\mathbf{p}^0) = \left[\frac{\partial F(\mathbf{p}^0)}{\partial p_1}, \frac{\partial F(\mathbf{p}^0)}{\partial p_2}, \ldots, \frac{\partial F(\mathbf{p}^0)}{\partial p_n}\right]^T \qquad (9.34)$$

and $\mathbf{H}(\mathbf{p}^0)$ is a Hessian matrix, defined as

$$\mathbf{H}(\mathbf{p}^0) \triangleq \left[\frac{\partial^2 F(\mathbf{p}^0)}{\partial p_i \, \partial p_j} \right], \quad i, j = 1, 2, \ldots, n \tag{9.35}$$

In the following considerations we will use the notation

$$F^0 \triangleq F(\mathbf{p}^0), \quad \nabla F \triangleq \nabla F(\mathbf{p}^0), \quad \mathbf{H} \triangleq \mathbf{H}(\mathbf{p}^0) \tag{9.36}$$

By taking the first two terms of a Taylor series expansion, assuming $\mathbf{p}^0 = E[\mathbf{p}] = \bar{\mathbf{p}}$, and using definitions of the expected value and variance (see Appendix 3), we obtain

$$E\big[F(\mathbf{p}) \big] \triangleq \bar{F}(\mathbf{p}) \approx E\big[F(\mathbf{p}^0) \big] = F(\mathbf{p}^0) \triangleq F^0 \tag{9.37}$$

because $E[\nabla^T F \cdot \Delta \mathbf{p}] = 0$, and

$$\text{var}\big[F(\mathbf{p}) \big] = E\big[F(\mathbf{p}) - F^0 \big]^2 \approx E\big[\nabla^T F^0 \cdot \Delta \mathbf{p} \big]^2$$

$$= E\big[\nabla^T F \cdot \Delta \mathbf{p} \cdot \Delta \mathbf{p}^T \cdot \nabla F \big] \tag{9.38}$$

Because ∇F is a vector of constant values, we obtain from (8.38) (see Appendix 3):

$$\text{var}\big[F(\mathbf{p}) \big] \approx \nabla^T F \mathbf{K}^\mathbf{p} \nabla F \tag{9.39}$$

where $\mathbf{K}^\mathbf{p} \triangleq [\text{cov}(p_i, p_j)] = [K_{ij}^\mathbf{p}]$ is a covariance matrix of the random variables $\mathbf{p}$. The equation (9.39) may be also written in the form:

$$\text{var}\big[F(\mathbf{p}) \big] \approx \sum_{i=1}^{n} \sum_{j=1}^{n} \frac{\partial F}{\partial p_i}\bigg|_{\mathbf{p}=\mathbf{p}^0} \cdot \frac{\partial F}{\partial p_j}\bigg|_{\mathbf{p}=\mathbf{p}^0} \cdot K_{ij}^\mathbf{p} \tag{9.40}$$

This relation is used to compute the variance of a circuit function $F(\mathbf{p})$. The partial derivatives of $F(\mathbf{p})$ are the sensitivities of a network readily obtained from adjoined network sensitivity analysis.

Relation (9.40) may be generalized for a case when F is multidimensional random variable $\mathbf{F}$, $\mathbf{F} = [F_1(\mathbf{p}), F_2(\mathbf{p}), \ldots, F_m(\mathbf{p})]$. The Taylor series expansion cut to two terms will have the form:

$$\mathbf{F} \approx \begin{bmatrix} F_1 \\ F_2 \\ \vdots \\ F_m \end{bmatrix} = \begin{bmatrix} F_1^0 \\ F_2^0 \\ \vdots \\ F_m^0 \end{bmatrix} + \begin{bmatrix} \nabla^T F_1^0 \\ \nabla^T F_2^0 \\ \vdots \\ \nabla^T F_m^0 \end{bmatrix} \cdot \Delta\mathbf{p} = \mathbf{F}^0 + \mathbf{J}^0 \, \Delta\mathbf{p} \qquad (9.41)$$

where $\mathbf{J}^0$ is the Jacobian matrix given as

$$\mathbf{J} \triangleq \frac{\partial \mathbf{F}}{\partial \mathbf{p}} = \begin{bmatrix} \dfrac{\partial F_1}{\partial p_1} & \dfrac{\partial F_1}{\partial p_2} & \cdots & \dfrac{\partial F_1}{\partial p_k} \\ \dfrac{\partial F_2}{\partial p_1} & \dfrac{\partial F_2}{\partial p_2} & \cdots & \dfrac{\partial F_2}{\partial p_k} \\ \vdots & \vdots & & \vdots \\ \dfrac{\partial F_m}{\partial p_1} & \dfrac{\partial F_m}{\partial p_2} & \cdots & \dfrac{\partial F_m}{\partial p_k} \end{bmatrix} \qquad (9.42)$$

and taken at $\mathbf{p} = \mathbf{p}^0$.

From (9.41) we have

$$E[\mathbf{F}] = \mathbf{F}^0 \qquad (9.43)$$

The covariance matrix $\mathbf{K}^{\mathbf{F}}$ of the circuit functions $\mathbf{F}$ is defined as

$$\mathbf{K}^{\mathbf{F}} = E[\Delta\mathbf{F} \cdot \Delta\mathbf{F}^T] \qquad (9.44)$$

where $\Delta\mathbf{F} = \mathbf{F} - \mathbf{F}^0$.

By using (9.41), we have $\Delta\mathbf{F} = \mathbf{J} \, \Delta\mathbf{p}$ and

$$\mathbf{K}^{\mathbf{F}} = E[\mathbf{J} \, \Delta\mathbf{p} \, \Delta\mathbf{p}^T \mathbf{J}^T] = \mathbf{J}\mathbf{K}^{\mathbf{P}}\mathbf{J}^T \qquad (9.45)$$

This equation determines the relation between the covariance matrix $\mathbf{K}^{\mathbf{P}}$ of the circuit parameters $\mathbf{p}$ and the covariance matrix $\mathbf{K}^{\mathbf{F}}$ of circuit functions. The terms of the Jacobian matrix $\mathbf{J}$ are obtained from the network sensitivity analysis described in chapter 4.

In practical applications, we often use the normalized sensitivities $S_p^f =$

$\partial(\ln f)/\partial(\ln p)$. In subsequent derivations, we use the following notations:

$$\delta\mathbf{F} \triangleq \left[\frac{\Delta F_1}{E[F_1]}, \frac{\Delta F_2}{E[F_2]}, \ldots, \frac{\Delta F_m}{E[F_m]}\right]^T \qquad (9.46)$$

$$\delta\mathbf{p} \triangleq \left[\frac{\Delta p_1}{E[p_1]}, \frac{\Delta p_2}{E[p_2]}, \ldots, \frac{\Delta p_k}{E[p_k]}\right]^T \qquad (9.47)$$

$$\mathbf{S}^{F_j} \triangleq \left[S_{p1}^{F_j}, S_{p2}^{F_j}, \ldots, S_{pk}^{F_j}\right]^T, \quad j = 1, 2, \ldots, m \qquad (9.48)$$

$$\tilde{\mathbf{S}} \triangleq \begin{bmatrix} (\mathbf{S}^{F_1})^T \\ (\mathbf{S}^{F_2})^T \\ \vdots \\ (\mathbf{S}^{F_m})^T \end{bmatrix} = \begin{bmatrix} S_{p1}^{F_1} & S_{p2}^{F_1} & \cdots & S_{pk}^{F_1} \\ S_{p1}^{F_2} & S_{p2}^{F_2} & \cdots & S_{pk}^{F_2} \\ \vdots & \vdots & & \vdots \\ S_{p1}^{F_m} & S_{p2}^{F_m} & \cdots & S_{pk}^{F_m} \end{bmatrix} \qquad (9.49)$$

Then, using relations (9.33) and (9.41), we have

$$\frac{\Delta F_j}{E[F_j]} \approx (\mathbf{S}^{F_j})^T \cdot \delta\mathbf{p} \qquad (9.50)$$

$$\frac{\text{var}[F_j(\mathbf{p})]}{E[F_j]} = (\mathbf{S}^{F_j})^T \cdot \mathbf{K}^{\delta\mathbf{p}} \cdot \mathbf{S}^{F_j} \qquad (9.51)$$

$$\delta\mathbf{F} = \tilde{\mathbf{S}}\, \delta\mathbf{p} \qquad (9.52)$$

$$\mathbf{K}^{\delta\mathbf{F}} = \tilde{\mathbf{S}}\mathbf{K}^{\delta\mathbf{p}}\tilde{\mathbf{S}}^T \qquad (9.53)$$

where $\mathbf{K}^{\delta\mathbf{F}}$ and $\mathbf{K}^{\delta\mathbf{p}}$ are, respectively, the covariance matrices for elements of vector $\delta\mathbf{F}$ and $\delta\mathbf{p}$. The elements of these matrices are

$$[\mathbf{k}^{\delta f}]_{ij} = \frac{\text{cov}[f_i, f_j]}{E[f_i]E[f_j]} \quad \text{and} \quad [k^{\delta p}]_{ij} = \frac{\text{cov}[p_i, p_j]}{E[p_i]E[p_j]} \qquad (9.54)$$

These relations may be used for statistical parameter computation (variances and covariances) of the magnitude and phase of a complex transmission function $F(\mathbf{p})$ $|F|e^{j\arg\{F\}}$.

By using relations (4.6) and (4.7), we have

$$\frac{\Delta|F|}{E[|F|]} \approx \left[\mathrm{Re}\{\mathbf{S}^F\}\right]^T \cdot \delta\mathbf{p} \qquad (9.55)$$

$$\Delta\arg\{F\} \approx \left[\mathrm{Im}\{\mathbf{S}^F\}\right]^T \cdot \delta\mathbf{p} \qquad (9.56)$$

where $\mathrm{Re}\{\mathbf{S}^F\}$ and $\mathrm{Im}\{\mathbf{S}^F\}$ are vectors, respectively, of real and imaginary parts of sensitivities $S_{p_i}^F$, $i = 1, 2, \ldots, n$.

From (9.51), we get

$$\frac{\mathrm{var}[|F|]}{E[|F|]} \approx \left[\mathrm{Re}\{\mathbf{S}^F\}\right]^T \mathbf{K}^{\delta\mathbf{p}}\left[\mathrm{Re}\{\mathbf{S}^F\}\right] \qquad (9.57)$$

$$\mathrm{var}\left[\arg\{F\}\right] \approx \left[\mathrm{Im}\{\mathbf{S}^F\}\right]^T \mathbf{K}^{\delta\mathbf{p}}\left[\mathrm{Im}\{\mathbf{S}^F\}\right] \qquad (9.58)$$

All sensitivities are computed at $\mathbf{p} = \mathbf{p}^0 = E[\mathbf{p}]$.

By defining $\delta\mathbf{F}$ as a two-element vector:

$$\delta\mathbf{F} \triangleq \left[\frac{\Delta|F|}{E[|F|]}, \arg\{F\}\right] \qquad (9.59)$$

and applying relation (9.53), we are able to determine a covariance between $|F|$ and $\arg\{F\}$,

$$\mathbf{K}^{\delta\mathbf{F}} = \begin{bmatrix} \left[\mathrm{Re}\{\mathbf{S}^F\}\right]^T \\ \left[\mathrm{Im}\{\mathbf{S}^F\}\right]^T \end{bmatrix} \mathbf{K}^{\delta\mathbf{p}}\left[\mathrm{Re}\{\mathbf{S}^F\}, \mathrm{Im}\{\mathbf{S}^F\}\right]$$

$$= \begin{bmatrix} \left[\mathrm{Re}\{\mathbf{S}^F\}\right]\mathbf{K}^{\delta\mathbf{p}}\,\mathrm{Re}\{\mathbf{S}^F\}, \left[\mathrm{Re}\{\mathbf{S}^F\}\right]^T \mathbf{K}^{\delta\mathbf{p}}\,\mathrm{Im}\{\mathbf{S}^F\} \\ \left[\mathrm{Im}\{\mathbf{S}^F\}\right]^T \mathbf{K}^{\delta\mathbf{p}}\,\mathrm{Re}\{\mathbf{S}^F\}, \left[\mathrm{Im}\{\mathbf{S}^F\}\right]^T \mathbf{K}^{\delta\mathbf{p}}\,\mathrm{Im}\{\mathbf{S}^F\} \end{bmatrix} \qquad (9.60)$$

The elements $[k^{\delta f}]_{12} = [k^{\delta f}]_{21}$ are equal to the required covariance between $|F|$ and $\arg\{F\}$:

$$\frac{\mathrm{cov}[|F|, \arg\{F\}]}{E[|F|]} = [k^{\delta\mathbf{F}}]_{12} = \left[\mathrm{Re}\{\mathbf{S}^F\}\right]^T \mathbf{K}^{\delta\mathbf{p}}\,\mathrm{Im}\{\mathbf{S}^F\} \qquad (9.61)$$

9.3.2 Computation of the Yield by Using the Method of Statistical Moments

Application of the moment method for the yield computation is rather restricted because of large errors related to such approach. Only when the considered problem satisfies conditions of the central limit theorem, may computation of the yield by using the moment method be acceptable. In such cases, the m-dimensional random variable $\mathbf{F} = \mathbf{F}(\mathbf{p})$ (circuit functions) has a normal distribution whose pdf defines the relation:

$$h_F(\mathbf{F}) = \frac{1}{(2\pi)^{m/2}\sqrt{|\mathbf{K}|}}\, e^{-1/2\, \Delta\mathbf{p}^T \mathbf{K}^{-1}\, \Delta\mathbf{p}} \qquad (9.62)$$

where $\Delta\mathbf{p} = \mathbf{p} - E[\mathbf{p}]$.

The covariance matrix of this distribution is $\mathbf{K} \triangleq \mathbf{K}^{\mathbf{F}} = \mathbf{JK}^{\mathbf{P}}\mathbf{J}^T$ (9.45) and the expected value $E[\mathbf{F}] = \mathbf{F}(\mathbf{p}^0)$. Evaluation of the yield Y requires computation of the m-dimensional integral (9.15), which is a rather computationally complicated task. For $m > 3$ numerical integration is not computationally effective.

Instead of computing integrals, the yield of the normal distribution may be estimated by using relations determining lower and upper boundary of the yield [8]:

$$Y^L \leq Y \leq Y^U \qquad (9.63)$$

where

$$Y^L = a + b - c, \quad Y^U = a + b \qquad (9.64)$$

$$a = 1 - \sum_{j=1}^{m} W_j, \quad b = \sum_{j=1}^{m-1}\sum_{k=j+1}^{m} W_{jk},$$

$$c = \sum_{j=1}^{m-2}\sum_{k=j+1}^{m-1}\sum_{l=j+1}^{m} W_{jkl} \qquad (9.65)$$

and

$$W_j = 1 - \mathrm{pr}\{S_j^l \leq f_j \leq S_j^u\}, \quad j = 1, 2, \ldots, m \qquad (9.66)$$

$$W_{jk} - 1 - \mathrm{pr}\{S_j^l \leq f_j \leq S_j^u \text{ and } S_k^l \leq f_k \leq S_k^u\}, \quad j, k = 1, 2, \ldots, m \qquad (9.67)$$

$$W_{jkl} = 1 - \mathrm{pr}\{S_j^l \leq f_j \leq S_j^u \text{ and } S_k^l \leq f_k \leq S_k^u \text{ and } S_l^l \leq f_l \leq S_l^u\},$$

$$j, k, l = 1, 2, \ldots, m \qquad (9.68)$$

The probabilities in (9.66), (9.67), and (9.68) may be calculated by using the tables of probability values [9], or they can be computed numerically. Both ways are more attractive than numerical computation of m-dimensional integrals.

Upper and lower bounds for the yield may be calculated by using less accurate relations:

$$Y^L = a, \quad Y^U = a + b \tag{9.69}$$

We thus avoid the expensive computation of W_{ijk}.

9.3.3 Monte Carlo Method for Tolerance Analysis

In the Monte Carlo method circuit parameters are assumed to be n-dimensional random variables $\mathbf{p} = [p_1, p_2, \ldots, p_n]^T$ with known n-dimensional joint pdf $h_p(\mathbf{p})$. The procedure consists of generating sample networks using randomly selected circuit parameters according to the given pdf $h_p(\mathbf{p})$ and compiling statistics of the circuit function $\mathbf{F}$ or the yield of the circuit from the resulting analyses [10, 11].

The basic or primitive Monte Carlo method is defined by the following algorithm:

Step 1 Set $i = 0$ (the sample index). Set $N_A = 0$ (the number of sample circuits satisfying circuit specifications).

Step 2 Set $i = i + 1$. Generate the ith realization of a vector of circuit parameters $\mathbf{p}^{(i)}$ according to the given pdf $h_p(\mathbf{p})$.

Step 3 Analyze the ith sample network and compute the ith realization of a vector of circuit functions $\mathbf{F}^{(i)}$.

Step 4 Check whether the ith sample circuit satisfies design specifications. When the ith sample circuit with parameter values $\mathbf{p}^{(i)}$ meets all specifications, $\mathbf{F}^{(i)} \in R_A$, set $N_A = N_A + 1$.

Step 5 If $i < N$ (N is the maximum number of sample circuits), return to step 2.

Step 6 Compute statistical parameters of the distribution $h_F(\mathbf{F})$ of circuit functions (statistical moments, correlation coefficients, histograms).

Step 7 Determine the estimate $\hat{Y}$ of the yield as

$$\hat{Y} = \frac{N_A}{N} = \frac{\text{Number of passing circuits}}{\text{Total number of circuits}} \tag{9.70}$$

The procedure simulates sets of circuit parameter values according to the prescribed distribution. For this purpose, we must generate random numbers according to the prescribed distribution. General methods for random number generation are discussed in literature [10]. The applications of these methods can be found in many papers [12–16].

The Monte Carlo method is the most accurate approach for circuit parameter tolerance analysis. It allows us to estimate the yield and accuracy of this estimation. The most important point is that the number of circuit samples necessary to achieve a given accuracy does not depend on the number of circuit parameters. Typically between 100 and 300 circuit simulations are performed.

The Monte Carlo analysis is used to verify tolerance design algorithms and also for worst-case problems when sensitivity sign changes make results of analysis unreliable. This is achieved by using the parameter distribution shown in Figure 9.3. In Monte Carlo worst-case analysis, circuits satisfying design requirements are of no interest. Using the parameter distributions as in Figure 9.3 we search for worst-case circuits at the vertices of the tolerance region.

9.3.4 Generation of Pseudorandom Parameter Values

The Monte Carlo requires the random generation of parameter values. General methods for pseudorandom number generation are described in literature [10]. Subroutines to generate uniformly distributed random numbers in the range 0 to 1 can be used to generate nearly all other distributions. The random numbers are generated with a computer by using the power-residue concept. The basic expression used is a recurrence relation, often in the form:

$$u_{i+1} = (au_i + c) \bmod(2^d) \qquad (9.71)$$

where u_i is the previous random number, d represents the number of significant bits in a computer word, and a and c are nonnegative integer numbers in the range 0 to $2^d - 1$.

The notation $(g) \bmod m$ means $\mathrm{int}(g/m) \cdot m$, where $\mathrm{int}(g/m)$ is integer part of the division inside brackets.

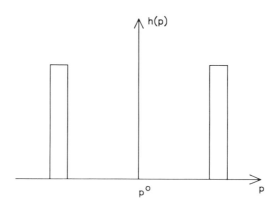

Figure 9.3 Circuit parameter distribution used for simulation of worst-case conditions.

Formula (9.71) generates a sequence of random numbers that repeat themselves with the period less than or equal to 2^d. To ensure a large repetition period, for computers with 32-bit word length d is usually chosen to be 31. In this case, the relationship (9.71) gives integer numbers in the range 0 to 2^{31}, which have to be divided by 2^{31}.

In Monte Carlo analysis, we need random number generators that will allow us to generate random numbers of required distributions. The uniformly distributed number sequences can be used to generate a new sequence having required distribution.

For example, the formula:

$$z_1 = \frac{\sum\limits_{i=1}^{N} x_i}{N} - \frac{1}{2}$$

$$z_2 = \frac{\sum\limits_{i=1}^{N} x_{N+i}}{N} - \frac{1}{2}, \quad 0 \le x_i \le 1$$

$$z_3 = \frac{\sum\limits_{i=1}^{N} x_{2N+i}}{N} - \frac{1}{2}$$

$$z_k = \frac{\sum\limits_{i=1}^{N} x_{(k-1)N+i}}{N} - \frac{1}{2}$$

(9.72)

will transform a pseudorandom uniform distribution $x_1, x_2, x_3, \ldots$ to the pseudorandom normal (Gaussian) distribution $z_1, z_2, z_3 \ldots$ According to the central limit theorem, if N (number of summed numbers x_i) is sufficiently large, the distribution of a new random variable z_k defined by (9.72) will approach the Gaussian distribution.

A more general procedure that may be used for generation sequences of pseudorandom numbers with any desired probability density function $h(x)$ is based on the *cumulative distribution* (or *density*) *function* (cdf), defined as

$$g(x) = \int_{x_1}^{x} h(x')\, dx' \tag{9.73}$$

and illustrated in Figure 9.4. In (9.73) and in the figure, x_1 and x_2 define an interval in which pdf(x) is considered.

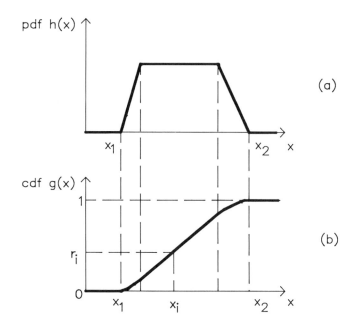

Figure 9.4 Generation of arbitrary distribution from uniform distribution using cumulative distribution function: (a) arbitrary probability distribution function; (b) cumulative distribution function.

The procedure is the following:
1. Integrate pdf $h(x)$ to get cdf $g(x)$.
2. Generate a pseudorandom value r_i uniformly distributed in the range 0 to 1.
3. Set $r_i = g(x)$ and find corresponding to it value x, as in Figure 9.4(b).

In practice, the cdf values are usually given as piecewise linear approximations or tables.

9.3.5 Accuracy of the Monte Carlo Method and Required Number of Samples

The procedure used in the Monte Carlo method to draw circuit samples for yield estimation corresponds to the Bernoulli drawing scheme, in which the probability of success equals $Y = N_A / N$ and the probability of failure equals $1 - Y = 1 - N_A / N$. The probability, that in N drawings (a number of sample points in a set of Monte Carlo analysis) occur exactly N_A successful, pass samples $(\mathbf{p} \in R_A)$, defines binomial (Bernoulli) distribution, and equals

$$\text{pr}\{x = N_A\} = \binom{N}{N_A} Y^{N_A}(1 - Y)^{N - N_A} \tag{9.74}$$

where x is a random variable whose values are equal to the number of success drawings, and

$$\binom{N}{N_A} \triangleq \frac{N!}{(N - N_A)!N_A!} \tag{9.75}$$

is the Newton symbol.

The binomial distribution has the expectation:

$$E[x] = N \cdot Y \tag{9.76}$$

and the variance:

$$\mathrm{var}[x] = \sigma^2 = N \cdot Y \cdot (1 - Y) \tag{9.77}$$

Our goal is to estimate the parameter Y (the yield) of this distribution. The "natural" estimator for Y is $\hat{Y} = N_A/N$. We get this estimator performing N Monte Carlo analyses of a circuit. This is a single simulation of circuit parameter dispersion. If the Monte Carlo analysis of a circuit is carried out a number of times, on each occasion with a new set of random samples, then a number of different estimates $\hat{Y}$ of yield will be generated. The estimator $\hat{Y}$ of the yield Y is the random variable with some probability distribution related to the distribution (9.74). For a large N, a binomial distribution may be approximated by a normal (Gaussian) distribution with the expectation and variance equal to the expectation (9.76) and variance (9.77) of a binomial distribution. We can show that the distribution of the yield estimator $\hat{Y}$ is asymptotically normal (for large N) with the expected value [4, 5]:

$$E[\hat{Y}] = Y \tag{9.78}$$

and the variance:

$$\mathrm{var}[\hat{Y}] = \sigma^2 = \frac{Y(1 - Y)}{N} \tag{9.79}$$

The property (9.78) indicates that the average of the estimates will tend toward the true value of the yield Y. This property is referred to as the *unbiasedness* of the estimator. Figure 9.5 presents the probability density function $h_Y(\hat{Y})$ of the estimator $\hat{Y}$ obtained for an infinite number of sets of Monte Carlo analyses.

By using the above-mentioned properties of the yield estimator pdf, $h_Y(\hat{Y})$, we are able to find some interesting and useful conclusions from the outcome y of a single set of Monte Carlo analysis with N samples. We consider the pdf, $h_Y(\hat{Y})$, the mean of which is the true but unknown value Y of the yield. The probability that any particular

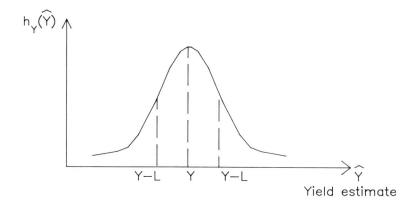

Figure 9.5 The probability distribution function $h_Y(\hat{Y})$ of yield estimates for an infinite number of sets of Monte Carlo analyses.

single estimate y lies between two limits symmetrically located about Y, $Y - L$ and $Y + L$ (see Figure 9.5):

$$\text{pr}\{Y - L < y < Y + L\} = \alpha \qquad (9.80)$$

is the area α under the pdf between these two limits. Interval $2L$ is called the *confidence interval* for a given confidence level, α. In a Gaussian distribution to compute $2L$, the size of confidence interval, we have to know σ^2, the variance of the distribution. The most commonly used value of the confidence level α is 0.95 (95%), which corresponds to a value of $L = 2\sigma$. (For other values of the confidence level, we have, for $\alpha = 0.85$, $L/\sigma = 1.44$; for $\alpha = 0.9$, $L/\sigma = 1.65$; and for $\alpha = 0.99$, $1/\sigma = 2.6$.) By rearranging the inequality inside the bracket of (9.80), we obtain a useful statement:

$$\text{pr}\{y - L < Y < y + L\} = \alpha \qquad (9.81)$$

which indicates that the probability that the interval $y \mp L$ contains the true value Y of the yield is also α.

According to (9.81), calculation of σ requires knowledge of the true value Y of yield, which is not known. Because the confidence interval has to be established for one set of Monte Carlo analysis, we calculate an adequate approximation to σ from the equation (9.81), by substituting for Y the estimate $\hat{Y}$ of this quantity. Having σ, we multiply it by appropriate value of L/σ and thereby obtain the confidence interval $y \mp L$. For example, for $N = 100$ (number of Monte Carlo circuit analyses) the estimated yield of 80% is found. From (9.81), we calculate $\sigma = 0.04$ (4%). This gives a 95% confidence level that the actual yield lies between $(80 - 2 \cdot 4)\% = 72\%$ and $(80 + 2 \cdot 4)\% = 88\%$.

The interpretation of the confidence interval is as follows. If a large number of sets of Monte Carlo analyses were performed, on average, 95% of the confidence intervals would embrace the true value of the yield.

The discussion just presented may be summarized as follows:

1. the greater the required confidence in the yield estimate, the wider must be the confidence interval;
2. the size of the confidence interval is inversely proportional to the square root of a Monte Carlo sample analysis (9.81);
3. the standard deviation σ that determines the uncertainty in the yield estimate does not depend on the number of circuit parameters.

The third property is of great significance. It makes the Monte Carlo methods superior to deterministic techniques when the number of circuit parameters exceeds about five to eight.

REFERENCES

[1] J.W. Bandler, P.C. Liu, and H. Tromp, "A Nonlinear Programming Approach to Design Centering, Tolerancing and Tuning," *IEEE Trans. Circuits Syst.*, Vol. CAS-23, 1976, pp. 155–165.

[2] J.W. Bandler, P.C. Liu, and H. Tromp, "Integrated Approach to Microwave Design," *IEEE Trans. Microwave Theory Tech.*, Vol. MTT-24, 1976, pp. 584–591.

[3] H.L. Abdel-Malek and J.W. Bandler, "Yield Optimization for Arbitrary Statistical Distributions: Part I—Theory," *IEEE Trans. Circuits Syst.*, Vol. CAS-27, 1980, pp. 245–253.

[4] K.S. Miller, *Multidimensional Gaussian Distributions*, John Wiley and Sons, New York, 1964.

[5] J.E. Freund, *Mathematical Statistics*, Prentice-Hall, Englewood Cliffs, NJ, 1962.

[6] A.R. Thorbjornsen and S. W. Director, "Computer-Aided Tolerance Assignment for Linear Circuits with Correlated Elements," *IEEE Trans. Circuit Theory*, Vol. CT-20, 1973, pp. 518–524.

[7] A.R. Thorbjornsen and M. Malek-Daaboul, "Computer-Aided Tolerance and Correlation Coefficient Assignment: A First Attempt," *Proc. 14th Allerton Conf. on CST*, 1976, pp. 578–586.

[8] B.J. Karafin, "The General Component Tolerance Assignment Problem in Electrical Networks," Ph.D. Dissertation, University of Pennsylvania, Philadelphia, 1974.

[9] A. Abramovitz and I.A. Stegun, *Handbook of Mathematical Functions*, National Bureau of Standards, Applied Mathematical Series, No. 55, Washington, DC, 1964.

[10] J.M. Hammerslaey and D.C. Handscomb, *Monte Carlo Methods*, Methuen and Company, New York, 1964.

[11] R. Spence and R.S. Soin, *Tolerance Design of Electronic Circuits*, Addison-Wesley, London, 1988.

[12] J. Logan, "Characterization and Modeling for Statistical Design," *BSTJ*, Vol. 50, No. 4, April 1971, pp. 1105–1146.

[13] I.A. Cernak and D.B. Kirby, "Nonlinear Circuits and Statistical Design," *BSTJ*, Vol. 50, No. 4, April 1971, pp. 1173–1195.

[14] P. Balaban and J.J. Golembeski, "Statistical Analysis for Practical Circuit Design," *IEEE Trans. Circuits Syst.*, Vol. CAS-22, February 1975, pp. 100–108.

[15] D.A. Diverak, R.W. Dutton, and W.J. Mc Calla, "Experimental Study of Gummel-Poon Model Parameter Correlations for Bipolar Junction Transistors," *IEEE J. Solid State Circuits*, Vol. SC-12, October 1977, pp. 552–559.

[16] M. Styblinski, "Factor Analysis Model of Resistor Correlations in Monolithic Integrated Circuits," *IEEE Proc. ISCAS-79*, Tokyo, Japan, 1979.

Chapter 10
Tolerance Design of Microwave Circuits

During the manufacture of mass-produced microwave networks, their parameters are subjected to statistical variations due to manufacturing tolerances. Environmental factors such as temperature or humidity are sources of circuit parameter variations. Because of these effects, practical circuits have to be designed in the face of parameter variations. The objective of parameter tolerance assignment is to attain full use of allowable design margins to minimize manufacturing costs by incorporating *design centering* (a change in nominal parameter values with the tolerances held fixed), *tolerance assignment* (a change in tolerances with constant nominal parameters of a circuit), or a combination of these two approaches. Figure 10.1 illustrates these methods for increasing the manufacturing yield of a circuit. Starting with the original circuit design (Figure 10.1(a)), by adjusting nominal values of the circuit parameters, but leaving their tolerances fixed, we may shift the tolerance region to locate it more centrally within the region of acceptability R_A (Figure 10.1(b)). This procedure, called *design centering*, leads to an increase in the production yield of the circuit. By changing tolerances of the optimally centered circuit we may get a design corresponding to 100% production yield (Figure 10.1(c)) or a design leading to the minimum production cost of the circuit (Figure 10.1(d)).

There are two approaches to tolerance assignment. In the worst-case design the objective is to assign tolerances such that no rejects on the production line could be attributed to parameter tolerance limits (Figure 10.1(c)). In the broader view of the worst-case design, every network will satisfy all required specifications in all expected operational environments. The yield is not considered in the worst-case design; a 100% yield is implied by a feasible solution. A worst-case design is not always appropriate or a 100% yield may not even be practically realizable. In either case, a worst-case design with 100% yield may lead to unrealistically tight tolerances and, as a result, high production cost.

In the second method to tolerance assignment we assume that the more economical approach is to eliminate some circuit units that do not satisfy the specified

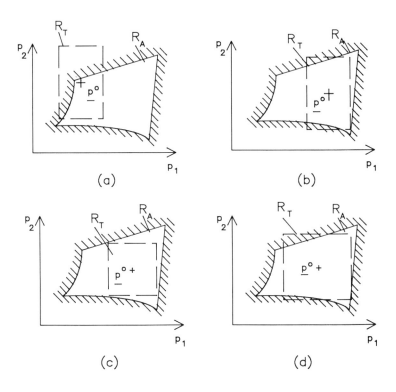

Figure 10.1 Tolerance design: (a) original circuit design with yield less than 100%; (b) tolerance centering leading to maximum yield; (c) tolerance assignment leading to 100% manufacturing yield; (d) tolerance assignment leading to lower cost of circuit manufacturing.

requirements. We know that the cost of a circuit component is typically an inverse function of its tolerance (Figure 10.2(a)). Because of this, the minimum production cost of a circuit understandably may not occur at 100% yield, as shown in Figure 10.2(b) [1]. The objective then is to minimize the cost C_s of circuits that satisfy all design requirements (specifications), such that

$$C_s = C_u \frac{\text{Total number of manufactured circuits}}{\text{Number of circuits satisfying all specifications}}$$

$$= \frac{C_u}{Y} \tag{10.1}$$

where C_u is per-circuit manufacturing cost, and Y is the yield. Both C_u and Y depend on nominal parameter values p^0 and their tolerances. Thus, product cost minimization may be achieved by adjusting nominal values to maximize the yield Y.

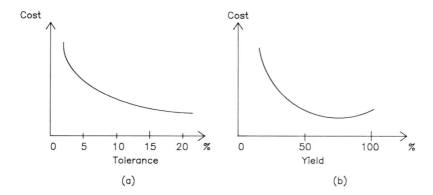

Figure 10.2 Typical relations between circuit cost and (a) tolerance; (b) production yield.

Methods for tolerance design fall into two categories. First, we have methods that use techniques of the optimization theory. This is the deterministic approach to tolerance centering and assignment [2–5]. The second group of methods uses the Monte Carlo technique to evaluate the yield. The information obtained from Monte Carlo trials may then be used for yield maximization [6–11].

10.1 BASIC CONSIDERATIONS

The discussion presented in Section 9.1 gives basic definitions used in statistical circuit design problems.

The constraints on circuit performance, defined as

$$\mathbf{g}(\mathbf{p}) = \mathbf{F}(\mathbf{p}) - \mathbf{S} \geq 0 \tag{10.2}$$

determine the acceptable region:

$$R_A = \{\mathbf{p} \,|\, \mathbf{g}(\mathbf{p}) \geq 0\} \tag{10.3}$$

in the parameter space.

In the practical design of a physical circuit, we must account for variations of circuit parameters related to the manufacturing process (manufacturing tolerances), the lifetime of the circuit (aging of the circuit elements), or environmental factors (temperature, humidity). These processes are usually characterized by determining a statistical distribution for each circuit parameter. For circuit parameters that vary independently of one another, we define the pdf, $h_i(p_i)$, such that

$$\int_a^b h_i(p_i) \, dp_i = \mathrm{pr}\{a \leq p_i \leq b\} \tag{10.4}$$

where $i = 1, 2, \ldots, n$, and n is a number of circuit parameters.

If two or more parameters vary in a correlated way, we define a joint probability density function (see Appendix 3). Figure 10.3 presents examples of level lines of joint pdf values of two variables. The centers of the pdf values are the means $\mathbf{p}^0 = [\,p_1^0,\, p_2^0\,]$ or nominal values.

In many applications the circuit parameters are statistically independent and joint pdf is the product of the individual pdf values.

The pdf depends on the nominal values $\mathbf{p}^0$ of circuit parameters. By changing the nominal values $\mathbf{p}^0$ of circuit parameters we alter the location of the mean in the parameter space, as well as level lines by the same amount. The shape of the level lines is not assumed to depend on $\mathbf{p}^0$. The yield of the design at a given nominal point $\mathbf{p}^0$ with a given probability density function $h(\mathbf{p}, \mathbf{p}^0)$ and acceptable (feasible) region, R_A, is given as (9.15):

$$Y(\mathbf{p}^0) = \int_{R_T \cap R_A} h(\mathbf{p}, \mathbf{p}^0)\, d\mathbf{p} \tag{10.5}$$

The yield is the probability that the circuit will work; that is, it will satisfy the design specifications given by (10.2).

The yield maximization problem is to adjust the nominal point, $\mathbf{p}^0$, to maximize the yield; that is,

$$\max_{\mathbf{p}^0} Y(\mathbf{p}^0) \tag{10.6}$$

The effect of the location of the nominal point $\mathbf{p}^0$ on the yield is illustrated in

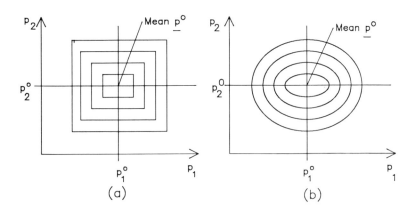

(a) (b)

Figure 10.3 Level lines of joint probability density functions (pdf) of two variables: (a) uniform; (b) Gaussian.

Figure 10.1. The primary difficulty with (10.5) and (10.6) is the evaluation of the yield: the objective function of the optimization. One difficulty is because the region R_A is not known. The acceptable region is not presented as in Figure 10.1, but always is known only implicitly through the solution of the circuit equations and the circuit response functions. The other difficulty is the requirement for the evaluation of the multidimensional integral given by (10.5). A method commonly used for computing integrals in large dimension is the Monte Carlo technique. If $h(\mathbf{p}, \mathbf{p}^0)$ is the joined pdf of circuit parameters, then the yield $Y(\mathbf{p}^0)$ defined by (10.5) is the expected value of the real variable $z(\mathbf{p})$

$$ Y = \iint_{R_T} z(\mathbf{p}) h(\mathbf{p}, \mathbf{p}^0) \, d\mathbf{p} \tag{10.7} $$

where

$$ z(\mathbf{p}) = \begin{cases} 1, & \text{if } \mathbf{p} \in R_A \text{ (the circuit with parameter values } \mathbf{p} \text{ meets} \\ & \text{all design specifications)} \\ 0, & \text{otherwise (the circuit does not meet design} \\ & \text{specifications).} \end{cases} \tag{10.8} $$

In (10.7) the integration is taken over the entire n-dimensional tolerance region.

The simplest yield estimation technique is the Monte Carlo method based on the *Strong law of large numbers* (see Appendix 3). The Monte Carlo evaluation of the yield at a particular nominal value of $\mathbf{p}^0$ consists of generating N random samples $\mathbf{p}^{(1)}, \mathbf{p}^{(2)}, \ldots, \mathbf{p}^{(N)}$ according to the given pdf, $h(\mathbf{p}, \mathbf{p}^0)$, analyzing the circuit for each trial set of parameters $\mathbf{p}^{(i)}$ and determining $z(\mathbf{p}^{(i)})$ for each $\mathbf{p}^{(i)}$. The estimate $\hat{Y}$ of Y is given as

$$ \hat{Y} = \frac{1}{N} \sum_{i=1}^{N} z(\mathbf{p}^{(i)}) $$

$$ = \frac{1}{N} \sum_{i=1}^{N} z_i = \frac{\text{Number of acceptable circuits}}{\text{Total number of circuits}} \tag{10.9} $$

10.2 DETERMINISTIC APPROACH TO TOLERANCE DESIGN

An initial approach to design centering (calculation of yield and yield maximization) involves forming a "*simplicial*" approximation of the acceptable region defined in (10.3). Several methods have been proposed for approximating R_A [2, 12, 13].

Director and Hachtel [2] have devised a simplicial approximation method based on the proposal to approximate the acceptable region R_A by a piecewise region

$\hat{R}_A \subset R_A$. Region R_A is assumed to be convex. The method begins by determining points $\mathbf{p}^{(k)}$ that lie on the boundary of R_A given by $\Omega_A = \{\mathbf{p} \mid \mathbf{g}(\mathbf{p}) = \mathbf{0}\}$. To find these points we start with a design of the circuit that satisfies all performance specifications. This circuit corresponds to the point lying inside R_A. Points belonging to Ω_A are sought by changing a value of one parameter while keeping all others constant. Points lying on the boundary Ω_A of R_A are then searched for along all axes of the n-dimensional parameter space. The convex hull of these points forms a polyhedron $\hat{R}_A$. If R_A is convex, then $\hat{R}_A$ is contained in R_A, $\hat{R}_A \subseteq R_A$. Figure 10.4 presents such approximation for a two-dimensional problem. The first approximation to the region of acceptability is a triangle. In the n-dimensional case, the boundaries of $\hat{R}_A$ are portions of hyperplanes intersecting each other and forming a polyhedron. Each hyperplane Π_i, being a wall of $\hat{R}_A$, can be defined by the outward direction of the plane normal, $\boldsymbol{\eta}_i$, and a constant b_i:

$$\Pi_i = \left\{ \mathbf{p} \mid \boldsymbol{\eta}_i^T \mathbf{p} = b_i \right\} \tag{10.10}$$

Using (10.10) we can define the region $\hat{R}_A$ as

$$\hat{R}_A = \{\mathbf{p} \mid \mathbf{C}\mathbf{p} \leq \mathbf{b}\} \tag{10.11}$$

where

$$\mathbf{C} = \left[\boldsymbol{\eta}_1, \boldsymbol{\eta}_2, \ldots, \boldsymbol{\eta}_\nu \right] \tag{10.12}$$

$$\mathbf{b} = \left[b_1, b_2, \ldots, b_\nu \right] \tag{10.13}$$

In (10.12) and (10.13) ν is a number of hyperplanes creating walls of the region

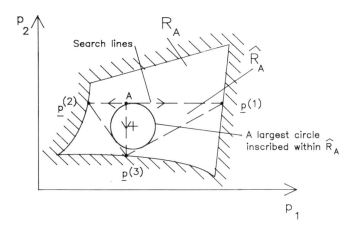

Figure 10.4 Piecewise approximation $\hat{R}_A$ of the convex of the acceptable region R_A.

$\hat{R}_A$. Algorithms are available for computing the hyperplane normals η_i and the constants b_i, $i = 1, 2, \ldots, \nu$ for a given set of points $\{\mathbf{p}^{(k)}\}$ [14].

The approximation $\hat{R}_A$ of R_A, given by (10.11), (10.12), and (10.13), is useful and convenient in computing the yield $Y(\mathbf{p}^0)$ of the design. The yield $Y(\mathbf{p}^0)$ can be computed by evaluating an approximation to $z(\mathbf{p})$ defined by (10.8):

$$\hat{z}(\mathbf{p}) = \begin{cases} 1, & \text{if } \mathbf{p} \in \hat{R}_A \\ 0, & \text{otherwise} \end{cases} \tag{10.14}$$

and applying (10.9). The test that the given sample point $\mathbf{p}$ belongs to $\hat{R}_A$, $\mathbf{p} \in \hat{R}_A$, is simple. We have to check only whether the given trial point $\mathbf{p}$ satisfies the inequality $\mathbf{C} \cdot \mathbf{p} \leq \mathbf{b}$ (relation (10.11)). Using (10.14) and (10.9) we are able to evaluate the yield estimation by generating Monte Carlo sample points with the known probability distribution without having to perform N circuit analyses. We should mention here that $\hat{z}(\mathbf{p}) \leq z(\mathbf{p})$ as $\hat{R}_A \subseteq R_A$. Consequently, the integral (10.7) with $\hat{z}(\mathbf{p})$ (10.14) instead of $z(\mathbf{p})$ (10.8) is the lower bound for the yield $Y(\mathbf{p}^0)$. The main computational effort of the method is related to a number of circuit analyses involved in forming a good simplicial approximation of the acceptable region R_A. Figure 10.4 presents an initial approximation $\hat{R}_A$ of a two-dimensional region of acceptability. The approximation may be improved by using the following algorithm. First, the largest possible circle is embedded into the polygon (see Figure 10.5). Next, the largest side of the polygon that touches the circle is selected. We now want to locate a new direction and base point

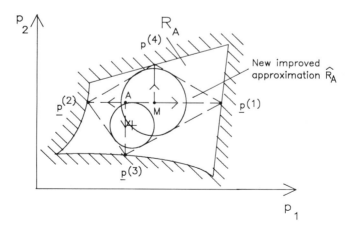

Figure 10.5 Improvement of the piecewise approximation by expanding $\hat{R}_A$: a new point ($\mathbf{p}^4$) is found and included in $\hat{R}$.

from which to search. Starting from the midpoint of the largest side of the polygon, perpendicular to the side and directed away from the interior of $\hat{R}_A$, we locate a point on the boundary of R_A. This new point is used to expand and improve the approximation R_A. The largest possible circle is inscribed within the new polygon. This process can be repeated many times. The approximation is assumed to be sufficient, when the difference of locations of two consecutive largest circles inscribed into the approximations is smaller than a given small number. For tolerance problems with many circuit parameters, the approximation $\hat{R}_A$ of R_A is a polyhedron. The improvement of the approximation is based on the expansion of the polyhedron about the side of greatest area tangential to the largest hypersphere embedded into the approximation [2, 15, 16].

A deterministic approach to design centering requires computationally expensive circuit simulations in generating the approximation to the acceptable region, R_A. As the number of the toleranced parameters increases the computational effort and the amount of information required to maintain a good simplicial approximation to the region of acceptability tend to grow exponentially. Thus, as a practical method, simplicial approximation should be applied to circuits having no more than five to eight toleranced parameters [17].

In addition to these problems, this method requires that the region of acceptability R_A be a convex body. There is a danger of significant error in the method if the feasible region R_A is not convex. This restriction may be relaxed at the expense of greater complexity to the algorithm and greater computational cost [3, 4].

10.3 STATISTICAL APPROACH TO TOLERANCE DESIGN

To avoid the great computational cost associated with design centering for many toleranced parameters ($n \geq 8$), a statistical exploration approach may be used to tolerance design [6–11]. In the statistical methods, the basic problem of yield optimization, yield estimation, is dealt with by using Monte Carlo techniques. In Monte Carlo methods, the actual circuit manufacturing process is simulated by drawing sample nominal points (nominal values of circuit parameters) from the parameter pdf and then evaluating, by means of a circuit analysis procedure, the performance of each sample circuit. The circuit simulations are used to find how many circuits pass the specifications. This information is used to find an estimate of the circuit yield. The most important point of such Monte Carlo analysis is that the number of circuit samples necessary for a given accuracy (estimated) does not depend on the number of designable parameters. However, the accuracy of the estimate depends on the number of samples used in the Monte Carlo analysis. Typically, between 100 and 400 circuit simulations are performed.

For a small number of toleranced parameters, a statistical exploration approach is inefficient compared to design centering. However, as the number of designed parameters increases, Monte Carlo methods become superior.

10.3.1 The Gravity Method

Based on a Monte Carlo analysis of the designed circuit, the centers of gravity of the failed and passed sample points in the parameter space are respectively determined as

$$\mathbf{CF} = \frac{\sum_{k \in J} \mathbf{p}^{(k)}}{N_f} \qquad (10.15)$$

$$\mathbf{CP} = \frac{\sum_{k \notin J} \mathbf{p}^{(k)}}{N_p} \qquad (10.16)$$

where J is the index set identifying the failed sample points, and N_f and N_p are the numbers of failed and passed sample circuits, respectively. The centers of gravity of passed (**CP**) and failed (**CF**) sample circuits are illustrated in Figure 10.6. The gravity center of passed circuits is located in the overlap ($R_T \cap R_A$) between the tolerance region, R_T, and the acceptable region, R_A. The gravity center of failed circuits (**CF**) is located in the remainder of R_T. We can show that, as the number of sample points tends to infinity, the centers of gravity tend to the actual centers of gravity of the two regions when all sample points are weighted according to the pdf of the circuit parameters. We also can show that the nominal and true centers of gravity are collinear, independent of the pdf values of the circuit parameters [18].

A better design is found by moving the nominal point, $\mathbf{p}^0$, along the direction $\mathbf{s} = \mathbf{CF} - \mathbf{CP}$. This algorithm is simple but also heuristic. How the gravity centers are related to the yield in the general multidimensional problems is not clear. Consequently, how far the move of the nominal point $\mathbf{p}^0$ along $\mathbf{s}$ should be is not obvious. The authors of the method, R.S. Soin and R. Spence, suggest that the length of the move should be identical to the separation of the centers of gravity [19]. There are other suggestions concerning the move length. If the new circuit design can be expressed as

$$\mathbf{p}^0_{new} = \mathbf{p}^0_{old} + \lambda(\mathbf{CP} - \mathbf{CF}) \qquad (10.17)$$

then the optional choices of the move length may be expressed as $\lambda = 1$ [19], $\mathbf{p}^0_{new} = \mathbf{CP}$ [19], or $\lambda = (1 - \hat{Y})$ [20].

10.3.2 The Parametric Sampling Method

The results of Monte Carlo analysis used to estimate the yield of the design may also be employed to estimate the gradients of the yield with respect to nominal values and tolerances of the designable parameters. The estimation of yield gradients does not require any additional circuit analysis. A family of tolerance design methods uses the information on the yield gradients [21–24].

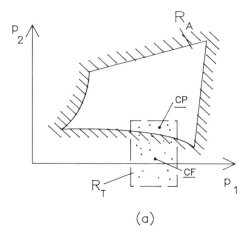

(a)

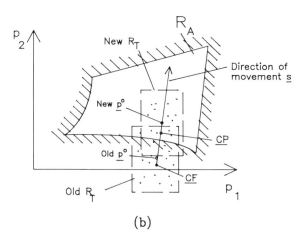

(b)

Figure 10.6 (a) The centers of passed (CP) and failed (CF) sample points; (b) the direction of movement of a nominal point leading to increased yield.

Parametric Estimators

The yield of circuit design defines the integral (10.7):

$$Y(\mathbf{p}^0) = \int_{R_T} \cdots \int z(\mathbf{p}) h(\mathbf{p}, \mathbf{p}^0) \, d\mathbf{p} \qquad (10.18)$$

where $\mathbf{p}^0 = [p_1^0, p_2^0, p_3^0, \ldots, p_n^0]^T$ is a vector of nominal values of circuit parameters. Differentiating (10.18) with respect to p_k^0, we have

$$\frac{\partial Y(\mathbf{p}^0)}{\partial p_k^0} = \int_{RT} \cdots \int z(\mathbf{p}) \frac{\partial h(\mathbf{p}, \mathbf{p}^0)}{\partial p_k^0} \, d\mathbf{p} \qquad (10.19)$$

Assuming that the pdf, $h(\mathbf{p}, \mathbf{p}^0)$, is differentiable and nonzero in the region of acceptability, (10.19) may be rewritten in the form:

$$\frac{\partial Y(\mathbf{p}^0)}{\partial p_k^0} = \int_{RT} \cdots \int \frac{z(\mathbf{p})}{h(\mathbf{p}, \mathbf{p}^0)} \frac{\partial h(\mathbf{p}, \mathbf{p}^0)}{\partial p_k^0} h(\mathbf{p}, \mathbf{p}^0) \, d\mathbf{p} \qquad (10.20)$$

By using the Monte Carlo technique, we generate N sample circuits with parameter values $\mathbf{p}^{(1)}, \mathbf{p}^{(2)}, \ldots, \mathbf{p}^{(N)}$ randomly selected according to the assumed pdf, $h(\mathbf{p}, \mathbf{p}^0)$. The yield of a design may be estimated by using relation (10.9):

$$\hat{Y} = \frac{1}{N} \sum_{i=1}^{N} z(\mathbf{p}^{(i)}) \qquad (10.21)$$

where $z(\mathbf{p}^{(i)})$ is the testing function defined in (10.8) and evaluated from the results of the Monte Carlo analysis.

By comparing (10.18) and (10.21), we note that the integral (10.20) can be estimated as

$$\widehat{\frac{\partial Y}{\partial p_k^0}} = \frac{1}{N} \sum_{i=1}^{N} \frac{z(\mathbf{p}^{(i)})}{h(\mathbf{p}^{(i)}, \mathbf{p}^0)} \frac{\partial h(\mathbf{p}^{(i)}, \mathbf{p}^0)}{\partial p_k^0} \qquad (10.22)$$

As is obvious from the preceding relation, the results of the Monte Carlo analysis of the designed circuit, which are used for computing yield estimation $\hat{Y}$, may be applied for computing yield sensitivity estimation $\partial Y / \partial p_k^0$. No additional Monte Carlo analysis is required. Higher-order sensitivities of $\hat{Y}$ may be evaluated with very similar formulas but still using the results of the same Monte Carlo analysis as for yield evaluation:

$$\widehat{\frac{\partial^2 Y}{\partial p_k^0 \partial p_l^0}} = \frac{1}{N} \sum_{i=1}^{N} \frac{z(\mathbf{p}^{(i)})}{h(\mathbf{p}^{(i)}, \mathbf{p}^0)} \frac{\partial^2 h(\mathbf{p}^{(i)}, \mathbf{p}^0)}{\partial p_k^0 \partial p_l^0} \qquad (10.23)$$

The yield and its sensitivities are estimated via Monte Carlo analysis and used to find a new trial circuit design. Thanks to the knowledge of the yield gradients, gradient-based optimization procedures may be used in tolerance design [22]. At each iteration of an optimization procedure (for each new set of circuit variables $\mathbf{p}^0$), the yield and its gradients have to be evaluated. Normally, for each new set of variables $\mathbf{p}^0$ we have a new pdf, $h(\mathbf{p}, \mathbf{p}^0)$; and therefore, the sampling and circuit analysis have to be repeated.

The method of parametric sampling is based on the concept of importance sampling according to which the samples $p_k^{(i)}$ are generated from pdf, $h_s(\mathbf{p})$, which is different from the designable parameter pdf, $h(\mathbf{p}, \mathbf{p}^0)$. The integral (10.18) defining the yield of the design is now

$$Y(\mathbf{x}) = \int_{R_S} \cdots \int \left\{ z(\mathbf{p}) \frac{h(\mathbf{p}, \mathbf{x})}{h_s(\mathbf{p})} \right\} h_s(\mathbf{p}) \, d\mathbf{p} \qquad (10.24)$$

where $h_s(\mathbf{p})$ is called the *sampling density function*. The pdf of circuit parameters is denoted by $h(\mathbf{p}, \mathbf{x})$, which depends on the vector $\mathbf{x}$ denoting nominal values, tolerances, and correlations of circuit parameters $\mathbf{p}$.

Differentiation of (10.24) with respect to x_i, a component of $\mathbf{x}$, gives the yield sensitivity:

$$\frac{\partial Y(\mathbf{x})}{\partial x_k} = \int_{R_S} \cdots \int \left\{ \frac{z(\mathbf{p})}{h_s(\mathbf{p})} \frac{\partial h(\mathbf{p}, \mathbf{x})}{\partial x_k} \right\} h_s(\mathbf{p}) \, d\mathbf{p} \qquad (10.25)$$

An estimator of the yield is given as

$$\hat{Y} = \frac{1}{N} \sum_{i=1}^{N} z(\mathbf{p}^{(i)}) \frac{h(\mathbf{p}^{(i)}, \mathbf{x})}{h_s(\mathbf{p}^{(i)})} = \frac{1}{N} \sum_{i=1}^{N} z(\mathbf{p}^{(i)}) W(\mathbf{p}^{(i)}, \mathbf{x}) \qquad (10.26)$$

Similarly, an estimator of the yield sensitivity is given as

$$\widehat{\frac{\partial Y}{\partial x_k}} = \frac{1}{N} \sum_{i=1}^{N} \frac{z(\mathbf{p}^{(i)})}{h_s(\mathbf{p}^{(i)})} \frac{\partial h(\mathbf{p}^{(i)}, \mathbf{x})}{\partial x_k}$$

$$= \frac{1}{N} \sum_{i=1}^{N} z(\mathbf{p}^{(i)}) \frac{\partial W(\mathbf{p}^{(i)}, \mathbf{x})}{\partial x_k} \qquad (10.27)$$

The ratio of the pdf values, $h(\mathbf{p}, \mathbf{x})/h_s(\mathbf{p})$, denoted by $W(\mathbf{p}, \mathbf{x})$ is the weight of the point $\mathbf{p}$, compensating the use of the sampling pdf, which is different from the parameter pdf.

The parametric sampling approach has two important features. First, once the testing function $z(\mathbf{p}^{(i)})$ is calculated using Monte Carlo analysis with samples generated according to the sampling pdf, $h_s(\mathbf{p})$, no more circuit simulations are required when $\mathbf{x}$ is changed. Furthermore, if the parameter pdf, $h(\mathbf{p}, \mathbf{x})$, is a differentiable function, then sensitivities of the estimated yield are readily available without any excessive computing cost. Hence, powerful gradient optimization methods may be used for yield maximization. In practice, the procedure starts with the large number of base points generated according to the sampling pdf, $h_s(\mathbf{p})$ and located in a large region of exploration, R_s. This is shown in Figure 10.7. The results of the analysis of the sample circuits create an initial data bank. In the next step, an initial design $\mathbf{p}^{0(1)}$ with initial tolerance region $R_T^{(1)}$ is located within R_s. The results of the analysis of sample circuits corresponding to samples located within $R_T^{(1)}$ are treated as conventional Monte Carlo data. They are used for computations of the yield and yield gradient estimates. Using an optimization algorithm, a new nominal point and tolerances are then found and located in R_s as $\mathbf{p}^{0(2)}$ and $R_T^{(2)}$. This iterative process is repeated until a local maximum of the yield is achieved. Also, a new base point located in a new region of exploration R_s may be required to achieve optimal solution of a given tolerance design problem. More detailed

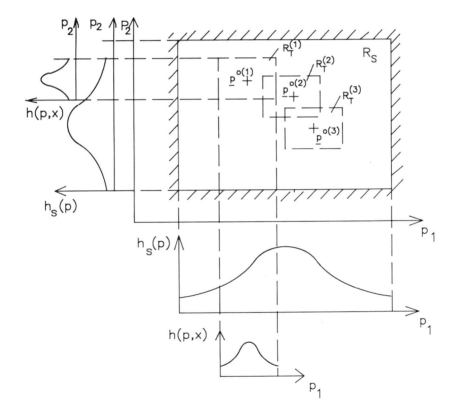

Figure 10.7 Parametric sampling method (R_s = region of exploration, R_{T1} = tolerance region of the initial trial design, R_{T3} = tolerance region of the final, optimal design).

discussions on the method and its practical implementation may be found in [17, 21, 23].

10.4 WORST-CASE DESIGN

As we know, a goal of the worst-case design is a 100% yield solution. A worst-case design procedure leads to establishing a set of nominal circuit parameters such that all manufactured circuit samples pass all design specifications. One of many approaches to the worst-case design is the cut method [25], which is applicable for convex and nonconvex problems. The cut method guarantees convergence when a solution exists. It is the iterative method. Each iteration starts with selecting a nominal design and initiating a Monte Carlo analysis. After detecting, during the Monte Carlo analysis, a violation of design specifications for the actual nominal design point, the analysis can be halted because the 100% yield goal will not be reached. The result of such exploration of parameter space within the orthotope of the tolerance region R_T indicates that the worst-case design cannot lie inside the orthotope having the same shape as R_T and centered at the fail point $\mathbf{p}_F$. (Here we assume fixed absolute tolerances of the design parameters.) This orthotope is called a *cut region* because it can be removed from further consideration. In the next iteration the new nominal point generated by the algorithm must be located outside the actual cut region defined as the union of all orthotopes surrounding points obtained outside the acceptability region during the nominal point search or during the Monte Carlo analysis. The essence of the cut method is the sequential connection of cuts that define a region of impossible nominal worst-case designs. At the same time the number of pass circuit samples generated before the single fail may be used as indication of where to locate a new nominal design.

If at the ith iteration, the jth sample point $\mathbf{p}^{(i,\,j)}$ does not satisfy design specifications, the Monte Carlo analysis is stopped and the cut region $C^{(i)}$ is expanded with

$$C^{(i)} = \{\mathbf{p} \mid \mathbf{p} = \mathbf{p}^{(i,\,j)} + \mathbf{E}\boldsymbol{\mu}\} \tag{10.28}$$

where

$$\mathbf{E} = \begin{bmatrix} \epsilon_1 & & & \\ & \epsilon_2 & & 0 \\ & 0 & \ddots & \\ & & & \epsilon_n \end{bmatrix} \tag{10.29}$$

$$\boldsymbol{\mu} = [\mu_1, \mu_2, \ldots, \mu_n] \tag{10.30}$$

In (10.30) ϵ_k, $k = 1, 2, \ldots, n$ are absolute tolerances of the circuit parameters and μ_k, $k = 1, 2, \ldots, n$ are randomly generated numbers in the range $\{-1, 1\}$.

The cut region $C^{(i)}$ is an n-dimensional orthotope, centered at $\mathbf{p}^{(i,\,j)}$ with sides of length $2\epsilon_k$, $k = 1, 2, \ldots, n$.

The new cut region is the union of $C^{(i-1)}$ and $C^{(i)}$:

$$C^{(i+1)} = C^{(i)} \cup C^{(i-1)} \qquad (10.31)$$

Figure 10.8 presents incrementation of the cut region following the discovery of a fail point in the current tolerance region being explored.

The search of the new nominal design point is begun in the design region but outside the cut region $C^{(i+1)}$. The method of selecting a new nominal design may be described as follows. As a starting point we select the best of the previous nominal designs, having the largest number of pass network samples that were generated before the fail sample was discovered. Next a random vector of proper length is added to the previous best nominal design. This new point is the new trial nominal design.

The application of the cut method requires determining the maximum number of sample networks to be tested within a tolerance region of a single trial design. This number is usually between 300 and 400. If all trial circuits satisfy design specifications, we assume that a satisfactory worst-case design has been achieved.

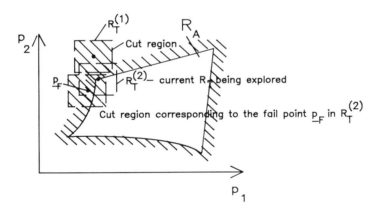

Figure 10.8 The cut method and its incrementation.

REFERENCES

[1] J.F. Pinel, "Tolerance Assignment and Network Alignment of Linear Networks in the Frequency Domain," *Computer Aided Network Design*, IEEE Short Course, Course Notes, 73-SC-06.

[2] S.W. Director and G.D. Hachtel, "The Simplicial Approach to Design Centering," *IEEE Trans. Circuits and Systems*, Vol. CAS-24, July 1977, pp. 363–372.

[3] L.M. Vidigal and S.W. Director, "A Design Centering Algorithm for Nonconvex Regions of Acceptability," *IEEE Trans. CAD of Int. Circuits and Systems*, Vol. CAD-1, No. 1, January 1982, pp. 13–24.

[4] R.K. Brayton, S.W. Director, G.D. Hachtel, and L.M. Vidigal, "A New Algorithm for Statistical Circuit Design Based on Quasi-Newton Methods and Function Splitting," *IEEE Trans. Circuits and Systems*, Vol. CAS-26, September 1979, pp. 784–794.

[5]   E. Polak and A. Sangiovanni-Vincentelli, "Theoretical and Computational Aspects of the Optimal Design Centering, Tolerancing, and Tuning Problem," *IEEE Trans. Circuits and Systems*, Vol. CAS-26, September 1979, pp. 795-813.

[6]   R.S. Soin and R. Spence, "Statistical Exploration Approach to Design Centering," *Proc. IEE*, Vol. 127, Part G, No. 6, 1980, pp. 260-262.

[7]   K.S. Tahim and R. Spence, "A Radial Exploration Approach to Manufacturing Yield Estimation and Design Centering," *IEEE Trans. Circuits and Systems*, Vol. CAS-26, September 1979, pp. 768-774.

[8]   K.J. Antreich and R. K. Koblitz, "Design Centering by Yield Prediction," *IEEE Trans. Circuits and Systems*, Vol. CAS-29, February 1982, pp. 88-95.

[9]   K. Singhal and J.F. Pinel, "Statistical Design Centering and Tolerancing Using Parametric Sampling," *IEEE Trans. Circuits and Systems*, Vol. CAS-28, July 1981, pp. 692-702.

[10] P. Baloban and J.L. Golembeski, "Statistical Analysis for Practical Circuit Design," *IEEE Trans. Circuits and Systems*, Vol. CAS-22, February 1975, pp. 100-108.

[11] J.P. Spoto, W.T. Coston, and C.P. Hernandez, "Statistical Integrated Circuit Design and Characterization," *IEEE Trans. CAD of Int. Circuits and Systems*, Vol. CAD-5, January 1986, pp. 90-103.

[12] H.L. Abdel-Malek and J.W. Bandler, "Yield Optimization for Arbitrary Statistical Distributions, Part I; Theory," *IEEE Trans. Circuits and Systems*, Vol. CAS-27, 1980, pp. 245-253.

[13] R.M. Biernacki and M.A. Styblinski, "Statistical Circuit Design with a Dynamic Constrained Approximation Scheme," *Proc. IEEE Int. Symp. Circuits and Systems*, San Jose, CA, 1986, pp. 976-979.

[14] D.R. Chand and S.S. Kapur, "An Algorithm for Convex *Polyhtopes*," *J. of ACM*, Vol. 17, No 7, 1970, pp. 78-86.

[15] S.W. Director and G.D. Hachtel, "The Simplicial Approximation Approach to Design Centering and Tolerance Assignment," *Proc. IEEE Int. Symp. Circuits and Systems*, Munich, 1976, p. 706.

[16] S.W. Director, G.D. Hachtel, and L.M. Vidigal, "Computationally Efficient Yield Estimation Procedures Based on Simplicial Approximation," *IEEE Trans. Circuit and Systems*, Vol. CAS-25 March 1978, pp. 121-130.

[17] R. Spence and R. G. Soin, *Tolerance Design of Electronic Circuits*, Addison-Wesley, London 1988.

[18] K.J. Antreich and R.K. Koblitz, "A New Approach to Design Centering Based on a Multiparameter Yield Prediction Formula," *Proc. IEEE Int. Symp. Circuits and Systems*, Houston, 1980, pp 270-277.

[19] R.S. Soin and R. Spence, "Statistical Exploration Approach to Design Centering," *Proc. IEE*, Vol 127, Part G, 1980, pp. 260-269.

[20] I.R. Ibbotson, E. Compton, and D. Boardman, "Improved Statistical Design Centering for Electrical Networks," *Electronics Letters*, Vol. 20, pp. 757-758.

[21] K. Singhal and J.F. Pinel, "Statistical Design Centering and Tolerancing Using Parametric Sampling," *IEEE Trans. Circuits and Systems*, Vol. CAS-28, July 1981, pp. 692-702.

[22] B.V. Batalov, Y.N. Belyakov, and F.A. Karmaev, "Some Methods for Statistical Optimization Integrated Microcircuits with Statistical Relations among the Parameters of the Components," *Soviet Microelectronics* (U.S.), Vol. 7, pp. 228-238.

[23] J.F. Pinel and K. Singhal, "Efficient Monte Carlo Computation of Circuit Yield Using Importance Sampling," *Proc. IEEE Int. Symp. Circuits and Systems*, Phoenix, April 1977, pp. 575-578.

[24] P.W. Becker and F. Jensen, *Design of Systems and Circuits for Maximum Reliability & Maximum Production Yield*, McGraw-Hill, New York, 1977.

[25] E. Wehrhahn, "A Cut Algorithm for Circuit Centering," *Proc. IEEE Int. Symp. Circuits and Systems*, Montreal, 1984, pp. 970-973.

Chapter 11
Optimization Techniques for Microwave
Circuit Design

A very important class of problems in the CAD of microwave circuits concerns techniques for adjusting circuit parameters to minimize the deviation between the circuit performance achieved at some stage of the design and the desired specifications. For example, the design of a microwave transistor amplifier can have as performance functions gain, bandwidth, noise figure, and perhaps input and output impedance as well. Such techniques are used to refine iteratively an initial design, usually conceived by intuitive methods, until specifications are met. This approach yields results otherwise unobtainable and is particularly attractive economically in situations involving excluding specifications. The refinement process is an optimization process involving the minimization of a performance criterion that measures the difference between existing and desired response characteristics of the designed circuit. The following sections will cover some of the basic concepts of optimization theory, including descriptions of several algorithms and the discussion of some practical aspects of their implementation and use.

Emphasis will be put on the methods suitable for use in the design of linear microwave circuits in the frequency domain, and some examples will be given from this area. This does not mean that application of these methods is restricted to the linear microwave circuit design or that the particular techniques discussed are the only ones suitable for work in this field. Only a very small fraction of the optimization theory will be considered in this chapter. For a broader view of this subject, the reader is referred to the literature.

1.1 BASIC CONCEPTS AND DEFINITIONS

1.1.1 Definition of the Optimization Problem

Optimization in general means determining the extreme value (maximum or, more often, minimum) of a mathematical function. The optimization problem is to minimize the scalar function [1–5]:

$$U = U(\mathbf{x}) \tag{11.1}$$

subject to the inequality constraints:

$$\mathbf{c}(\mathbf{x}) \leq \mathbf{0} \tag{11.2}$$

and equality constraints:

$$\mathbf{h}(\mathbf{x}) = \mathbf{0} \tag{11.3}$$

Here $\mathbf{c}(\mathbf{x})$ and $\mathbf{h}(\mathbf{x})$ are, in general, nonlinear vector-valued functions of $\mathbf{x}$; for example,

$$\mathbf{c}(\mathbf{x}) = \begin{bmatrix} c_1(\mathbf{x}) \\ c_2(\mathbf{x}) \\ \vdots \\ c_{m_c}(\mathbf{x}) \end{bmatrix} \tag{11.4}$$

$$\mathbf{h}(\mathbf{x}) = \begin{bmatrix} h_1(\mathbf{x}) \\ h_2(\mathbf{x}) \\ \vdots \\ h_{m_h}(\mathbf{x}) \end{bmatrix} \tag{11.5}$$

In (11.1) to (11.5), $\mathbf{x}$ is a vector of n independent variables or parameters:

$$\mathbf{x} = \begin{bmatrix} x_1 \\ x_2 \\ \vdots \\ x_n \end{bmatrix} \tag{11.6}$$

Thus, the vector $\mathbf{x}$ is an element of an n-dimensional parameter space $\mathbf{R}^n$.

An acceptable region R_A (or a feasible, a constrained region) is given by

$$R_A = \{\mathbf{x} \,|\, \mathbf{c}(\mathbf{x}) \leq \mathbf{0}, \mathbf{h}(\mathbf{x}) = \mathbf{0}\} \tag{11.7}$$

Any point belonging to R_A is said to be feasible.

The objective function, which also can be called a *cost function* or an *error criterion*, generates a response *hypersurface*.

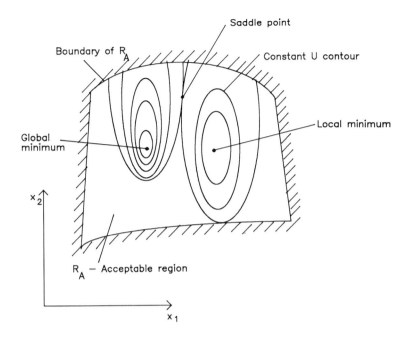

igure 11.1 Example of a two-dimensional optimization problem.

Figure 11.1 illustrates some features encountered in optimization problems. It is a plot of constant U contours for the case of two-dimensional $\mathbf{x}$ vector. The goal of optimization is to find location of $\mathbf{x}^*$ in R_A such that

$$U(\mathbf{x}^*) \leq U(\mathbf{x}), \quad \text{for all } \mathbf{x} \in R_A \tag{11.8}$$

This is a global minimum of $U(\mathbf{x})$. A local minimum of $U(\mathbf{x})$ occurs at a point $\mathbf{x}^*$ n R_A with respect to a suitably defined neighborhood $N(\mathbf{x})$ if

$$U(\mathbf{x}^*) \leq U(\mathbf{x}), \quad \text{for all } \mathbf{x} \in R_A \cap N(\mathbf{x}) \tag{11.9}$$

n the optimization problems with continuous parameters $\mathbf{x}$ the neighborhood is taken as n open hypersphere containing $\mathbf{x}^*$.

The necessary conditions satisfied by an unconstrained minimum of a continuously differentiable function U of a single variable x are

$$\frac{dU}{dx} = 0 \tag{11.10}$$

$$\frac{d^2U}{dx} \geq 0 \tag{11.11}$$

(Strong inequality (11.11) is a sufficient, but not necessary condition for local minimum of a single variable function.)

For a function in n variables (parameters) $\mathbf{x}$, the condition (11.10) becomes

$$\frac{\partial U}{\partial x_i} = 0, \quad i = 1, 2, \ldots, n \tag{11.12}$$

or in vector notation:

$$\nabla U(\mathbf{x}^*) = \mathbf{0} \tag{11.13}$$

The point $\mathbf{x}^*$ satisfying the above condition is termed a *stationary point* of U.

Condition (11.11) becomes a requirement that the matrix of second partial derivatives of U be semipositive definite:

$$\Delta\mathbf{x}^T \left[\frac{\partial^2 U(\mathbf{x}^*)}{\partial x_i \partial x_j} \right] \Delta\mathbf{x} \geq 0 \tag{11.14}$$

where $\Delta\mathbf{x}$ is any deviation vector from $\mathbf{x}^*$.

We can prove that (11.13) and (11.14) are necessary conditions for $\mathbf{x}^*$ to be a strong local minimum of the unconstrained minimization problem in n dimensions:

$$\min_{\mathbf{x} \in \mathbf{R}^n} U(\mathbf{x}) \tag{11.15}$$

The nature of the problem and the algorithm employed determines a result of a minimum search. The global minimum, a local minimum, several local minima, or no minimum can be expected. As more information is provided about U, $\mathbf{c}$, and $\mathbf{h}$ in an optimization problem, more can be said about the problem's solution and more efficient algorithms can be used. The processes of determining the existence of an optimal solution, characterizing it, and searching for it fall into the domain of mathematical programming.

Optimization problems encountered in the CAD of microwave circuits may be divided into the following classes:

linear programming: U, $\mathbf{c}$, $\mathbf{h}$ linear function of $\mathbf{x}$;

quadratic programming: U quadratic, $\mathbf{c}$, $\mathbf{h}$ linear;

nonlinear programming: at least one of U, $\mathbf{c}$, $\mathbf{h}$ is a nonlinear function of $\mathbf{x}$.

Although, in general, microwave circuit functions are not linear or quadratic in the design parameters, the techniques used for circuit optimization are iterative, and each step, the problem is approximated by a linear or quadratic function. Each step of the iteration can be done very rapidly, because linear or quadratic programs can be solved very efficiently.

Note that, in microwave circuit design problems, the dependence of the function

$U(\mathbf{x})$, $\mathbf{c}(\mathbf{x})$, and $\mathbf{h}(\mathbf{x})$ on the design parameters $\mathbf{p} = \mathbf{x}$ is implicit through the circuit equations. Thus the objective function evaluations required to run an optimization algorithm involve an often expensive simulation of the designed microwave circuit.

11.1.2 Convexity

A function is said to be convex in a region if a linear interpolation between every two points $\mathbf{x}^a$ and $\mathbf{x}^b$ on its hypersurface never underestimates the function. Thus, for $\mathbf{x}^a \neq \mathbf{x}^b$:

$$U\left[\mathbf{x}^a + \lambda(\mathbf{x}^b - \mathbf{x}^a)\right] \leq U(\mathbf{x}^a) + \lambda\left[U(\mathbf{x}^b) - U(\mathbf{x}^a)\right]$$

$$\text{for } 0 \leq \lambda \leq 1 \quad (11.16)$$

Strictly convex functions are defined in the same way, but in (11.16) strong inequalities must be used. A concave (strictly concave) function is one whose negative is convex (strictly convex). The origin of the term *convexity* can be seen from the one-dimensional examples shown in Figure 11.2. Any two points on the curve of Figure 11.2(a) and (c) can be joined by straight lines that never fall below the curve. Figure 11.3 presents other examples of convex and nonconvex functions.

Strict convexity is a sufficient condition that a local minimum be a global minimum of a function. However, as can be observed from the nonconvex function examples in Figure 11.2(c) and Figure 11.3(a), convexity is not a necessary condition for uniqueness. It is too restrictive.

The convexity idea also is applicable to regions. A region (set of points or a domain) is said to be convex if, for any two points $\mathbf{x}^a$ and $\mathbf{x}^b$ in the region, $\mathbf{x}^a \neq \mathbf{x}^b$, and all points defined as

$$\mathbf{x} = \mathbf{x}^a + \lambda(\mathbf{x}^a - \mathbf{x}^b), \quad \text{for all } 0 \leq \lambda \leq 1 \quad (11.17)$$

lie in the region. In accordance with this definition a region R is convex if

$$\mathbf{x}^a, \mathbf{x}^b, \mathbf{x}^a + \lambda(\mathbf{x}^a - \mathbf{x}^b) \in R \quad \text{for all } 0 \leq \lambda \leq 1 \quad (11.18)$$

Figure 11.4 shows examples of convex and nonconvex regions.

11.1.3 Constraints

Conditions or limitations that may be imposed on the minimization problem take the general forms given by (11.2) and (11.3). For example, they can represent bounds on parameter values or performance specifications. Inequality constraints may take the

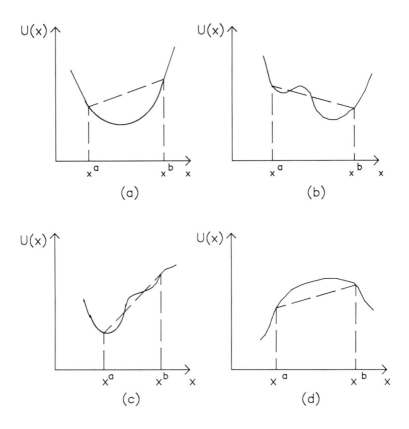

Figure 11.2 Illustration of one-dimensional convex and concave functions: (a) strictly convex; (b) nonconvex; (c) nonconvex; (d) strictly concave.

following particular forms:

$$c_i(\mathbf{x}) \le c_{ui} \tag{11.19}$$

$$c_i(\mathbf{x}) \ge c_{li} \tag{11.20}$$

$$c_{li} \le c_i(\mathbf{x}) \le c_{ui} \tag{11.21}$$

where c_{li} and c_{ui} are constants. Note that the constraint in the form (11.21) can easily be broken into two, each of the forms given by (11.19) and (11.20). Further, multiplication of both sides of (11.21) by minus one converts it into the form of (11.20). Then it is obvious that any inequality constraint, by proper transformations, can be put into the general form given by (11.2).

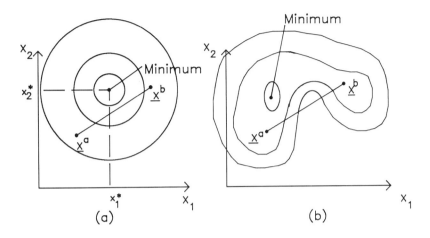

Figure 11.3 Illustration of two-dimensional (a) convex and (b) nonconvex functions.

As mentioned earlier, equality and inequality constraints can be, in general, linear or nonlinear functions of the decision variables $\mathbf{x}$. A form of a constraint encountered very often in the optimization problems of microwave circuits is a linear function:

$$l(\mathbf{x}) = \mathbf{a}^T\mathbf{x} + b \tag{11.22}$$

where $\mathbf{a}^T$ is the constant row vector, and b is a scalar. Thanks to the linearity of this relation vector $\mathbf{a}$ is the (constant) gradient of $l(\mathbf{x})$.

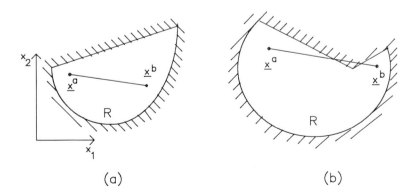

Figure 11.4 Examples of (a) convex and (b) nonconvex regions.

As in the general case of constraints, linear constraints can be divided into two types:
equality constraints

$$\mathbf{a}^T\mathbf{x} + b = 0 \qquad (11.23)$$

and inequality constraints

$$\mathbf{a}^T\mathbf{x} + b \leq 0 \qquad (11.24)$$

A particularly simple form of linear constraint occurs when $l(\mathbf{x})$ involves only one variable. In such case the three possible constraint forms are

$$x_i = b_i \quad (x_i \text{ is fixed at } b_i) \qquad (11.25)$$

$$x_i \geq b_i \quad (b_i \text{ is a lower bound for } x_i) \qquad (11.26)$$

$$x_i \leq b_i \quad (b \text{ is the upper bound for } x_i) \qquad (11.27)$$

The last two forms of the linear constraints are called *simple* (or *box*) *bounds* on the variable x_i.

Parameter constraints of the form

$$x_{li} \leq x_i \leq x_{ui} \qquad (11.28)$$

can be rewritten as

$$c_1(\mathbf{x}) = x_i - x_{ui} \leq 0 \qquad (11.29)$$

$$c_2(\mathbf{x}) = x_{li} - x_i \leq 0 \qquad (11.30)$$

This is a special class of inequality constraints, which involve linear expression in $\mathbf{x}$. Algorithms that exploit the linearity of this type of constraint will be discussed in Section 11.4.

For equality constraints in the form (11.3), an explicit solution should be used, if possible, because it reduces the number of parameters. Some $h_i(\mathbf{x}) = 0$ can be written as $x_j = f(x_1, x_2, \ldots, x_{j-1}, x_{j+1}, \ldots, x_n)$. Thus, we can optimize with $n -$ parameters.

Example 11.1

This simple example illustrates the way in which the problem of an amplifier design may be transformed into an optimization problem. Let us consider a design of a narrow band amplifier. Given that its gain at some frequency f_0 is to be maximized and that a 3 dB bandwidth of $2\,\Delta f$ is required, the problem might be stated as follows. Minimize:

$$U(\mathbf{x}, \mathbf{f}) = -G(\mathbf{x}, f_0)$$

subject to

$$c_1(\mathbf{x}, \mathbf{f}) = 10 \log\big[G(\mathbf{x}, f_0) - G(\mathbf{x}, f_0 + \Delta f)\big] - 3 \le 0$$

$$c_2(\mathbf{x}, \mathbf{f}) = 10 \log\big[G(\mathbf{x}, f_0) - G(\mathbf{x}, f_0 - \Delta f)\big] - 3 \le 0$$

where

$G =$ the gain,
$\mathbf{x} =$ the set of network parameters whose values are permitted to change during the optimization,
$\mathbf{f} =$ the set of frequency points $\{f_0 - \Delta f, f_0, f_0 + \Delta f\}$.

11.1.4 Continuous Functions and Their Derivatives

The ability to use information about a function at a particular point to analyze its general behavior determines methods used for solving optimization problems. We consider now some basic properties of multivariable functions that allow such conclusions to be drawn.

Continuity is the most important among such properties. To define the continuity of a multivariable function, we first must introduce the concept of a neighborhood in n-dimensional space. A δ-neighborhood of the point $\mathbf{x}$ is understood as the set of all points $\hat{\mathbf{x}}$ such that $\|\hat{\mathbf{x}} - \mathbf{x}\| \le \delta$, where δ is a positive number. The choice of the norm is not usually significant; the Euclidean norm is the most frequently used (see Appendix 1).

The multivariable function $U(\mathbf{x})$ is continuous at $\mathbf{x}$ if, for given any $\epsilon > 0$, there is $\delta > 0$ such that if $\|\hat{\mathbf{x}} - \mathbf{x}\| \le \delta$ (i.e., $\hat{\mathbf{x}}$ is in the δ-neighborhood of $\mathbf{x}$), then $|U(\hat{\mathbf{x}}) - U(\mathbf{x})| < \epsilon$.

A vector-valued function is continuous if each of its component functions is continuous. Illustrations of continuous and discontinuous functions are given in Figure 11.5.

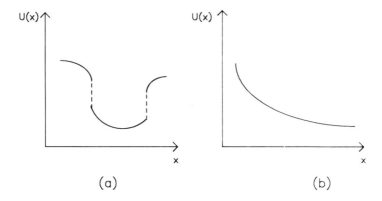

Figure 11.5 (a) Discontinuous and (b) continuous one-dimensional functions.

Let us now consider continuous scalar $U(\mathbf{x})$ and vector-valued $\mathbf{y}(\mathbf{x})$ functions with continuous first and second partial derivatives. Vectors $\mathbf{x}$ and $\mathbf{y}(\mathbf{x})$ are defined as

$$\mathbf{x} = \begin{bmatrix} x_1 \\ x_2 \\ \vdots \\ x_n \end{bmatrix} \quad \text{and} \quad \mathbf{y}(\mathbf{x}) = \begin{bmatrix} y_1(\mathbf{x}) \\ y_2(\mathbf{x}) \\ \vdots \\ y_l(\mathbf{x}) \end{bmatrix} \tag{11.31}$$

Let the n-dimensional partial derivative operator with respect to $\mathbf{x}$ be

$$\boldsymbol{\nabla} = \begin{bmatrix} \dfrac{\partial}{\partial x_1} \\[2mm] \dfrac{\partial}{\partial x_2} \\[2mm] \vdots \\[2mm] \dfrac{\partial}{\partial x_n} \end{bmatrix} \tag{11.32}$$

Then, $g(x) = \nabla U$ is the gradient vector of $U(x)$ with respect to x:

$$g(x) = \nabla U = \begin{bmatrix} \dfrac{\partial U(x)}{\partial x_1} \\[2ex] \dfrac{\partial U(x)}{\partial x_2} \\[1ex] \vdots \\[1ex] \dfrac{\partial U(x)}{\partial x_n} \end{bmatrix} \tag{11.33}$$

If the gradient of $U(x)$ is a constant vector, $U(x)$ is said to be a linear function of x. In this case, $U(x)$ is of the form

$$U(x) = \beta^T x + \alpha \tag{11.34}$$

where β is a fixed vector, $\beta = \nabla U(x)$, and α is a scalar.

Higher-order derivatives of a multivariable function are defined as in the univariate case. The "first derivative" of an n-variable function is an n-vector; the "second order derivative" of an n-variable function is defined by the n^2 partial derivatives of the n first derivatives with respect to the n variables:

$$\frac{\partial}{\partial x_i} \left(\frac{\partial U}{\partial x_j} \right) \quad i = 1, 2, \ldots, n, \quad j = 1, 2, \ldots, n \tag{11.35}$$

This quantity is usually written as

$$\frac{\partial^2 U}{\partial x_i \, \partial x_j} \quad \text{for } i \neq j, \text{ and } \quad \frac{\partial^2 U}{\partial x_i^2} \quad \text{for } i = j \tag{11.36}$$

If the partial derivatives $\partial U / \partial x_i$, $\partial U / \partial x_j$ and $\partial^2 U / \partial x_i \, \partial x_j$ are continuous, then $\partial^2 U / \partial x_j \, \partial x_i$ exists and $\partial^2 U / \partial x_i \, \partial x_j = \partial^2 U / \partial x_j \, \partial x_i$.

A square, symmetric matrix, termed the *Hessian matrix*, of $U(x)$ represents these n^2 "second partial derivatives." It is given by

$$\mathbf{H}(\mathbf{x}) = \nabla^T \nabla U(\mathbf{x}) = \begin{bmatrix} \dfrac{\partial^2 U(\mathbf{x})}{\partial x_1^2} & \dfrac{\partial^2 U(\mathbf{x})}{\partial x_1 \partial x_2} & \cdots & \dfrac{\partial^2 U(\mathbf{x})}{\partial x_1 \partial x_n} \\[2ex] \dfrac{\partial^2 U(\mathbf{x})}{\partial x_2 \partial x_1} & \dfrac{\partial^2 U(\mathbf{x})}{\partial x_2^2} & \cdots & \dfrac{\partial^2 U(\mathbf{x})}{\partial x_2 \partial x_n} \\[2ex] \vdots & \vdots & & \vdots \\[2ex] \dfrac{\partial^2 U(\mathbf{x})}{\partial x_n \partial x_1} & \dfrac{\partial^2 U(\mathbf{x})}{\partial x_n \partial x_2} & \cdots & \dfrac{\partial^2 U(\mathbf{x})}{\partial x_n^2} \end{bmatrix} \quad (11.37)$$

For a vector-valued function, derivatives can be defined by simply differentiating each component function separately. The Jacobian matrix of a vector function $\mathbf{y}(\mathbf{x}) = [y_1(\mathbf{x}), \ldots, y_l(\mathbf{x})]^T$ is defined as the $l \times n$ matrix whose (i, j) element is the first derivative of y_i with respect to x_j:

$$\mathbf{J}^T = \frac{\partial \mathbf{y}^T}{\partial \mathbf{x}} = \begin{bmatrix} \nabla y_1 & \nabla y_2 & \cdots & \nabla y_m \end{bmatrix}$$

$$= \begin{bmatrix} \dfrac{\partial y_1}{\partial x_1} & \dfrac{\partial y_2}{\partial x_1} & \cdots & \dfrac{\partial y_m}{\partial x_1} \\[2ex] \dfrac{\partial y_1}{\partial x_2} & \dfrac{\partial y_2}{\partial x_2} & \cdots & \dfrac{\partial y_m}{\partial x_2} \\[2ex] \vdots & \vdots & & \vdots \\[2ex] \dfrac{\partial y_1}{\partial x_n} & \dfrac{\partial y_2}{\partial x_n} & \cdots & \dfrac{\partial y_m}{\partial x_n} \end{bmatrix} \quad (11.38)$$

By C^k we denote the class of functions with continuous derivatives of order 1 through k. The class C^2 is called the set of *twice differentiable functions*. "Smooth" functions are those with high degrees of differentiability.

The multidimensional Taylor series of a general function $U(\mathbf{x})$ about point $\mathbf{x}$ allows us to construct simple approximations to the functions in the neighborhood of $\mathbf{x}$, which are very frequently used in optimization. For instance, ignoring all but the linear term of the Taylor series, we have

$$U(\mathbf{x} + \Delta\mathbf{x}) \approx U(\mathbf{x}) + \nabla U(\mathbf{x})^T \Delta\mathbf{x} \quad (11.39)$$

This expression defines a linear function of the n-vector Δx, the displacement from x:

$$\Delta x = \begin{bmatrix} \Delta x_1 \\ \Delta x_2 \\ \vdots \\ \Delta x_n \end{bmatrix} \tag{11.40}$$

Similarly, including one additional term from the Taylor series produces the approximation:

$$U(x + \Delta x) \approx U(x) + \nabla U(x)^T \Delta x + \frac{1}{2} \Delta x^T H(x) \Delta x \tag{11.41}$$

Now it is a quadratic function of Δx.

At a stationary point of $U(x)$, $\nabla U(x^*) = 0$ and (11.41) reduces to

$$U(x + \Delta x) = U(x^*) + \frac{1}{2} \Delta x^T H(x^*) \Delta x \tag{11.42}$$

The product $\Delta x^T H(x^*) \Delta x$ determines behavior of $U(x)$ at x^*. If the product is larger than zero for any choice of the vector Δx (in such case the matrix H is positive definite), then the stationary point x^* must be a minimum. If the product $\Delta x^T H \Delta x$ is smaller than zero for any Δx (the matrix H is negative definite), the point x^* has to be a maximum. If the product takes either a positive or negative value depending on the choice of Δx (the matrix H is indefinite) the stationary point x^* is a saddle point (see Figure 11.1).

11.1.5 Conjugate Directions

Our further discussion will refer to conjugate directions, in which properties are exploited by certain gradient optimization methods. Before discussing them let us define conjugate directions with respect to a positive definite matrix H. If

$$s^{(i)T} H s^{(j)} = \begin{cases} 0, & \text{for } i \neq j \\ k_j > 0, & \text{for } i = j \end{cases} \tag{11.43}$$

then the vectors $s^{(j)}$, $j = 0, 1, \ldots, n - 1$ are said to be conjugate with respect to an $n \times n$ positive definite matrix H. Conjugate directions are linearly independent.

We can prove that the minimum of a quadratic function can be found in, at most, n iterations. As is evident, one linear minimization along each direction in turn locates the minimum. This assumes perfect minimization in each direction.

11.2 VARIABLES AND FUNCTIONS

In this section our purpose is to discuss some basic concepts related to practical circuit optimization. We shall introduce and consider such terms as physical system and its simulation model, hierarchy of models and the sets of designable parameters and response functions associated with these models. Some specifications for circuit functions, error functions, and objective functions for microwave circuit optimization also will be presented.

11.2.1 The Physical System and Its Simulation Models

In microwave circuit design, the physical system can be a network, a device, or a measurement or manufacturing process. The element types and a topology of the system are assumed fixed. As an example, the physical system may be a bandpass waveguide filter, MESFET transistor amplifier in MIC technology, a procedure for noise parameter measurements, and so on. The system parameters will be referred to as the column vector $\mathbf{p}^M$, where the superscript M identifies concepts related to the physical system. Examples of physical, adjustable parameters are geometrical dimensions such as the width and height of the rectangular waveguide, the width of a strip, the metallization thickness, the length of the section of the stripline, *et cetera*. In the procedure of noise parameter measurements, hot and cold state temperatures of noise sources are other examples of possible components of $\mathbf{p}^M$.

For the physical system some measurable quantities describe its performance and characteristics, such as the frequency and time transient responses. $\mathbf{F}^M$ denote these measurable characteristics.

The responses of the system, whether in frequency domain or time domain, depend on the system parameters. Some suitable models are used to simulate these dependencies. In circuit optimization, such models may be defined at many levels [6, 7].

Typically four levels of hierarchy are used in circuit modeling [5]:

$$\mathbf{p}^L = \mathbf{p}^L(\mathbf{p}^M) \tag{11.44}$$

$$\mathbf{p}^H = \mathbf{p}^H(\mathbf{p}^L) \tag{11.45}$$

$$\mathbf{F}^L = \mathbf{F}^L(\mathbf{p}^H) \tag{11.46}$$

$$\mathbf{F}^H = \mathbf{F}^H(\mathbf{F}^L) \tag{11.47}$$

In these definitions $\mathbf{p}^L$ is a set of low-level parameters. The vector $\mathbf{p}^L$ should represent, as closely as possible, the set of adjustable parameters $\mathbf{p}^M$ of the actual physical system under consideration. The vector $\mathbf{p}^H$ represents a set of parameters of a higher-level model of the optimized system. This model is typically an equivalent circuit used in a computer-aided analysis subroutines. The set (11.45) represents the relationships between the higher-level and lower-level parameters, which may be derived from theoretical or experimental considerations of the system.

Furthermore, the model responses are defined, typically, at two levels of the hierarchy. The frequency-dependent complex scattering parameters $\mathbf{S}$, the four noise parameters F_{min}, $\mathrm{Re}\{\Gamma_{opt}\}$, $\mathrm{Im}\{\Gamma_{opt}\}$, and N, or noise correlation matrix elements $\mathbf{C}$ described with respect to the external ports of the model, undetermined y-parameters, *et cetera* are the examples of low-level model responses, denoted $\mathbf{F}^L$. Some of these quantities may not be directly measurable. To the higher-level responses $\mathbf{F}^H$ belong the actually measured system responses. In the case of microwave networks they may be, for example, the noise figure, transducer power gain, input port SWR, or insertion loss, return loss, and group delay of a circuit.

For a particular circuit design, we can choose a certain section or combination of this model hierarchy to formulate an optimization problem. For instance we can choose either $\mathbf{p}^L$ or $\mathbf{p}^H$, or select a combination of both, as the optimization parameters. In the same way, we can select the response functions of the designed system. They may be selected as $\mathbf{F}^H$ or $\mathbf{F}^L$, or a suitable combination of both. In the following discussions, we will just denote as $\mathbf{p}$ the designable parameters and as $\mathbf{F}$ the response functions of the designed system.

Example 11.2

Figure 11.6 presents a low-pass filter structure as it might appear if designed on microstrip line. The filter consists of a cascade of connection of microstrip lines of low characteristic impedance (the broad strips) and lines of high impedance (the narrow strips). Like the low-level model parameters, we can define 11 physical parameters of the microstrip filter structure. They are w_0, w_1, w_2, w_3, w_4 = strip widths of the succeeding microstrip line sections; l_1, l_2, l_3 = their lengths; ϵ_r = the dielectric constant of a substrate; b = the substrate thickness; and t = the strip metallization thickness. The vector $\mathbf{p}^L$ containing these parameters is

$$\mathbf{p}^L = \mathbf{p}^M = \left[\, w_0, w_1, w_2, w_3, w_4, l_1, l_2, l_3, \epsilon_r, b, t\,\right]^T$$

Electrically, the discussed microstrip low-pass filter structure can be thought of as a cascade of transmission line segments, each with a different characteristic impedance, phase shift, line loss, and guide wavelength. Figure 11.6(c) depicts the equivalent circuit of the discussed low-pass filter structure, taking into account effects related to the discontinuities associated with the abrupt change of strip width. It is a higher-level

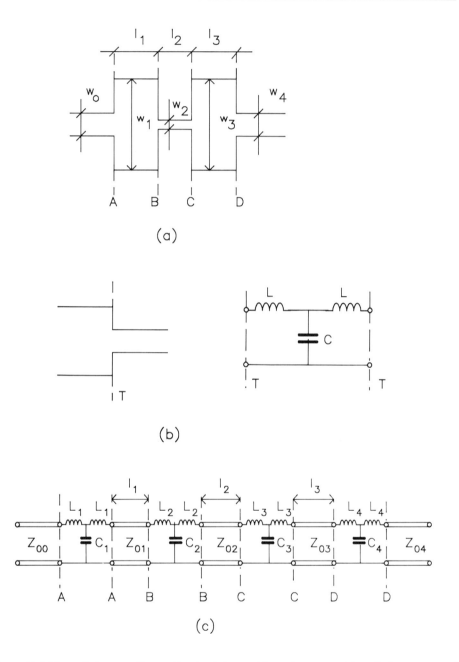

Figure 11.6 Microstrip low-pass filter and its simulation model: (a) the pattern of the strip-line structure of the filter; (b) lumped equivalent circuit of an abrupt step in width of the microstrip line; (c) equivalent circuit of the low-pass microstrip filter.

model, and the vector $\mathbf{p}^H$ containing its 16 parameters is

$$\mathbf{p}^H = \left[Z_{00}, Z_{01}, Z_{02}, Z_{03}, Z_{04}, l_1, l_2, l_3, L_1, L_2, L_3, C_1, C_2, C_3, C_4 \right]^T$$

The relationships between the high-level parameters and the physical, low-level parameters of the system are available from the theory of the microstrip line and its discontinuities [8].

In practice we can assume the symmetry of the filter structure. Thus, the number of high-level design parameters may be reduced to nine. Because $Z_{00} = Z_{04}$, $Z_{01} = Z_{03}$, $l_1 = l_3$, $L_1 = L_4$, $L_2 = L_3$, $C_1 = C_4$, $C_2 = C_3$, we have

$$\mathbf{p} = \left[Z_{00}, Z_{01}, Z_{02}, l_1, l_2, L_1, L_2, C_1, C_2 \right]^T$$

The responses $\mathbf{F}$ of the designed low-pass filter structure can be the frequency characteristics of the transmission loss $L(\omega) = 20 \log(\mid S_{21}(j\omega) \mid^{-1})$, and of the group delay $\tau = -(\partial / \partial \omega)(\arg\{S_{21}\})$. These two quantities, computed with respect to the idealized, matched terminations, are the elements of the low-level circuit functions $\mathbf{F}^L$. The same functions measured by the network analyzer at the actual complex terminations could be high-level responses $\mathbf{F}^H$ of the discussed design problem.

11.2.2 Design Specifications and Error Functions

The desired performance of a system may be expressed by a set of specifications, which are usually functions of some independent variables such as frequency, time, or temperature. Because the response functions can be computed at a finite number of values of one or more independent variables, we have to consider that discrete set of the samples of independent variables which satisfies the specifications at these points and guarantees satisfying them between the points as well. In general, we can simultaneously consider more than one type of response, and we can define more than one specification for each of the responses.

For example, a circuit can be designed to meet desired specifications in both the frequency and time domains. Very often in practice the design problem has to be defined by lower and upper specifications. Figure 11.7 presents an example of the lower and upper specifications for the insertion loss characteristic of a bandpass filter and the response function that violates these specifications. In this example ω_l and ω_u denote, respectively, the lower bound and upper bound of the independent variable interval in which the problem is being considered [3, 4, 9].

In Figure 11.8, we show an example of response specifications for a directional coupler design problem. This time the response function under consideration is the coupling coefficient frequency characteristic of the designed directional coupler. This example shows lower and upper specifications defined for the same interval of the

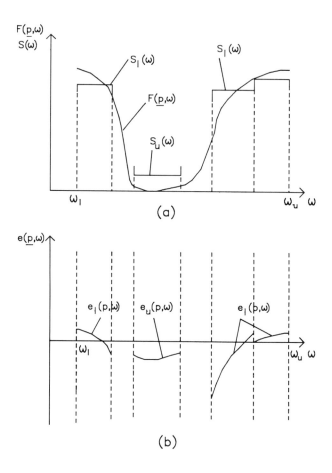

Figure 11.7 An example of lower $S_l(f)$ and upper $S_u(f)$ specifications imposed on a response function $F(\mathbf{p}, f)$ of a bandpass filter.

independent variable. The specification of this type is a "window" specification. The upper and lower specifications are one-sided specifications.

Another type of specification is a single specification, which is a two-sided specification. Many different design problems may be defined using single specification, which may be interpreted as a "window" specification having zero width. An example of a single specification for the design problem of an amplifier is shown in Figure 11.9. It describes the requirements on the shape and magnitude of the transducer power gain of the amplifier. In the amplifier design, the transducer power gain characteristic, the magnitude of the input port reflection coefficient, and a value of the noise figure in the frequency band of interest are considered simultaneously.

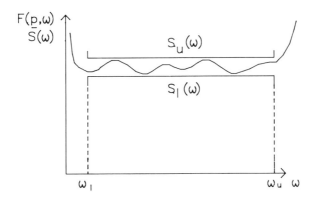

Figure 11.8 An example of design specifications and the response function for a directional coupler design problem.

In the discrete case, a set of calculated response functions and the corresponding set of sampled specifications are denoted as, respectively,

$$F_j(\mathbf{p}), \quad j \in J \tag{11.48}$$

and

$$S_j, \quad j \in J \tag{11.49}$$

In (11.48) and (11.49) the subscript j refers to quantities (already defined) evaluated at discrete values of independent variables (as for example, discrete frequency points).

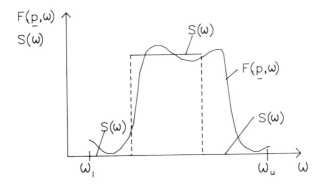

Figure 11.9 The single specification $S(f)$ and a response function for an amplifier design problem.

The error function must arise from the difference between the actually calculated responses and the given specifications.

In general, when we have lower and upper specifications, we define the error functions as [5]

$$e_{uj}(\mathbf{p}) = w_{uj}\big(F_j(\mathbf{p}) - S_{uj}\big), \quad j \in J_u \tag{11.50}$$

$$e_{lj}(\mathbf{p}) = w_{lj}\big(F_j(\mathbf{p}) - S_{lj}\big), \quad j \in J_l \tag{11.51}$$

where $F_j(\mathbf{p})$ is the response function, $\mathbf{p}$ represents the network parameters, S_{uj} is an upper specification, S_{lj} is a lower specification, w_{uj} is a nonnegative weighting factor for S_{uj}, and w_{lj} is a nonnegative weighting factor for S_{lj}.

Weighting factors w_{uj} and w_{lj} can be thought of as giving the relative importance of $e_{u,lj}(\mathbf{p})$. Simultaneously, each $w_{u,lj}$ must take into account the relative magnitudes or scales associated with $e_{uj,lj}$, which may be defined for different kinds of responses and in different units of measure.

The index sets J_u and J_l defined as

$$J_u = \{j_1, j_2, \ldots, j_k\} \tag{11.52}$$

$$J_l = \{j_{k+1}, j_{k+2}, \ldots, j_m\} \tag{11.53}$$

are not necessarily disjoint.

By redefining the error functions as [5]

$$e_i(\mathbf{p}) = \begin{cases} e_{uj}(\mathbf{p}), & j = j_i, \quad i = 1, 2, \ldots, k \\ -e_{lj}(\mathbf{p}), & j = j_i, \quad i = k+1, k+2, \ldots, m \end{cases} \tag{11.54}$$

we obtain a set of uniformly indexed error functions.

The response functions $F(\mathbf{p})$ corresponding to these defined error functions can be only real-valued functions. Obviously, a positive value of error function e_i indicates a violation of the corresponding specifications.

If we define

$$M_e(\mathbf{p}) = \max_{i \in I} e_i(\mathbf{p}), \quad I = \{1, 2, \ldots, m\} \tag{11.55}$$

then the sign of $M_e(\mathbf{p})$ indicates whether all upper and lower specifications are satisfied or violated. That is, if

$$M_e(\mathbf{p}) = \begin{cases} > 0, & \text{the specifications are violated,} \\ = 0, & \text{the specifications are met exactly,} \\ < 0, & \text{the specifications are satisfied with excess.} \end{cases}$$

In design problems with single specifications, the definition of error functions is much simpler:

$$e_i = w_i \,|\, F_i(\mathbf{p}) - S_i \,|, \quad i = 1, 2, \ldots, m \qquad (11.56)$$

where w_i is a nonnegative weighting factor.

In (11.56) the response functions and the specifications can be real or complex.

The error functions e_i, $i = 1, 2, \ldots, m$ can be treated as the components of a vector defined as

$$\mathbf{e}(\mathbf{p}) = \begin{bmatrix} e_1(\mathbf{p}) \\ e_2(\mathbf{p}) \\ \vdots \\ e_m(\mathbf{p}) \end{bmatrix} \qquad (11.57)$$

11.2.3 Objective Functions in CAD of Microwave Circuits

Many microwave circuit design problems are best formulated in terms of minimizing a scalar objective function $U(\mathbf{p})$, which is closely related to the error functions $e_i(\mathbf{p})$ defined in (11.54). Typically, we define the objective function $U(\mathbf{p})$ as a l_p norm or a generalized l_p function of $e_i(\mathbf{p})$.

The l_p Norm

The l_p norm of $\mathbf{e}(\mathbf{p})$ is defined as [10]

$$\|\mathbf{e}\|_p = \left\{ \sum_{i=1}^{m} |e_i(\mathbf{p})|^p \right\}^{1/p}, \quad \begin{array}{l} i \in I \\ 1 \le p \le \infty \end{array} \qquad (11.58)$$

l_p norm is a scalar measure of the deviations between the desired and actual responses of the designed circuit.

The choice of the value of the parameter p in the l_p norm has an essential significance. By taking a small value for p in effect we emphasize those error functions e_i that have smaller values. When the parameter p has a large value, about the value of the l_p norm, we emphasize those error functions e_i that have larger values. Three cases of the l_p norm, $p = 1$, $p = 2$, and $p = \infty$, are best known and most widely used in the circuit optimization.

The l_1 Norm (p = 1)

The l_1 norm, defined as

$$\|\mathbf{e}\|_1 = \sum_{i=1}^{m} |e_i(\mathbf{p})| \tag{11.59}$$

emphasizes the importance of the error functions e_i that are close to zero. The l_1 norm is used as the objective function in the approximation problems defined as the minimization:

$$\min_{\mathbf{p}} \|\mathbf{e}(\mathbf{p})\|_1 = \sum_{i=1}^{m} |e_i(\mathbf{p})| \tag{11.60}$$

Compared to the other norms l_p where $p > 1$, the l_1 norm, has the distinctive property that some components of $\mathbf{e}$ with large values are ignored; that is, at the solution of (11.60) a few values of e_i may be much larger than the others. The solution of (11.60) is usually a point where one or more e_i values are equal to zero whereas some large e_i values are completely ignored. This property is very important in problems of approximating data, which might contain some points weighted with large errors. It means that l_1 norm is preferable in problems of approximating a measured response by a response of a device, a network, or a system formulated as an optimization problem with respect to the equivalent circuit parameters of a proposed model [11, 12]. Because $\|\mathbf{e}\|_1$ is not differentiable in the ordinary sense, their minimization requires sophisticated algorithms [13–17], some of which are going to be discussed in Sections 11.3.2 and 11.3.5.

The l_2 Norm (p = 2)

The l_2 norm defined as

$$\|\mathbf{e}\|_2 = \left[\sum_{i=1}^{m} |e_i(\mathbf{p})|^2 \right]^{1/2} \tag{11.61}$$

is perhaps the best-known and most widely used norm [18]. It is called the *least-squares norm*. The l_2 norm used as the objective function has found various applications in design problems formulated as optimization problems [19–21]. A very important property of the l_2 norm is its differentiability. Its gradient vector can be obtained very easily from the partial derivatives of $\mathbf{e}$.

The l_∞ Norm (p = ∞)

By letting $p = \infty$ in (11.58) we have $\|\mathbf{e}\|_\infty$, which is called the *Chebyschev* or *uniform norm*. Because

$$\lim_{p \to \infty} \left\{ \sum_{i=1}^{m} |e_i|^p \right\}^{1/p} = \max_i |e_i|, \quad i \in I \tag{11.62}$$

as a consequence, the uniform norm

$$\|\mathbf{e}\|_\infty = \max_i |e_i|, \quad i \in I \tag{11.63}$$

takes a value of the error function e_i with the largest value (the worst case) and the other errors in effect are ignored.

"*Minimax*" optimization defined as

$$\min_{\mathbf{p}} \max_i |e_i(\mathbf{p})| \tag{11.64}$$

typically leads to the optimal solution $\tilde{\mathbf{p}}$ for which values of some error functions are the same:

$$e_{i_1}(\tilde{\mathbf{p}}) = e_{i_2}(\tilde{\mathbf{p}}) = \cdots = e_{i_k}(\tilde{\mathbf{p}}) \tag{11.65}$$

with the other values of $e_i(\tilde{\mathbf{p}})$ less than this value. This equality at the final answer is equivalent to the equal-ripple error characteristic of the circuit function. Such a solution is very often required in microwave filter design problems. Minimax optimization is widely employed in the microwave circuit design [22–29]. Some of the algorithms specially developed for minimax optimization will be presented in Sections 11.5.1 and 11.5.2.

Generalized l_p Objective Function

When we compute error functions with respect to the upper and lower specifications, they may have positive as well as negative values. A negative value of $e_i(\mathbf{p})$ means that at this point the specification has been fulfilled to excess. This consideration leads to the most general case, in which some of the e_i for $i \in I$ nonnegative (positive-real) or all of the e_i for $i \in I$ are negative. If I^+ is a set of i corresponding to nonnegative error functions:

$$I^+ \{i \mid e_i \geq 0, \, i \in I\} \tag{11.66}$$

then the generalized l_p function has a form [30, 31]:

$$U_p(\mathbf{p}) = \begin{cases} U_p^+(\mathbf{p}), & \text{if the set } I^+ \text{ is not empty} \\ U_p^-(\mathbf{p}), & \text{otherwise} \end{cases} \tag{11.67}$$

where

$$U_p^+(\mathbf{p}) = \left[\sum_{i \in I^+} (e_i(\mathbf{p}))^p \right]^{1/p} \tag{11.68}$$

$$U_p^-(\mathbf{p}) = -\left[\sum_{i=1}^m (-e_i(\mathbf{p}))^{-p} \right]^{-1/p} \tag{11.69}$$

In other words, $U_p^+(\mathbf{p})$ is formed from only the nonnegative error functions, and $U_p^-(\mathbf{p})$ is formed from all error functions when all the values of e_i are negative. This permits the minimization of the objective function even after all the specifications have been met, so that the response function may be further improved.

Because, for $e_i > 0$, $i \in I$,

$$\max_{i \in I} \{e_i\} = \lim_{p \to \infty} \left[\sum_{i \in I} (e_i)^p \right]^{1/p} \tag{11.70}$$

$$\min_{i \in I} \{e_i\} = \lim_{p \to \infty} \left[\sum_{i \in I} (e_i)^{-p} \right]^{-1/p} \tag{11.71}$$

we can understand that by minimizing the generalized l_p function given by (11.67) to (11.69) with very large values of p we obtain results very close to the minimax optimum.

The generalized l_p function without some modification would not be applicable in practice due to ill-conditioning, resulting from the numerical evaluation of $(\mp e_i)^{\mp p}$ for very large values of p. This ill-conditioned problem may be solved simply by dividing all error functions e_i by a value equal to the largest magnitude of an error function M_e.

We use this idea to define the objective function in the following way [30, 31]:

$$U(\mathbf{p}) = \begin{cases} M_e \left[\displaystyle\sum_{i \in K} \left(\dfrac{e_i(\mathbf{p})}{M_e} \right)^q \right]^{1/q}, & \text{for } M_e \neq 0 \\ 0, & \text{for } M_e = 0 \end{cases} \qquad (11.72)$$

where

$$q = \frac{M_e}{|M_e|} p \begin{cases} 1 < p < \infty, & \text{for } M_e > 0 \\ 1 \leq p < \infty, & \text{for } M_e < 0 \end{cases} \qquad (11.73)$$

$$K = \begin{cases} I^+, & \text{if } M_e > 0 \\ I, & \text{if } M_e < 0 \end{cases} \qquad (11.74)$$

In (11.72) to (11.74):

$$M_e = \begin{cases} \lim_{p \to \infty} U_p^+(\mathbf{p}), & \text{if the set } I^+ \text{ is not empty} \\ \lim_{p \to \infty} U_p^-(\mathbf{p}), & \text{otherwise} \end{cases} \qquad (11.75)$$

The gradient vector of the objective function is given by

$$\nabla U_p(\mathbf{p}) = \left[\sum_{i \in K} \left(\frac{e_i(\mathbf{p})}{M_e} \right)^q \right]^{1/q - 1}$$

$$\cdot \sum_{i \in K} \left(\frac{e_i(\mathbf{p})}{M_e} \right)^{q-1} \nabla e_i(\mathbf{p}), \quad \text{for } M_e \neq 0 \qquad (11.76)$$

Observe from (11.72) and (11.76) that if $e_i(\mathbf{p})$ for $i \in I$ is a set of continuous functions with continuous first partial derivatives, then the objective function U and its first partial derivative also are continuous. The discontinuities in the derivative of the objective function of (11.72) occur only at a point $\mathbf{p}$ where $M_e(\mathbf{p}) = 0$, and two or more maxima are equal. The minimization of the objective function defined by (11.72) to (11.75) can be performed by using gradient optimization methods that are more efficient than direct search methods.

Example 11.3

We present a microwave circuit design problem that can be solved by using optimization methods to analyze and design microwave oscillators with semiconductor devices, such as Gunn diodes.

The fundamental frequency of oscillations of a Gunn diode oscillator equals the resonant frequency of the oscillator circuit loaded with the Gunn diode admittance Y_d. At steady state oscillations of the circuit, a Gunn diode may be represented as a parallel connection of negative conductance G_d and positive susceptance B_d. Both G_d and B_d depend on the dc supply voltage V_{dc}, on the amplitude $\overline{V}$ of the sinusoidal voltage at diode terminals, and on the fundamental frequency f of oscillations. At a constant dc supply voltage, the steady state oscillation frequency and the amplitude of oscillations determines the equation

$$Y_d(f, \overline{V}) + Y_c(f) = 0$$

where $Y_d = G_d(f, \overline{V}) + jB_d(f, \overline{V})$ is the Gunn diode admittance and $Y_c(f)$ is the admittance of the microwave circuit interacting with the diode.

One possible configuration of a MIC Gunn diode oscillator is shown in Figure 11.10(a). A resonant circuit interacting with Gunn diode is a section of a microstrip line $L1$ with length $\approx \lambda/4$. The other section of microstrip line $L2$ parallel to $L1$ on a length l forms a circuit coupling the load impedance Z_L with the resonant circuit of the oscillator. A three-section low-pass MIC filter is used as a dc voltage supply circuit. Figure 11.10(b) presents an equivalent circuit of the oscillator. Conditions for steady state oscillations discussed earlier should be defined at the reference plane 0–0′ of the circuit in Figure 11.10(b).

The design of the MIC oscillator consists in finding the length l and gap spacing s of coupled striplines for given frequency of oscillation. The objective is to find a minimum of the function:

$$U(\mathbf{p}) = |G_d + G_c| + |B_d + B_c|$$

where

$$\mathbf{p} = \begin{bmatrix} l \\ s \end{bmatrix}$$

A minimum of $U(\mathbf{p}) = 0$ fulfills the steady state oscillation conditions.

Example 11.4

Transferred electron devices (TEDs) exhibit negative small signal resistance in a broad band of frequencies. They may be used for broadband reflection amplifiers. Because the input impedance of a circulator as well as small-signal TED impedance depend on frequency, the design of the amplifier with a given power gain in a given frequency band requires application of the optimization methods. Figure 11.11(a) presents a cross section of the physical structure of the amplifier. The amplifier design corresponds to

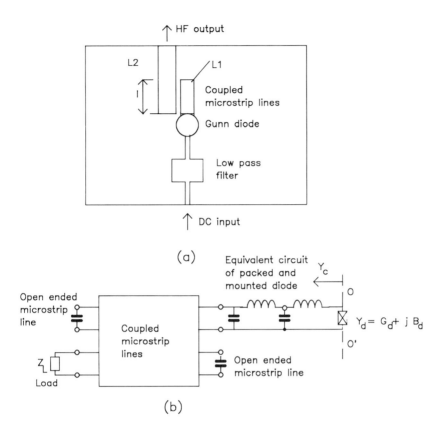

Figure 11.10 (a) Microstrip line configuration of MIC Gunn-diode oscillator and (b) its equivalent circuit.

designing a multiple-section coaxial line impedance transformer, the equivalent circuit of which is presented in Figure 11.11(b). The packed TED is represented by its measured impedance $Z_d(f)$. The term $Z_L(f)$ is the input impedance of the circulator port; Z_L is also described by data received from measurements. Capacitances C_1 and C_2 represent coaxial line discontinuity reactances.

The problem is to minimize a function:

$$U(\mathbf{p}) = \left\{ \sum_{i=1}^{m} |e_i(\mathbf{p})|^p \right\}^{1/p}$$

where

$$e_i(\mathbf{p}) = |\Gamma_i(\mathbf{p})|^2 - |\Gamma_0|^2$$

$$\mathbf{p} = [Z_{01}, Z_{02}, l_1, l_2]$$

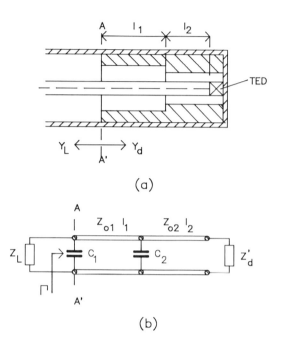

Figure 11.11 Coaxial line reflection amplifier with transferred electron device: (a) physical structure; (b) equivalent circuit of the amplifier.

$\Gamma_i(\mathbf{p})$ is the reflection coefficient at the ith frequency, given as

$$\Gamma_i(\mathbf{p}) = \frac{Z_d(\omega_i) - Z_L^*(\omega_i)}{Z_d(\omega_i) + Z_L(\omega_i)}, \quad i = 1, 2, \ldots, m$$

the squared magnitude of which equals the power gain of the amplifier, and $|\Gamma_0|^2$ is the specified constant power gain at m equidistantly spaced sample frequencies located in the desired frequency band of the amplifier [28].

Example 11.5

Let us consider a single-stage transistor amplifier design. Figure 11.12 presents an equivalent circuit of the amplifier. Two simple matching networks with shunt short-circuited stubs are used at the input and output of the transistor. Characteristic impedances

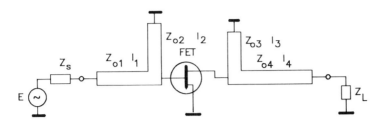

Figure 11.12 Single-stage transistor amplifier.

and lengths of transmission lines creating matching circuits are design parameters:

$$\mathbf{p} = \left[Z_{01}, l_1, Z_{02}, l_2, Z_{03}, l_3, Z_{04}, l_4 \right]^T$$

The scattering matrix parameters related to the input and output ports of the amplifier characterize the whole circuit. They are functions of the design parameters. For Z_s and Z_L equal the reference impedances, respectively, of the input and output port of the amplifier, $|S_{21}|^2$ represents transducer power gain of the amplifier, S_{11} and S_{22} are the input and output reflection coefficients, and $|S_{12}|^2$ is the backward transducer power gain of the amplifier.

The resulting parameters are used to create error functions and, ultimately, the objective function. We define the following error functions:

$$e_i(\mathbf{p}) = \left| S_{21}(\mathbf{p}, \omega_i) \right|^2 - \left| S_{21}^s(\omega_i) \right|^2$$

$$e_i(\mathbf{p}) = \left| S_{11}(\mathbf{p}, \omega_i) \right|$$

$$e_i(\mathbf{p}) = \left| S_{22}(\mathbf{p}, \omega_i) \right|$$

where $i \in I \triangleq \{1, 2, \ldots, m\}$ is a set of integers corresponding to a number of discrete frequency points ω_i in a frequency band of interest Ω, and $|S_{21}^s(\omega_i)|^2$ is the specified objective for the power gain in the frequency band Ω.

The objective function can be made to stress various aspects of the amplifier design: gain flatness, maximum power transfer, minimum reflection on the input and output of the amplifier, a minimum noise figure, and various combinations of these features.

Using the objective function in the form:

$$U(\mathbf{p}) = \sum_{i=1}^{m} \left\{ w_{21i} \left| \left| S_{21}(\mathbf{p}, \omega_i) \right|^2 - \left| S_{21}^s(\omega_i) \right|^2 \right|^p + w_{11i} \left| S_{11}(\mathbf{p}, \omega_i) \right|^p \right.$$

$$\left. + w_{22i} \left| S_{22}(\mathbf{p}, \omega_i) \right|^p \right\}$$

and minimizing it with respect to circuit parameters **p** are a compromise between power gain and a good match at the input and output of amplifier. In $U(\mathbf{p})$ the w_{21i}, w_{11i}, and w_{22i} are positive weighting factors.

When designing the amplifier with optimum noise performance, we can utilize a two-step procedure. In the first step we design the input matching network to provide the best noise performance of the transistor. The appropriate objective function has a form:

$$ U(\mathbf{p}) = \sum_{i=1}^{m} w_i \big(F(\mathbf{p}, \omega_i) \big)^p $$

where $F(\mathbf{p}, \omega_i)$ is a noise figure of the amplifier, and **p** is a vector of design parameters given as

$$ \mathbf{p} = [Z_{01}, l_1, Z_{02}, l_2] $$

In the second step, the output matching network is designed to provide maximum constant gain in the frequency band of interest.

Example 11.6

In general-purpose computer programs, the objective functions usually have a form:

$$ U(\mathbf{p}) = \sum_{j=1}^{k} \left\{ \sum_{i=1}^{l_j} \sum_{m=1}^{6} \left[w_{ij}^{(m)} \big(F_{ij}^{(m)}(\mathbf{p}) - S_{ij}^{(m)} \big) \right]^p \right\} $$

where

$j =$ optimization frequency band 1 to k,
$i =$ frequency points 1 to l_j in a frequency band j,
$m =$ circuit function being optimized,
$w_{ij}^{(m)} =$ weight for circuit function m at a frequency point i in band j,
$F_{ij}^{(m)}(\mathbf{p}) =$ a value of circuit function m at a frequency point i in band j,
$S_{ij}^{(m)} =$ desired value of circuit function m at a frequency point i in band j,
$\mathbf{p} =$ circuit parameters.

The six circuit functions usually are four overall scattering parameters referred t the input and output ports of a circuit, noise figure, and a group delay. Becaus scattering parameters of a circuit are complex, their real and imaginary parts have to b controlled separately. To do so, components of the objective function $U(\mathbf{p})$ correspond ing to overall scattering parameters should be computed as

$$\left| w_{ij}^{(m)} \, \mathrm{Re}\left\{ F_{ij}^{(m)}(\mathbf{p}) - S_{ij}^{(m)} \right\} \right|^{p} + \left| w_{ij}^{(m)} \, \mathrm{Im}\left\{ F_{ij}^{(m)}(\mathbf{p}) - S_{ij}^{(m)} \right\} \right|^{p}$$

Instead of controlling the real and imaginary parts of complex scattering parameters, we can control their magnitudes and phases. Here, components of the sum forming the objective function will have the form:

$$\left| w_{ij}^{(m)} \left(\left| F_{ij}^{(m)}(\mathbf{p}) \right| - \left| S_{ij}^{(m)} \right| \right) \right|^{p} + \left| w_{ij}^{(m)} \left(\arg\left\{ F_{ij}^{(m)}(\mathbf{p}) \right\} - \arg\left\{ S_{ij}^{(m)} \right\} \right) \right|^{p}$$

The choice depends on the problem to be solved. If the sign of the real or imaginary part of the circuit function is essential, the first approach should be used.

The values of scattering parameters differ in magnitude. For passive circuits they range from zero to unity, and for active circuits they are allowed to be greater than one. In a general optimization problem, there may be a wide variance in values of different circuit functions or in values of one function for different frequency bands. So that all circuit functions will contribute adequately to the objective function, weights $w_{ij}^{(m)}$, which may be different for each circuit function and each frequency band, are allowed.

11.3 BASIC GRADIENT-BASED METHODS FOR UNCONSTRAINED FUNCTION MINIMIZATION

Most optimization methods are iterative and based on determining a sequence of vectors $\mathbf{x}^{(k)}$, $k = 0, 1, 2, \ldots, n$, such that

$$U\left(\mathbf{x}^{(0)}\right) > U\left(\mathbf{x}^{(1)}\right) > \cdots > U\left(\mathbf{x}^{(k)}\right) \tag{11.77}$$

To proceed from an initial starting point $\mathbf{x}^{(0)}$ to other points we need to determine a direction that in n-dimensional space is an n-dimensional vector $\mathbf{s} = s_1, s_2, \ldots, s_n]^{T}$. For the kth iteration, the new point $\mathbf{x}^{(k)}$ is

$$\mathbf{x}^{(k+1)} = \mathbf{x}^{(k)} + d^{(k)}\mathbf{s}^{(k)} \tag{11.78}$$

and the increment of $\mathbf{x}$

$$\Delta\mathbf{x}^{(k+1)} = \mathbf{x}^{(k+1)} - \mathbf{x}^{(k)} = d^{(k)}\mathbf{s}^{(k)} \tag{11.79}$$

In (11.78) and (11.79) $d^{(k)}$ is a real constant determining the length of a step in the direction $\mathbf{s}^{(k)}$.

Before starting the discussion on iterative minimization methods, we present a general algorithm valid for most methods.

Given: $U(\mathbf{x})$ is an objective function; accuracy of computations ϵ_1, ϵ_2.

Step 1. Set $k = 0$. Select the starting point $\mathbf{x}^{(0)}$ (the initial guess).

Step 2. Calculate $U(\mathbf{x}^{(k)})$ and $\nabla U(\mathbf{x}^{(k)})$ (if the method uses the gradient information).

Step 3. Determine the search direction $\mathbf{s}^{(k)}$ and normalize it to the unit length.

Step 4. Find a step length $d^{(k)}$ in the direction $\mathbf{s}^{(k)}$ such that either

$$U(\mathbf{x}^{(k)} + d^{(k)}\mathbf{s}^{(k)}) < U(\mathbf{x}^{(k)}), \quad \text{or}$$

$$U(\mathbf{x}^{(k)} + d^{(k)}\mathbf{s}^{(k)}) = \min_{d^{(k)}} U(\mathbf{x}^{(k)} + d^{(k)}\mathbf{s}^{(k)})$$

Step 5. Obtain the new solution $\mathbf{x}^{(k+1)}$:

$$\Delta \mathbf{x}^{(k)} = d^{(k)}\mathbf{s}^{(k)}$$

$$\mathbf{x}^{(k+1)} = \mathbf{x}^{(k)} + d^{(k)}\mathbf{s}^{(k)}$$

Step 6. Check, if convergence criterion is met. If $|U(x^{(k+1)}) - U(\mathbf{x}^{(k)})| < \epsilon_1$ or $\|\Delta \mathbf{x}^{(k)}\| < \epsilon_2$, then stop.

Step 7. Increment k by one, $k = k + 1$ and return to step 2.

Before discussing different optimization methods we want to mention here an important observation valid for any differentiable function $U(\mathbf{x})$ whose knowledge i important for understanding many optimization methods. If we assume that $U(\mathbf{x})$ i minimized in a given search direction $\mathbf{s}^{(k)}$, then at the minimum point $\mathbf{x}^{(k)}$ in thi direction the gradient vector and the search direction vector are orthogonal to eac other; that is,

$$\left(\mathbf{s}^{(k)}\right)^T \nabla U^{(k+1)} = \left(\nabla U^{(k+1)}\right)^T \mathbf{s}^{(k)} = 0 \tag{11.80}$$

This is true because at the minimum point in search direction $\mathbf{s}^{(k)}$ the gradient vecto may not have a component in the search direction. Figure 11.13 illustrates (11.80).

To reach the optimum point we can use the derivatives of the objective functio $U(\mathbf{x})$. To get more information about the best direction of search, we can obtain secon derivatives or their approximations. As shown in Chapter 5, the adjoint network metho provides the gradient vector of circuit functions very effectively numerically. There fore, so-called direct search methods that do not require derivatives of the objectiv functions will be not discussed here.

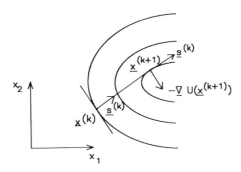

Figure 11.13 At a minimum point $\mathbf{x}^{(k+1)}$ in the direction $\mathbf{s}^{(k)}$ the gradient vector $\nabla U(\mathbf{x}^{(k+1)})$ of the objective function is orthogonal to the direction vector $\mathbf{s}^{(k)}$.

11.3.1 Steepest Descent Method

Optimization methods in which the minimum of the objective function is searched along the negative of the direction of the gradient vector

$$\mathbf{s}^{(k)} = -\frac{\nabla U(\mathbf{x})}{\|\nabla U(\mathbf{x})\|} \tag{11.81}$$

are called *steepest descent methods*.

The choice for $\mathbf{s}^{(k)}$ at $\mathbf{x}^{(k)}$ given by (11.81) is the most obvious. It results from 11.39). A first-order change in the objective function is given by

$$\Delta U = \nabla U(\mathbf{x})^T \Delta \mathbf{x} \tag{11.82}$$

and if $\Delta \mathbf{x} = d\mathbf{s}$, where $d > 0$ is fixed and $\|\mathbf{s}\| = 1$, then $\mathbf{s}$ minimizing ΔU is given by 11.81). Although a normalized gradient vector provides the greatest local change, effectiveness of the steepest descent methods depends strongly on scaling. If the minimum of a function is located in an elongated valley (see Figure 11.14), the method tends to oscillate, and the convergence is slow.

Several modifications have been proposed to reduce the possibility of oscillation. These methods do not use information from previous iteration steps to define new directions of search.

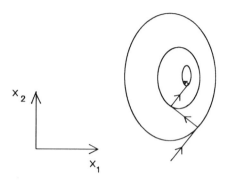

x_2

x_1

Figure 11.14 A steepest descent search and its oscillation.

11.3.2 Conjugate Gradient Methods

Faster convergence of function minimization may be expected if gradient data from previous iteration steps are used to determine the direction of search in the current iteration step. Such methods, called *conjugate gradient methods* [8, 32], are highly effective.

The first search direction $\mathbf{s}^{(0)}$ is taken as the steepest descent direction:

$$\mathbf{s}^{(0)} = -\nabla U(\mathbf{x}^{(0)}) \tag{11.83}$$

The subsequent search directions are linear combinations of the gradient vector a a current search point $\mathbf{x}^{(k)}$ and other previously used search directions:

$$\mathbf{s}^{(1)} = -\nabla U(\mathbf{x}^{(1)}) + k_{11}\mathbf{s}^{(0)}$$

$$\cdots \tag{11.84}$$

$$\mathbf{s}^{(k)} = -\nabla U(\mathbf{x}^{(k)}) + \sum_{i=1}^{k} k_{1k}\mathbf{s}^{(i-1)}$$

Because the search direction in the first iteration $\mathbf{s}^{(0)}$ is known, the search direction $\mathbf{s}^{(1)}$ for the second iteration is found by satisfying $\mathbf{H}$ conjugacy:

$$\left(\mathbf{s}^{(1)}\right)^T \mathbf{H}\mathbf{s}^{(0)} = \left(-\nabla U(\mathbf{x}^{(1)}) + k_{11}\mathbf{s}^{(0)}\right)^T \mathbf{H}\mathbf{s}^{(0)} = 0 \tag{11.85}$$

where $\mathbf{H}$ is the Hessian matrix.

Because the matrix $\mathbf{H}$ is generally not known, it should be replaced by a expression involving only objective function gradients at the respective search points. This may be done using the following relations.

For a general quadratic function in a form:

$$U(\mathbf{x}) = a + \mathbf{b}^T\mathbf{x} + \frac{1}{2}\mathbf{x}^T\mathbf{H}\mathbf{x} \tag{11.86}$$

the gradient vector is

$$\nabla \mathbf{U} = \mathbf{b} + \mathbf{H}\mathbf{x} \tag{11.87}$$

The difference of the gradient vectors at two subsequent points $\mathbf{x}^{(k)}$ and $\mathbf{x}^{(k+1)}$ is

$$\nabla U^{(k+1)} - \nabla U^{(k)} = \mathbf{H}\left(\mathbf{x}^{(k+1)} - \mathbf{x}^{(k)}\right) \tag{11.88}$$

As the difference of two successive points is

$$\mathbf{x}^{(k+1)} - \mathbf{x}^{(k)} = \varDelta\mathbf{x}^{(k)} = d^{(k)}\mathbf{s}^{(k)} \tag{11.89}$$

then (11.88) may be written as

$$\nabla U^{(k+1)} - \nabla U^{(k)} = \mathbf{H}\,\varDelta\mathbf{x}^{(k)} = d^{(k)}\mathbf{H}\mathbf{s}^{(k)} \tag{11.90}$$

and finally

$$\mathbf{H}\mathbf{s}^{(k)} = \frac{\nabla U^{(k+1)} - \nabla U^{(k)}}{d^{(k)}} \tag{11.91}$$

Substituting (11.91) into (11.85), we have

$$\left(-\nabla U(\mathbf{x}^{(1)}) + k_{11}\mathbf{s}^{(0)}\right)^T \frac{\nabla U^{(1)} - \nabla U^{(0)}}{d^{(0)}} = 0 \tag{11.92}$$

Assuming perfect minimization in the search direction (satisfaction of (11.80)) and using the fact that $\mathbf{s}^{(0)} = -\nabla U^{(0)}$, we have

$$\alpha_1^{(1)} = k_{11} = \frac{\left[\nabla U^{(1)}\right]^T \nabla U^{(1)}}{\left[\nabla U^{(0)}\right]^T \nabla U^{(0)}} \tag{11.93}$$

Applying the preceding result to the second step direction, the result for the second direction to the third direction, and so on, we have the result:

$$\alpha_1^{(k)} = k_{kk} = \frac{\left[\nabla U^{(k)}\right]^T \nabla U^{(k)}}{\left[\nabla U^{(k-1)}\right]^T \nabla U^{(k-1)}} \tag{11.94}$$

and $k_{ik} = 0$.

Finally, the direction of search $s^{(k)}$ is given by

$$s^{(k)} = -\nabla U^{(k)} + \alpha^{(k)} s^{(k-1)} \tag{11.95}$$

where $\alpha^{(k)}$ is calculated from (11.94) and, initially, $\alpha^{(0)} = 0$. Apart from round-off errors, the procedure will find a minimum of a quadratic function in at most n iterations. In nonquadratic functions, we recommended resetting the search directions after n steps to $s^{(n)} = -\nabla U^{(k)}$.

The method referred to as the *conjugate gradient method* of Fletcher and Reeves [8] is very effective for nonquadratic functions.

11.3.3 The Newton Method

An optimization method requiring knowledge of second-order derivatives of the objective function is called the *Newton method*. Let us consider a form of the Taylor series (11.41). Differentiating it with respect to Δx we have

$$\nabla U(x^{(k)} + \Delta x) = \nabla U(x^{(k)}) + H(x^{(k)}) \Delta x + \cdots \tag{11.96}$$

We want $x^{(k)} + \Delta x$ to be the minimum point, so that $\nabla U(x^{(k)} + \Delta x)$ will be zero:

$$\nabla U(x^{(k)}) + H(x^{(k)}) \Delta x = 0 \tag{11.97}$$

or

$$H^{(k)} \Delta x^{(k)} = -\nabla U^{(k)} \tag{11.98}$$

The latter relation is a system of linear equations that may be solved for Δx by using one of methods discussed in Chapter 6.

The search direction is set equal to Δx and written formally as

$$s^{(k)} = \Delta x^{(k)} = -\left(H^{(k)}\right)^{-1} \nabla U^{(k)} = -B^{(k)} \nabla U^{(k)} \tag{11.99}$$

where $B^{(k)}$ is the inverse of the Hessian matrix at the kth iteration.

In a quadratic function only one iteration with incremental change Δx of variables leads us to the minimum. However, this scheme has several disadvantages. For example, the Hessian matrix may not be positive definite, leading to divergence ($-\mathbf{H}^{-1}\nabla U^{(k)}$ might not point downhill). Finally, the computation of the Hessian and its inverse usually are very time consuming and expensive. To avoid this, instead of computing the matrix, its approximation is derived from first derivative information. A class of methods based on these ideas is called *quasi-Newton* or *variable metric* methods.

11.3.4 Quasi-Newton Methods

Quasi-Newton or variable metric methods are based on the computation of the gradient vector of the objective function and approximation of the Hessian matrix or its inverse. The simulation is made by continually updating a positive definite matrix in a way that preserves the positive definiteness. In all variable metric methods, the search direction is [33]

$$\mathbf{s}^{(k)} = -\mathbf{B}^{(k)}\nabla U^{(k)} \tag{11.100}$$

where $\mathbf{B}^{(k)}$ is the kth approximation to the inverse of the Hessian matrix. Initially, $\mathbf{B}^{(0)}$ is the unit matrix, and $\mathbf{s}^{(0)}$ is the steepest descent direction.

In each iteration, at a given point $\mathbf{x}^{(k)}$, we generate a direction $\mathbf{s}^{(k)} = -\mathbf{B}^{(k)}\nabla U^{(k)}$ and then perform minimization in this direction. In the succeeding iteration the matrices $\mathbf{B}^{(k)}$ are computed according to the following scheme.

We evaluate vectors:

$$\Delta\mathbf{x}^{(k)} = \mathbf{x}^{(k)} - \mathbf{x}^{(k-1)} \tag{11.101}$$

$$\mathbf{r}^{(k)} = \nabla U(\mathbf{x}^{(k)}) - \nabla U(\mathbf{x}^{(k-1)}) \tag{11.102}$$

and compute a new matrix $\mathbf{H}^{(k)}$ according to the formula:

$$\mathbf{B}^{(k)} = \mathbf{B}^{(k-1)} + \Delta\mathbf{B}^{(k)} \tag{11.103}$$

where $\Delta\mathbf{B}^{(k)}$ is the updated matrix. The new approximation to the inverse of the Hessian matrix takes into account newly accumulated curvature information. Computation of $\Delta\mathbf{B}^{(k)}$ in (11.103) may be performed in many different ways [34, 35]. In the

Davidon, Fletcher, and Powell (DFP) method [33], the updating formula for $\mathbf{B}^{(k)}$ has the form:

$$\mathbf{B}^{(k+1)} = \mathbf{B}^{(k)} + \frac{\Delta\mathbf{x}^{(k)}\left[\Delta\mathbf{x}^{(k)}\right]^T}{\left[\Delta\mathbf{x}^{(k)}\right]^T\mathbf{r}^{(k)}} - \frac{\mathbf{B}^{(k)}\mathbf{r}^{(k)}\left[\mathbf{r}^{(k)}\right]^T\mathbf{B}^{(k)}}{\left[\mathbf{r}^{(k)}\right]^T\mathbf{B}^{(k)}\mathbf{r}^{(k)}} \qquad (11.104)$$

As in the conjugate gradient method, a very important element of the variable metric method is reinitialization, understood as resetting $\mathbf{H}$ to the identity matrix. It is performed every nth iteration.

The updating formula (11.104) is generally known as the DFP formula. Derivation of the formula is beyond the scope of this text. Other updating formulas exist and are available in literature. The best known are the Broyden, Fletcher, Goldfarb, and Shanno (BFGS) and the Powell symmetric Broyden (PSB) formulas [36–38].

The merits of these formulas usually are compared in terms of their preservation of positive definiteness, numerical performance, and convergence to the true Hessian. Discussions are given in several books [34, 35].

11.3.5 Line Search

Almost all nonlinear optimization algorithms require the determination of the step length $d^{(k)}$ along the search direction $\mathbf{s}^{(k)}$. As mentioned in previous sections of this chapter, properties of many minimization algorithms depend strongly on an exact line search. Ideally, $d^{(k)}$ should be chosen to minimize $U(\mathbf{x})$ in the search direction $\mathbf{s}^{(k)}$ so that (11.80) is satisfied. In practice, however, a completely accurate line search is numerically very expensive, and therefore it is replaced by other methods. The line search may be performed in many different ways, such as bisection, curve fitting, or other methods. The evaluation of $U(\mathbf{x})$, $\nabla U(\mathbf{x})$, or both usually is limited to only a few points located along the search direction. In curve fitting methods, quadratic or cubic interpolation or extrapolation techniques are then incorporated [34, 35].

Figure 11.15 illustrates a principle of the line search with quadratic interpolation. The line search function given by

$$T(d) = U(\mathbf{x}^{(k)} + d\mathbf{s}^{(k)}) \qquad (11.105)$$

is interpolated by a parabola $T(d) = Ad^2 + Bd + C$. The first function value $T_0 = T(0) = U(\mathbf{x}^{(k)})$ is known: it is the result of the previous minimization step. Two arbitrarily selected step sizes d_1 and d_2 and the function values $T_1 = T(d_1)$ and $T_2 = T(d_2)$ provide additional points that are used to interpolate a parabola. Th

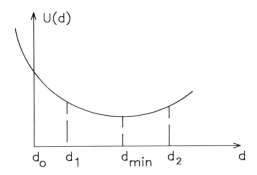

Figure 11.15 Quadratic interpolation.

method discussed here does not use gradient information. If the following conditions are satisfied:

$$d_0 < d_1 < d_2 \tag{11.106}$$

$$T(d_1) < T(d_0)$$

$$T(d_1) \le T(d_2) \tag{11.107}$$

function $T(d)$ has a minimum in the region $[d_0, d_2]$.

As we can easily find, the distance d_{min} to the minimum of the parabola (provided $A > 0$) is given by

$$d_{min} = -\frac{B}{2A} = \frac{T_2^2(T_1 - T_0) - T_1^2(T_2 - T_0)}{2[T_2(T_1 - T_0) - T_1(T_2 - T_1)]} \tag{11.108}$$

Evaluation of T at d_{min} gives the estimate of the minimum and completes one step of the search. A new step can be started and the procedure may be repeated until we find the true minimum with the given accuracy.

What is a function to minimize using the line search? An answer to this question is clear in cases of unconstrained problems. However, this is not the case in some constrained optimization methods, because in a line search we not only wish to reduce the objective function, but also to satisfy the constraints. This problem has led to development of special line search objective functions [36]. Some elements of this problem will be presented in the following sections.

11.4 GRADIENT-BASED METHODS FOR CONSTRAINED FUNCTION MINIMIZATION

Most optimization problems arising from microwave circuit design involve minimization of an objective function subject to certain constraints. Such a programming problem may be defined as the minimization of

$$\min_{\mathbf{x} \in R^n} U(\mathbf{x}) \tag{11.109}$$

subject to $c_i(\mathbf{x}) \leq 0$, $i = 1, 2, \ldots, m$. The constraints define an acceptable (feasible) region R_A in R^n given as $R_A \triangleq \{\mathbf{x} \mid c_i(\mathbf{x}) \leq 0, i = 1, 2, \ldots, m\}$.

In the past the penalty function methods were popular for solving such problems. These techniques replaced the constraint optimization problem with a sequence of parameterized unconstrained optimization, which within limits converted to the solution of constrained problem. Appropriate penalty or barrier functions were used to transform the constrained optimization problem into an unconstrained one [37, 38], but such methods are considered now computationally inefficient. In recent years, progress has been achieved in the area of constrained optimization. The basic optimization techniques are based on creating a sequence of models of the problem that are computationally simpler than the problem itself. These models usually are nonlinear programming problems without constraints, linear programming problems, or quadratic programming problems. We can show that the solutions to the simpler problems converge to a local solution of the original problem.

In this section we discuss only constrained quasi-Newton and penalty multiplier methods for scalar objective functions and constrained Gauss-Newton and quasi-Newton methods for multiple objective functions. These methods have proved to be most efficient tools for solving constrained optimization problems.

The key to these methods is the solution of the Kuhn-Tucker equations, which are necessary conditions for constrained minimum. We will consider them as a preliminary problem.

11.4.1 Kuhn-Tucker Conditions

In Section 11.1.1 we discussed requirements for a point $\mathbf{x}^*$ to be an unconstrained minimum of a function $U(\mathbf{x})$ of many variables.

The necessary conditions for the optimal solution of the nonlinear programming problem defined by (11.109) are called Kuhn-Tucker conditions [39].

We assume that $U(\mathbf{x})$ and $\mathbf{c}(\mathbf{x})$ are differentiable in the neighborhood of $\mathbf{x}^*$. Let $\mathbf{x}^*$ be a point that satisfies the constraints $\mathbf{c}(\mathbf{x}) \leq 0$, where just the first t of these constraints are satisfied as equalities. The nonlinear inequality constraints that are exactly zero at point $\mathbf{x}^*$ are said to be active at $\mathbf{x}^*$.

The necessary condition that $\mathbf{x}^*$ minimizes $U(\mathbf{x})$ subject to the constraints $\mathbf{c}(\mathbf{x}) \leq 0$ is that the gradient vector of the objective function $\nabla U(\mathbf{x})$ at $\mathbf{x} = \mathbf{x}^*$ can be expressed in the form:

$$\nabla U(\mathbf{x}^*) = - \sum_{i=1}^{t} \lambda_i^* \frac{\partial c_i(\mathbf{x}^*)}{\partial \mathbf{x}} \qquad (11.110)$$

$$\lambda_i^* \geq 0, \quad i = 1, 2, \ldots, t \qquad (11.111)$$

This relation says that, when the constraint gradient vectors are linearly independent, the gradient vector of the objective function $U(\mathbf{x})$ at $\mathbf{x} = \mathbf{x}^*$ is a negative linear combination of the gradients $\partial c_i(\mathbf{x}^*)/\partial \mathbf{x}$ of those constraints active at $\mathbf{x}^*$ ($i = 1, 2, \ldots, t$).

In the more usual notation, (11.110) would be expressed in the form:

$$\nabla U(\mathbf{x}^*) = - \sum_{i=1}^{m} \lambda_i \frac{\partial c_i(\mathbf{x}^*)}{\partial \mathbf{x}} \qquad (11.112)$$

$$\lambda_i \geq 0, \quad i = 1, 2, \ldots, m \qquad (11.113)$$

where the conditions $\lambda_i = 0$ for $i = t + 1, t + 2, \ldots, m$ are obtained by imposing the auxiliary restrictions:

$$\lambda_i c_i(\mathbf{x}^*) = 0, \quad i = 1, 2, \ldots, m \qquad (11.114)$$

The λ_i, $i = 1, 2, \ldots, m$ are termed the *Lagrange multipliers*. The graphical interpretation of these ideas is presented in Figure 11.16.

If the problem fulfills the conditions of convex programming, that is, if $U(\mathbf{x})$ and $\mathbf{c}(\mathbf{x})$ are convex functions, and R_A has a nonempty interior, the Kuhn-Tucker conditions (11.112) to (11.114) become both necessary and sufficient for $\mathbf{x}^*$ to be the constrained minimum. The Kuhn-Tucker sufficiency theorem may be stated as follows.

If $U(\mathbf{x})$ and $\mathbf{c}(\mathbf{x})$ are convex and continuously differentiable then the existence of $\lambda_i \geq 0$ multipliers such that the Kuhn-Tucker relations are satisfied is the sufficient condition that $\mathbf{x}^*$ is a global minimum of an inequality constrained optimization problem defined by (11.109).

More detailed considerations on the Kuhn-Tucker conditions indicate that the constrained qualification (convexity of R_A) must also hold [39].

Most problems we encounter in the microwave CAD practice are not convex. However, if the Kuhn-Tucker equations can be solved, the local optimum of the objective function can be found.

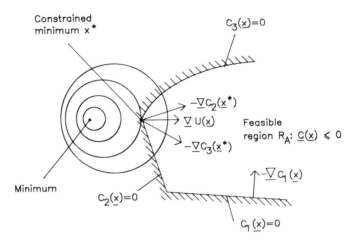

Figure 11.16 Graphic representation of the Kuhn-Tucker sufficient conditions for a constrained minimum of inequality constrained problem: $\lambda_1 = 0$, $\lambda_2 > 0$, $\lambda_3 = 0$.

In an optimization problem with inequality and equality constraints defined as the minimization of

$$\min_{x \in R^n} U(\mathbf{x}) \tag{11.115}$$

subject to

$$\mathbf{c}_i(\mathbf{x}) \le 0, \quad i = 1, 2, \ldots, m_c$$

$$h_i(\mathbf{x}) = 0, \quad i = 1, 2, \ldots, m_h$$

we can prove that the Kuhn-Tucker necessary conditions for $\mathbf{x}^*$ to be a minimum point are

$$\nabla U(\mathbf{x}^*) + \sum_{i=1}^{m_c} \lambda_i^* \nabla c_i(\mathbf{x}^*) + \sum_{i=1}^{m_h} \mu_i^* \nabla h_i(\mathbf{x}^*) = 0 \tag{11.116}$$

$$\lambda_i^* c_i(\mathbf{x}) = 0, \quad i = 1, 2, \ldots, m_c \tag{11.117}$$

where $\lambda^* \in R^{m_c}$, $\lambda^* \ge 0$ and $\mu^* \in R^{m_h}$.

We should mention here that the multipliers λ_i^*, $i = 1, 2, \ldots, m_c$, corresponding to inequality constraints, are nonnegative, whereas the multipliers μ^*, $i = 1, 2, \ldots, m_h$, corresponding to equality constraints, have any sign.

Discussing a constrained optimization problem with inequality constraints it is common to introduce the Lagrange function of the problem:

$$L(\mathbf{x}, \boldsymbol{\lambda}) = U(\mathbf{x}) + \sum_{i=1}^{m} \lambda_i c_i(\mathbf{x}) \tag{11.118}$$

Using the Lagrange function, the necessary conditions (11.112) to (11.114) may be written in the form:

$$\nabla_x L(\mathbf{x}^*, \boldsymbol{\lambda}^*) = 0 \tag{11.119}$$

$$[\boldsymbol{\lambda}^*]^T \nabla_\lambda L(\mathbf{x}^*, \boldsymbol{\lambda}^*) = 0 \tag{11.120}$$

$$\nabla_\lambda L(\mathbf{x}^*, \boldsymbol{\lambda}^*) \leq 0 \tag{11.121}$$

$$\boldsymbol{\lambda}^* \geq 0 \tag{11.122}$$

where ∇_x denotes the gradient vector of the Lagrange function with respect to $\mathbf{x}$, and ∇_λ is the gradient with respect to $\boldsymbol{\lambda}$.

Similarly, for an optimization problem with inequality and equality constraints, the Lagrange function has a form:

$$L(\mathbf{x}, \boldsymbol{\lambda}, \boldsymbol{\mu}) = U(\mathbf{x}) + \sum_{i=1}^{m_c} \lambda_i c_i(\mathbf{x}) + \sum_{i=1}^{m_h} \mu_i h_i(\mathbf{x}) \tag{11.123}$$

and the necessary conditions (11.116) and (11.117) are written as

$$\nabla_x L(\mathbf{x}^*, \boldsymbol{\lambda}^*, \boldsymbol{\mu}^*) = 0 \tag{11.124}$$

$$[\boldsymbol{\lambda}^*]^T \nabla_\lambda L(\mathbf{x}^*, \boldsymbol{\lambda}^*, \boldsymbol{\mu}^*) = 0 \tag{11.125}$$

$$\nabla_\lambda L(\mathbf{x}^*, \boldsymbol{\lambda}^*, \boldsymbol{\mu}^*) \leq 0 \tag{11.126}$$

$$\nabla_\mu L(\mathbf{x}^*, \boldsymbol{\lambda}^*, \boldsymbol{\mu}^*) = 0 \tag{11.127}$$

$$\boldsymbol{\lambda}^* \geq 0 \tag{11.128}$$

11.4.2 Constrained Quasi-Newton Methods

M.C. Biggs [40], S.P. Han [41], and M.J.D. Powell [36] have formulated a theory for the constrained quasi-Newton methods of a scalar objective function $U(\mathbf{x})$. We first present the constrained problem with equality constraints defined as

$$\min_{x \in R^n} U(x) \qquad (11.129)$$

subject to

$$h(x) = 0 \qquad (11.130)$$

where $h(x) = [h_1(x), h_2(x), \ldots, h_m(x)]^T$.

According to the Kuhn-Tucker equations, at an optimal point x^* we can always find Lagrange multipliers λ_i such that

$$m(x^*, \lambda) \equiv \frac{\partial U(x^*)}{\partial x} + \sum_{i=1}^{m} \lambda_i \frac{\partial h_i(x^*)}{\partial x} = 0 \qquad (11.131)$$

$m(x, \lambda) = 0$ is an n component function that, together with the equality constraints $h_i(x) = 0$, $i = 1, 2, \ldots, m$ form a set of $n + m$ nonlinear equations with $n + m$ unknowns x and λ.

A Newton method for the nonlinear system of nonlinear equations

$$f(x) = 0 \qquad (11.132)$$

can be defined as follows.

From the Taylor series expansion of f about $x^{(k)}$, we can obtain a linear approximation of f:

$$f(x^*) \approx f(x^{(k)}) + J(x^{(k)})(x^* - x^{(k)}) \qquad (11.133)$$

where x^* is the solution of (11.132).

The Newton step Δx is an approximation to $x^* - x^{(k)}$, and we define it by equating the right-hand side of (11.133) to zero. Thus, Δx satisfies the system of equations:

$$J(x^{(k)}) \Delta x = -f(x^{(k)}) \qquad (11.134)$$

and the next iteration is given by $x^{(k+1)} = x^{(k)} + \Delta x$.

Applying the Newton method to the set of $m + n$ equations given by (11.130) and (11.131), we have

$$\begin{bmatrix} H(x) + \sum \lambda_i H_{ci}(x) & -C_x^T \\ C_x & 0 \end{bmatrix} \begin{bmatrix} \Delta x \\ \Delta \lambda \end{bmatrix} = \begin{bmatrix} -m(x, \lambda) \\ -h(x) \end{bmatrix} \qquad (11.135)$$

In (11.135) $\mathbf{H}(\mathbf{x})$ is the Hessian of $U(\mathbf{x})$, $\mathbf{H}_{hi}(\mathbf{x})$ is the Hessian of $h_i(\mathbf{x})$, and

$$
\mathbf{C}_x = \begin{bmatrix}
\dfrac{\partial h_1(\mathbf{x})}{\partial x_1} & \dfrac{\partial h_1(\mathbf{x})}{\partial x_2} & \cdots & \dfrac{\partial h_1(\mathbf{x})}{\partial x_n} \\[2ex]
\dfrac{\partial h_2(\mathbf{x})}{\partial x_1} & \dfrac{\partial h_2(\mathbf{x})}{\partial x_2} & \cdots & \dfrac{\partial h_2(\mathbf{x})}{\partial x_n} \\[2ex]
\vdots & \vdots & & \vdots \\[2ex]
\dfrac{\partial h_m(\mathbf{x})}{\partial x_1} & \dfrac{\partial h_m(\mathbf{x})}{\partial x_2} & \cdots & \dfrac{\partial h_m(\mathbf{x})}{\partial x_n}
\end{bmatrix}
\tag{11.136}
$$

We would like to solve the preceding equation for $\Delta\mathbf{x}$ and $\Delta\lambda$ and use the result to update $\mathbf{x}$, λ by $\mathbf{x} = \mathbf{x}^{(k)} + \Delta\mathbf{x}^{(k)}$, $\lambda = \lambda^{(k)} + \Delta\lambda^{(k)}$.

According to the basics of the quasi-Newton approach, we assume a quadratic approximate for the Lagrangian function:

$$
L(\mathbf{x}, \lambda) \triangleq U(\mathbf{x}) - \sum_{i=1}^{m} \lambda_i h_i(\mathbf{x})
\tag{11.137}
$$

that is

$$
L(\mathbf{x} + \Delta\mathbf{x}, \lambda) \approx L(\mathbf{x}, \lambda) + \mathbf{m}(\mathbf{x}, \lambda)^T \Delta\mathbf{x} + \frac{1}{2} \Delta\mathbf{x}^T \mathbf{G} \, \Delta\mathbf{x}
\tag{11.138}
$$

where an $n \times n$ matrix $\mathbf{G}$ is an approximation to the Hessian of $L(\mathbf{x}, \lambda)$. Then, (11.135) is approximated by

$$
\begin{bmatrix} \mathbf{G} & -\mathbf{C}_x^T \\ \mathbf{C}_x & 0 \end{bmatrix}
\begin{bmatrix} \Delta\mathbf{x} \\ \Delta\lambda \end{bmatrix} =
\begin{bmatrix} -\mathbf{m}(\mathbf{x}, \lambda) \\ -\mathbf{h}(\mathbf{x}) \end{bmatrix}
\tag{11.139}
$$

from which we obtain the solution $\Delta\mathbf{x}$ and $\Delta\lambda$. Equation (11.139) can be also written in terms of $\lambda^* = \lambda + \Delta\lambda$ and $\Delta\mathbf{x}$; that is,

$$
\begin{bmatrix} \mathbf{G} & -\mathbf{C}_x^T \\ \mathbf{C}_x & 0 \end{bmatrix}
\begin{bmatrix} \Delta\mathbf{x} \\ \lambda^* \end{bmatrix} =
\begin{bmatrix} -\nabla U(\mathbf{x}) \\ -\mathbf{h}(\mathbf{x}) \end{bmatrix}
\tag{11.140}
$$

As we can see, a quasi-Newton method for constrained optimization at each step requires, first, the matrix $\mathbf{G}$ that is an approximation to the Hessian of the Lagrangian function:

$$
L(\mathbf{x}, \lambda) = U(\mathbf{x}) + \sum_{i=1}^{m} \lambda_i h_i(\mathbf{x})
$$

and, second, the solution of a set of linear equations (11.139) (or equivalently the solution of (11.140)). The approximate Hessian **G** depends on the current estimate of the Lagrange multipliers, λ. When the constraints are linear, the matrix **G** must approximate only **H(x)**, the Hessian of $U(\mathbf{x})$, and the Lagrange multipliers are not needed. The solution of (11.139) are vectors $\Delta\mathbf{x}$ and $\Delta\lambda$, which are used to update **x** and λ.

At each step the $\Delta\mathbf{x}$ obtained is the constrained quasi-Newton direction, which defines a search direction for the minimum of some function. If this function were the objective function $U(\mathbf{x})$, searching for its minimum along $\Delta\mathbf{x}$ might violate the constraints $\mathbf{h(x)} = \mathbf{0}$. It is suggested searching along $\Delta\mathbf{x}$ to minimize the function [41]:

$$\hat{U}(\mathbf{x}) = U(\mathbf{x}) + \sum_{i=1}^{m} \mu_i |h_i(\mathbf{x})| \qquad (11.141)$$

where $\mu_i > |\lambda_i + \Delta\lambda_i|$. This suggestion comes from the fact that, if the matrix **G** is positive definite, we can prove that there is such an $\alpha > 0$ that

$$U(\mathbf{x}) + \sum_{i=1}^{m} \mu_i |h_i(\mathbf{x})| > U(\mathbf{x} + \alpha\,\Delta\mathbf{x}) + \sum_{i=1}^{m} \mu_i |\mathbf{h}_i(\mathbf{x}) + \alpha\,\Delta\mathbf{x}| \quad (11.142)$$

which means that $\Delta\mathbf{x}$ defines a descent direction for the function $\hat{U}(\mathbf{x})$ defined by (11.141).

The inequality constrained problem defined as

$$\min_{\mathbf{x} \in R^n} U(\mathbf{x})$$

subject to $c_i(\mathbf{x}) \leq 0$, $i = 1, 2, \ldots, m$, can be solved iteratively using the ideas developed for the constrained problem with equality constraints. According to (11.131) and (11.137) $\mathbf{m(x}, \lambda)$ is the gradient with respect to **x** of the Lagrangian function $L(\mathbf{x}, \lambda)$. To update the current approximate Jacobian of $\mathbf{m(x}, \lambda)$ (it is the Hessian matrix of L), which we denote by **G**, change $\alpha\Delta\mathbf{x}$ in **x** and the change $\Delta\mathbf{m}$ in the gradient of $L(\mathbf{x}, \lambda)$ are used. If no additional information is available the initial Hessian approximation $\mathbf{G}^{(0)}$ is usually taken as the identity matrix. With such choice, the first iteration of a quasi-Newton method is equivalent to an iteration in the steepest descent method. After $\mathbf{x}^{(k)}$ has been computed, a new Hessian approximation $\mathbf{G}^{(k)}$ is obtained by updating $\mathbf{G}^{(k-1)}$. **G** can be updated by using any of the quasi-Newton updating methods. The best known are the Powell symmetric Broyden (PSB) update [42], the Davidon-Fletcher-Powell (DFP) update [33], and the Broyden-Fletcher-Goldfarb-Shanno (BFGS) update [32, 43–45].

Having the current approximate **G** we solve the constrained quadratic problem with inequality constraints:

$$\min_{\Delta\mathbf{x}} U(\mathbf{x}) + \nabla^T U(\mathbf{x})\,\Delta\mathbf{x} + \tfrac{1}{2}\,\Delta\mathbf{x}^T \mathbf{G}\,\Delta\mathbf{x} \qquad (11.143)$$

subject to

$$c_i(\mathbf{x}) + \nabla^T c_i\,\Delta\mathbf{x} \le 0, \quad i = 1, 2, \ldots, m \qquad (11.144)$$

This quadratic problem corresponds to (11.140) because the solutions $\Delta\mathbf{x}$ and $\boldsymbol{\lambda}^*$ are the same, except that, in the solution of (11.143) and (11.144), $\boldsymbol{\lambda}^* \ge 0$.

For inequality constraints, the function that controls the step-size α in the line search along $\Delta\mathbf{x}$ is given by

$$U(\mathbf{x}) + \sum_{i=1}^{m} \mu_i \left| \min\big(0, c_i(\mathbf{x})\big) \right| \qquad (11.145)$$

As has been found advantageous from numerical experimentation, instead of choosing $\mu_i = \lambda_i^*$, we let [39]

$$\mu_i = \max\left\{ \lambda_i^*, \tfrac{1}{2}\big(\bar{\mu}_i + \lambda_i^*\big) \right\} \qquad (11.146)$$

where $\bar{\mu}_i$ is the value of μ_i for the previous iteration. Thus, we allow a positive contribution from constraints that are currently inactive but have been active in previous iterations. The exact search along $\Delta\mathbf{x}$ to find the exact minimum is considered too expensive, and it is replaced by other methods such as interpolation and extrapolation techniques.

11.4.3 Penalty-Multiplier Methods (Augmented Lagrangian Methods)

The methods termed *augmented Lagrangian methods* are derived from the optimality conditions, and the role of Lagrange multipliers is crucial. Augmented Lagrangian methods can be applied to both equality and inequality constraints. We first discuss an optimization problem with equality constraints given by (11.129) and (11.130).

Assume that conditions are sufficient for a local minimum hold at $\mathbf{x}^*$; that is,

$$\nabla U(\mathbf{x}^*) - \mathbf{A}(\mathbf{x}^*)^T \boldsymbol{\lambda}^* = \mathbf{0} \qquad (11.147)$$

where $\mathbf{A}$ is a matrix with t rows that are the gradients of constraints active at $\mathbf{x}^*$. The vector of Lagrangian multipliers is unique when $\mathbf{A}(\mathbf{x}^*)$ holds full rank.

The Lagrangian function for the discussed problem is

$$L(\mathbf{x}, \boldsymbol{\lambda}) \triangleq U(\mathbf{x}) + \boldsymbol{\lambda}^T \mathbf{h}(\mathbf{x})$$

Equation (11.147) can be interpreted as a statement that $\mathbf{x}^*$ is a stationary point of the Lagrangian function when $\boldsymbol{\lambda} = \boldsymbol{\lambda}^*$. This might suggest that the Lagrangian function could be used as the objective function of an unconstrained subproblem equivalent to (11.129) and (11.130). Unfortunately, $\mathbf{x}^*$ is not necessarily a minimum of the Lagrangian function. Because of that, the Lagrangian function cannot be taken as the objective function of the subproblem.

The most popular choice for the objective function of the subproblem is the augmented Lagrangian function given by [46]

$$L_A(\mathbf{x}, \lambda, \alpha) = U(\mathbf{x}) + \lambda \mathbf{h}(\mathbf{x}) + \alpha^{-1}\left[\mathbf{c}(\mathbf{x})^T \mathbf{c}(\mathbf{x})\right] \qquad (11.148)$$

where $\alpha > 0$ is a penalty parameter.

The quadratic penalty term $\mathbf{c}(\mathbf{x})^T\mathbf{c}(\mathbf{x})$ of (11.148) and its gradient vanish at $\mathbf{x}^*$ (under the assumption that the vector $\mathbf{h}(\mathbf{x})$ contains the constraints active at $\mathbf{x}^*$). Then, if $\lambda = \lambda^*$, $\mathbf{x}^*$ is a stationary point of the optimization problem with equality constraints (11.129) and (11.130).

The objective function L_A can be minimized as a sequence of problems where the elements of λ will change according to some iteration formula. We can show that a finite α^* always exists such that, if we can choose the correct Lagrange multipliers λ_i, then $\mathbf{x}^*$ is an unconstrained minimum of $L_A(\mathbf{x}, \lambda^*, \alpha)$ for all α in $0 \leq \alpha \leq \alpha^*$. We can treat α^* as the value of α sufficient to make $L_A(\mathbf{x}, \lambda, \alpha)$ convex at $\mathbf{x}^*$. At such point the augmented Lagrangian method should work well, because the Kuhn-Tucker equations are necessary and sufficient conditions for $\mathbf{x}^*$ to be the constrained minimum.

The preceding discussion suggests that $\mathbf{x}^*$ could be computed by a single unconstrained minimization of the differentiable function (11.148). Unfortunately, in general, λ^* will not be available until the solution has been found. Therefore, an augmented Lagrangian method must include a procedure for establishing the Lagrange multipliers. This explains the second name for such an algorithm: the *penalty-multiplier method*.

The iteration suggested for the multiplier λ is

$$\lambda^{(k)} = \lambda^{(k-1)} - 2\alpha^{-1}\mathbf{h}\left[\mathbf{x}(\lambda^{(k-1)}, \alpha)\right] \qquad (11.149)$$

where $\mathbf{x}[\lambda^{(k-1)}, \alpha]$ is the minimum of (11.148) in the previous iteration.

The last problem left to be solved is that of finding α^*. It is suggested to let α be a sequence chosen in the following way [46]:

$$\alpha^{(k+1)} = \begin{cases} \beta\alpha^{(k)}, & \text{if } \mathbf{h}(\mathbf{x}^{(k)})^T\mathbf{h}(\mathbf{x}^{(k)}) > \gamma\mathbf{h}[\mathbf{x}^{(k+1)}]^T\mathbf{h}(\mathbf{x}^{(k+1)}) \\ \alpha^{(k)}, & \text{otherwise} \end{cases} \qquad (11.150)$$

where $\gamma < 1$ and $\beta < 1$. It happens in practice that near the solution a value of $\mathbf{h}^T\mathbf{h}$ decreases fast enough so that the iteration counter k greater than some K, $\alpha^{(k)} = \alpha^{(K)}$.

In an optimization problem with inequality constraints

$$\min_{\mathbf{x}} U(\mathbf{x})$$

subject to $c(x) \geq 0$, can be converted into a problem with equality constraints

$$\min_{x} U(x)$$

subject to $c(x) + z^2 = 0$, where $z = [z_1, z_2, \ldots, z_m]$ is a vector of slack variables z_i. It is possible to eliminate z conceptually. In such case the iteration formula (11.149) for the multipliers λ is modified to

$$\lambda^{(k)} = \max\{0, \lambda^{(k-1)} - 2\alpha^{-1}c(x^{(k-1)}, \alpha)\} \tag{11.151}$$

The approach discussed here generally follows that of D.F. Bertsekas [46]. Surveys of the properties of the augmented Lagrangian methods are discussed in [47–49].

11.5 MULTIPLE OBJECTIVE OPTIMIZATION

As shown in Section 11.2.2, many microwave circuit design problems may be formulated as minimization problems of a number of objective functions. This is the simultaneous optimization of several objectives, which in general may compete with each other.

A multiple objective optimization problem can be stated as follows:

$$\min_{x \in R^n} \begin{bmatrix} f_1(x) \\ f_2(x) \\ \vdots \\ f_l(x) \end{bmatrix} \tag{11.152}$$

subject to $c_i(x) \leq 0$, $i = 1, 2, \ldots, m_c$, and $h_i(x) = 0$, $i = 1, 2, \ldots, m_h$. The l objective functions f_i, $i = 1, 2, \ldots, l$ may be denoted as the l-vector:

$$f(x) = \begin{bmatrix} f_1(x) \\ f_2(x) \\ \vdots \\ f_l(x) \end{bmatrix} \tag{11.153}$$

representing a point in the l-dimensional objective function space.

If the components of $f(x)$ are competing, there is no unique solution to this problem, which means that no point simultaneously minimizes all $f_i(x)$, $i = 1, 2, \ldots, l$. In such a situation, the so-called Poreto or noninferior solution must be defined.

The noninferior solution of the multiobjective optimization problem is a point for which improvement in one objective requires a degradation in at least one of the other objective functions. This situation is illustrated in Figure 11.17 for a two objective minimization problem. Generally, there is an infinite number of Poreto points. Figure 11.17 also illustrates the concepts of global noninferior solutions, which are called the *noninferior solution set* or *noninferior surface*. Finding all points **x** of noninferior solutions is computationally infeasible. The most straightforward technique in finding a point on the noninferior surface is to combine the multiple objectives into a single one and use the classical techniques for minimizing single objective functions. We can show that a weighted minimax method is a general approach for converting a multiobjective optimization problem given by (11.152) into a scalar objective function problem. In this scheme a weighted vector of objective functions is chosen and the following scalar objective function problem is solved:

$$\min_{\mathbf{x} \in R^n} \max\left[w_1 f_1(\mathbf{x}), w_2 f_2(\mathbf{x}), \ldots, w_l f_l(\mathbf{x}) \right]$$

$$w_i \geq 0, \quad i = 1, 2, \ldots, l \tag{11.154}$$

subject to

$$c_i(\mathbf{x}) \leq 0, \quad i = 1, 2, \ldots, m_c \tag{11.155}$$

$$h_i(\mathbf{x}) = 0, \quad i = 1, 2, \ldots, m_h \tag{11.156}$$

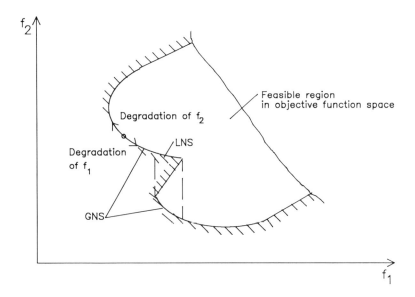

Figure 11.17 Feasible region, global noninferior solution (surface) GNS, and local noninferior (surface) LNS for a two-objective minimization problem.

By changing a set of weights w_i, $i = 1, 2, \ldots, l$, and solving (11.154) to (11.156) any point located on the noninferior surface may be found [50].

In the next two sections we will discuss special techniques that have been developed for multiple objective optimization.

We should mention here that the minimax problem defined by (11.154) to (11.156) may be substituted by an equivalent constrained optimization problem in a form:

$$\min x_{n+1} \tag{11.157}$$

subject to

$$w_i f_i(\mathbf{x}) - x_{n+1} \le 0, \quad i = 1, 2, \ldots, l \tag{11.158}$$

$$c_i(\mathbf{x}) \le 0, \quad i = 1, 2, \ldots, m_c \tag{11.159}$$

$$h_i(\mathbf{x}) = 0, \quad i = 1, 2, \ldots, m_h \tag{11.160}$$

x_{n+1} here is an additional independent variable. It may be interpreted as a level on the functions $w_i f_i(\mathbf{x})$, $i = 1, 2, \ldots, l$, which is forced down by minimization. At a minimum, at least one of the constraints $w_i f_i(\mathbf{x}) - x_{n+1} \le 0$, $i = 1, 2, \ldots, l$ must be an equality. Otherwise x_{n+1} could be decreased without violation.

The constrained optimization problem in the form (11.157) to (11.160) may be solved using any of standard constrained optimization methods discussed in previous section.

11.5.1 Constrained Gauss-Newton Methods for Multiple Objective Functions

The Gauss-Newton methods are widely used for microwave circuit optimization with multiple (vector-valued) objective functions. In these methods each function f_i is locally linearized, and linear or quadratic programming is used to solve the problem. A linearization of $\mathbf{f}$ functions is made as

$$\bar{\mathbf{f}}(\mathbf{x}^{(k)} + \Delta\mathbf{x}) = \mathbf{f}(\mathbf{x}^{(k)}) + \mathbf{J}(\mathbf{x}^{(k)}) \Delta\mathbf{x} \tag{11.161}$$

where $\mathbf{J}[\mathbf{x}^{(k)}]$ is the Jacobian matrix whose ith row is equal to the transposed gradient vector of the ith function f_i with respect to $\mathbf{x}$ at a current point $\mathbf{x}^{(k)}$:

$$\mathbf{g}_i[\mathbf{x}^{(k)}] = \frac{\partial f_i(\mathbf{x}^{(k)})}{\partial \mathbf{x}} \tag{11.162}$$

and $\Delta\mathbf{x}$ is a step used to find a better estimate of the solution.

The l_1 Optimization

Let us consider first linearly constrained l_1 problem. It is defined as

$$\min_{\mathbf{x} \in R^n} U(\mathbf{x}) = \sum_{i=1}^{l} |f_i(\mathbf{x})| \tag{11.163}$$

subject to linear equality and inequality constraints:

$$\mathbf{a}_j^T \mathbf{x} + b_j = 0, \quad j = 1, 2, \ldots, m_{eq} \tag{11.164}$$

$$\mathbf{a}_j^T \mathbf{x} + b_j \geq 0, \quad j = m_{eq} + 1, \ldots, m \tag{11.165}$$

where $\mathbf{a}_j, b_j, j = 1, \ldots, m$ are constants.

At the kth stage of the method, let $\mathbf{x}^{(k)}$ be the current best point. Each function $f_i(\mathbf{x})$ is linearized and approximated by (11.161). A step $\Delta\mathbf{x} = [\Delta x_1, \Delta x_2, \ldots, \Delta x_n]^T$, which is used to find a better estimate, is found by the solution of the following linearized problem [13, 14]:

$$\min_{\Delta\mathbf{x}} \overline{U}(\mathbf{x}^{(k)}, \Delta\mathbf{x}^{(k)}) = \sum_{i=1}^{l} \left| f_i(\mathbf{x}^{(k)}) + g_i(\mathbf{x}^{(k)})^T \Delta\mathbf{x}^{(k)} \right| \tag{11.166}$$

subject to

$$\|\Delta\mathbf{x}\|_\infty \leq \Lambda^{(k)} \tag{11.167}$$

$$\mathbf{a}_j^T(\mathbf{x}^{(k)} + \Delta\mathbf{x}^{(k)}) + b_j = 0, \quad j = 1, 2, \ldots, m_{eq} \tag{11.168}$$

$$\mathbf{a}_j^T(\mathbf{x}^{(k)} + \Delta\mathbf{x}^{(k)}) + b_j \geq 0, \quad j = m_{eq} + 1, \ldots, m \tag{11.169}$$

The solution of the problem defined by (11.166) to (11.169), $\Delta\mathbf{x}^{(k)}$, may be found by any standard linear programming method, but we suggest [14] using a very efficient algorithm of R.H. Bartels, A.R. Conn, and J.W. Sinclair [8].

Note that (11.166) is a linear approximation to (11.163). The value of the scalar $\Lambda^{(k)}$ controls the step size $\|\Delta\mathbf{x}^{(k)}\|_\infty$ so that $U(\mathbf{x})$ is decreased. If $\Lambda^{(k)}$ were small enough, the linear approximations to the $f_i(\mathbf{x})$ would be accurate and the decrease achieved. The point $(\mathbf{x}^{(k)} + \Delta\mathbf{x}^{(k)})$ is accepted as the next iterate provided that the decrease in U exceeds a prescribed fraction of the decrease described by the linear approximation. If [14, 15]

$$U(\mathbf{x}^{(k)}) - U(\mathbf{x}^{(k)} + \Delta\mathbf{x}^{(k)}) \geq \rho\left(\overline{U}(\mathbf{x}^{(k)}, 0) - \overline{U}(\mathbf{x}^{(k)}, \Delta\mathbf{x}^{(k)})\right) \tag{11.170}$$

with $\rho = 0.01$, then $\mathbf{x}^{(k+1)} = \mathbf{x}^{(k)} + \Delta\mathbf{x}^{(k)}$, otherwise $\mathbf{x}^{(k+1)} = \mathbf{x}^{(k)}$.

The value of the bound $\Lambda^{(k)}$ is adjusted at each step according to the following rules: if

$$U(\mathbf{x}^{(k)}) - U(\mathbf{x}^{(k)} + \Delta\mathbf{x}^{(k)}) \leq 0.25[\overline{U}(\mathbf{x}^{(k)}, 0) - \overline{U}(\mathbf{x}^{(k)}, \Delta\mathbf{x}^{(k)})] \quad (11.171)$$

then the bound is decreased $\Lambda^{(k+1)} = \Lambda^{(k)}/4$. Otherwise, if

$$U(\mathbf{x}^{(k)}) - U(\mathbf{x}^{(k)} + \Delta\mathbf{x}^{(k)}) \geq 0.75[\overline{U}(\mathbf{x}^{(k)}, 0) - \overline{U}(\mathbf{x}^{(k)}, \Delta\mathbf{x}^{(k)})] \quad (11.172)$$

we let $\Lambda^{(k+1)} = 2 \cdot \Lambda^{(k)}$. If neither (11.171) nor (11.172) holds, then the bound is left unchanged, $\Lambda^{(k+1)} = \Lambda^{(k)}$.

Minimax Optimization

The Gauss-Newton type methods also are used for nonlinear minimax optimization problems. The linearly constrained minimax problem is defined as

$$\min_{\mathbf{x} \in R^n} U(\mathbf{x}) = \max_i \{f_i(\mathbf{x})\} \quad (11.173)$$

subject to

$$\mathbf{a}_j^T \mathbf{x} + b_j = 0, \quad j = 1, 2, \ldots, m_{eq} \quad (11.174)$$

$$\mathbf{a}_j^T \mathbf{x} + b_j \geq 0, \quad j = m_{eq} + 1, \ldots, m \quad (11.175)$$

where $\mathbf{a}_j, b_j, j = 1, 2, \ldots, m$ are constants.

As in the l_1 problems, by using the linear approximation (11.161) of nonlinear functions $f_i(\mathbf{x})$, the linearly constrained nonlinear minimax problem defined by (11.173) to (11.175) can be transformed into a linear minimax problem [27, 28]:

$$\min_{\mathbf{x} \in R^n} \overline{U} = \max\{f_i(\mathbf{x}^{(k)}) + \mathbf{q}_i^{(k)}(\mathbf{x}^{(k)})^T \Delta\mathbf{x}^{(k)}\} \quad (11.176)$$

subject to

$$\|\Delta\mathbf{x}^{(k)}\|_\infty \leq \Lambda^{(k)} \quad (11.177)$$

$$\mathbf{a}_j^T(\mathbf{x}^{(k)} + \Delta\mathbf{x}^{(k)}) + b_j = 0, \quad j = 1, 2, \ldots, m_{eq} \quad (11.178)$$

$$\mathbf{a}_j^T(\mathbf{x}^{(k)} + \Delta\mathbf{x}^{(k)}) + b_j \geq 0, \quad j = m_{eq} + 1, \ldots, m \quad (11.179)$$

The solution of (11.176) to (11.179) denoted $\Delta\mathbf{x}^{(k)}$, is found by linear programming.

The conditions for acceptance of a new point $\mathbf{x}^{(k+1)} = \mathbf{x}^{(k)} + \Delta\mathbf{x}^{(k)}$ and the adjustment of the bound $\Lambda^{(k)}$ at each iteration step, discussed for the l_1 optimization problem, are also applicable to the method of the solution of the minimax problem.

Least-Squares Optimization

Normally, to find a minimum of the least-squares objective

$$U(\mathbf{x}) = \frac{1}{2} \sum_{i=1}^{l} f_i^2(\mathbf{x}) = \frac{1}{2} \left(\| f(\mathbf{x}) \|_2 \right)^2 \tag{11.180}$$

we can use a general optimization method, but in most circumstances the properties of (11.180) make it worthwhile to use methods designed especially for the least-squares optimization problems. These particular methods utilize special structures of the gradient vector and the Hessian matrix of the least-squares objective (11.180). Let the $l \times n$ Jacobian matrix of $\mathbf{f}(\mathbf{x}) = [f_1(\mathbf{x}), f_2(\mathbf{x}), \ldots, f_l(\mathbf{x})]^T$ be denoted by $\mathbf{J}(\mathbf{x})$, and let the matrix $\mathbf{H}_i(\mathbf{x})$ denote the Hessian matrix of $f_i(\mathbf{x})$. Then the gradient vector of (11.180) would equal

$$\mathbf{g}(\mathbf{x}) = \nabla U(\mathbf{x}) = \mathbf{J}(\mathbf{x})^T \mathbf{f}(\mathbf{x}) \tag{11.181}$$

and the Hessian

$$\mathbf{H}(\mathbf{x}) = \mathbf{J}(\mathbf{x})^T \mathbf{J}(\mathbf{x}) + \mathbf{Q}(\mathbf{x}) \tag{11.182}$$

where

$$\mathbf{Q}(\mathbf{x}) = \sum_{i=1}^{l} f_i(\mathbf{x}) \mathbf{H}_i(\mathbf{x}) \tag{11.183}$$

From (11.181) and (11.182) the Newton equation $\mathbf{H}^{(k)} \Delta\mathbf{x}^{(k)} = -\mathbf{g}^{(k)}$ becomes

$$\left(\mathbf{J}^{(k)}(\mathbf{x})^T \mathbf{J}^{(k)}(\mathbf{x}) + \mathbf{Q}^{(k)}(\mathbf{x}) \right) \Delta\mathbf{x}^{(k)} = -\mathbf{J}^{(k)}(\mathbf{x})^T \mathbf{f}^{(k)}(\mathbf{x}) \tag{11.184}$$

If $\| \mathbf{f}^{(k)} \|_2$ tends to zero as $\mathbf{x}^{(k)}$ approaches the solution, the matrix $\mathbf{Q}^{(k)}$ also tends to zero. Thus the Newton direction can be approximated by the solution of the equation:

$$\mathbf{J}^{(k)T} \mathbf{J}^{(k)} \Delta\mathbf{x}^{(k)} = -\mathbf{J}^{(k)T} \mathbf{f}^{(k)} \tag{11.185}$$

The solution of (11.185) is a solution of the linear least-squares problems defined as

$$\min \frac{1}{2} \left(\| \mathbf{J}^{(k)} \Delta\mathbf{x}^{(k)} + \mathbf{f}^{(k)} \|_2 \right)^2 \tag{11.186}$$

The vector $\Delta\mathbf{x}^{(k)}$ that solves (11.185) is called the *Gauss-Newton direction*.

In the Levenberg-Marquardt method [51, 52], the equation (11.185) is modified to the form:

$$\left(\mathbf{J}^{(k)T}\mathbf{J}^{(k)} + \Lambda^{(k)}\mathbf{I}\right)\Delta\mathbf{x}^{(k)} = -\mathbf{J}^{(k)T}\mathbf{f}^{(k)} \qquad (11.187)$$

where $\mathbf{I}$ is an identity matrix, and $\Lambda^{(k)}$ is a nonnegative scalar.

The Levenberg-Marquardt search direction $\Delta\mathbf{x}$ is found by the solution of the system of equations (11.187) by using, for example, LU or better QR factorization. The new iterand $\mathbf{x}^{(k+1)}$ is given by $\mathbf{x}^{(k)} + \Delta\mathbf{x}^{(k)}$. The parameter $\Lambda^{(k)}$ is very critical for this method. A "good" value for $\Lambda^{(k)}$ must be chosen to ensure descent. When $\Lambda^{(k)}$ is zero, $\Delta\mathbf{x}^{(k)}$ is the Gauss-Newton direction. As $\Lambda^{(k)} \to \infty$, $\|\Delta\mathbf{x}^{(k)}\| \to 0$ and $\mathbf{H}^{(k)}$ becomes parallel to the steepest descent direction.

11.5.2 Constrained Quasi-Newton Methods for Multiple Objective Functions

The constrained quasi-Newton methods for multiple (vector-valued) objective optimization applicable to the l_1 norm and minimax optimization problems have been developed by J. Hald and K. Madsen [14, 15, 27].

The l_1 Optimization

The method discussed now is a local method, which means that it works well only in the vicinity of the solution. In the linearly constrained l_1 problem defined by (11.163) and (11.164), zeros of the nonlinear functions $\mathbf{f}(\mathbf{x})$ play very important roles in the characteristics of the l_1 objective function.

Let $\mathbf{x}^*$ be a solution point of the problem. We define the set of functions f_i that are zero at $\mathbf{x}^*$

$$Z(\mathbf{x}^*) \triangleq \{i \mid f_i(\mathbf{x}^*) = 0\} \qquad (11.188)$$

and the set of active linear constraints:

$$C(\mathbf{x}^*) \triangleq \{j \mid \mathbf{a}_j^T\mathbf{x}^* + b_j = 0\} \qquad (11.189)$$

Because of continuity, $Z(\mathbf{x}) \subseteq Z(\mathbf{x}^*)$ for $\mathbf{x}$ in the neighborhood of $\mathbf{x}^*$; and therefore $U(x)$ can be partitioned into smooth (differentiable), and nonsmooth (nondifferentiable) parts:

$$U(\mathbf{x}) = \sum_{i \notin Z(\mathbf{x})} |f_i(\mathbf{x})| + \sum_{i \in Z(\mathbf{x})} |f_i(\mathbf{x})|$$

$$= g(\mathbf{x}) + \sum_{i \in Z(\mathbf{x})} |f_i(\mathbf{x})| \qquad (11.190)$$

In (11.190), $g(\mathbf{x})$ denotes the differentiable part of $U(\mathbf{x})$. The functions $f_i(x)$ whose values are zero at $\mathbf{x}^*$ contribute to the kinks of $U(\mathbf{x})$. Functions $f_i(\mathbf{x})$ that are nonzero at $\mathbf{x}^*$ form smooth part of $U(\mathbf{x})$, because for such functions $|f_i(\mathbf{x})|$ are smooth near $\mathbf{x}^*$.

Hald and Madsen proved [13] that the stationary point $\mathbf{x}^*$ of the linearly constrained l_1 problem must satisfy the equation:

$$\nabla g(\mathbf{x}) + \sum_{i \in Z^*} \delta_i^* \nabla f_i(\mathbf{x}^*) - \sum_{j \in C^*} \lambda_j^* a_j = \mathbf{0} \qquad (11.191)$$

in which $|\delta_i^*| \leq 1$, $\lambda_j \geq 0$ for $j > m_{\mathrm{eq}}$, $Z^* = Z(\mathbf{x}^*)$, and $C^* = C(\mathbf{x}^*)$.

If we define the l_1 Lagrangian function as

$$L(\mathbf{x}, \delta, \lambda) \equiv g(\mathbf{x}) + \sum_{i \in Z} \delta_i f_i(\mathbf{x}) - \sum_{j \in C} \lambda_j \left(a_j^T \mathbf{x} + b_j\right) \qquad (11.192)$$

then according to the equation (11.191) the constrained stationary point $\mathbf{x}^*$ (local minimum point) can be characterized by the following set of nonlinear equations:

$$\nabla_x L(\mathbf{x}, \delta, \lambda) = \mathbf{0} \quad Z = Z^*, C = C^*$$

$$f_i(\mathbf{x}) = \mathbf{0} \quad i = Z^* \qquad (11.193)$$

$$a_j^T \mathbf{x} + b_j = \mathbf{0} \quad j = C^*$$

These are the Kuhn-Tucker equations corresponding to the l_1 programming problem defined by (11.163) and (11.164). We see that in (11.192) the number of unknowns $\mathbf{x}$, δ, and λ equals the number of equations.

An approximate Newton method can be used for solving the nonlinear system (11.193) in the variables $(\mathbf{x}, \delta, \lambda)$. We can easily verify that a Newton step for the nonlinear system (11.192) is the solution of the following set of linear equations:

$$\begin{bmatrix} \nabla_{xx} L(\mathbf{x}, \delta, \lambda) & A(\mathbf{x}) \\ \hline A^T & 0 \end{bmatrix} \begin{bmatrix} \Delta \mathbf{x} \\ \hline \Delta \delta \\ -\Delta \lambda \end{bmatrix} = - \begin{bmatrix} \nabla_x L(\mathbf{x}, \delta, \lambda) \\ f(\mathbf{x}) \\ c(\mathbf{x}) \end{bmatrix} \qquad (11.194)$$

where $f(\mathbf{x})$ is the vector of active functions, $c(\mathbf{x})$ is the vector of active constraints, and A is the matrix, the columns of which are the gradients of the active functions and constraints:

$$A(\mathbf{x}) = \left[\{\nabla f_i(\mathbf{x}) \mid i \in Z(\mathbf{x})\}, \{a_j \mid j \in C(\mathbf{x})\} \right] \qquad (11.195)$$

The exact first derivatives are used to compute $\mathbf{A}(\mathbf{x})$, but the Hessian $\nabla_{xx} L(\mathbf{x}, \boldsymbol{\delta}, \boldsymbol{\lambda})$ is approximated by using a modified BFGS update [53], that keeps the approximation $\mathbf{G}^{(k)}$ of $\nabla_{xx} L$ positive definite. $\mathbf{G}^{(k)}$ is updated through

$$\mathbf{G}^{(k+1)} = \mathbf{G}^{(k)} - \frac{\mathbf{G}^{(k)} \Delta \mathbf{x}^T \mathbf{B}^{(k)}}{\Delta \mathbf{x}^T \mathbf{G}^{(k)} \Delta \mathbf{x}} + \frac{\mathbf{y}\mathbf{y}^T}{\mathbf{y}^T \mathbf{y}} \tag{11.196}$$

with

$$\Delta \mathbf{x} = \mathbf{x}^{(k+1)} - \mathbf{x}^{(k)}$$

$$\mathbf{y} = \nabla_x L\left(\mathbf{x}^{(k+1)}, \boldsymbol{\delta}, \boldsymbol{\lambda}\right) - \nabla_x L\left(\mathbf{x}, \boldsymbol{\delta}, \boldsymbol{\lambda}\right)$$

In this way, an approximate coefficient matrix of the system of linear equations (11.194) is obtained at the estimate $(\mathbf{x}^{(k)}, \boldsymbol{\delta}^{(k)}, \boldsymbol{\lambda}^{(k)})$ of the solution of (11.191). The next estimate is obtained by

$$\begin{bmatrix} \mathbf{x}^{(k+1)} \\ \boldsymbol{\delta}^{(k+1)} \\ \boldsymbol{\lambda}^{(k+1)} \end{bmatrix} = \begin{bmatrix} \mathbf{x}^{(k)} \\ \boldsymbol{\delta}^{(k)} \\ \boldsymbol{\lambda}^{(k)} \end{bmatrix} + \begin{bmatrix} \Delta \mathbf{x}^{(k)} \\ \Delta \boldsymbol{\delta}^{(k)} \\ \Delta \boldsymbol{\lambda}^{(k)} \end{bmatrix} \tag{11.197}$$

Details of this technique are presented in [13, 27] and an efficient implementation of the scheme in the form of a two-stage FORTRAN subroutine for linearly constrained l_1 optimization is given in [14].

Minimax Optimization

The approach discussed for the constrained l_1 problem also may be used to solve the linearly constrained minimax problem formulated as (11.173) to (11.175).

Let us assume that we know a point $\mathbf{x}$ close to the solution point $\mathbf{x}^*$, the set of active functions

$$A(\mathbf{x}^*) \equiv \left\{ i \mid f_i(\mathbf{x}^*) = U(\mathbf{x}^*) \right\} \tag{11.198}$$

and the set of active constraints

$$C(\mathbf{x}^*) \equiv \left\{ j \mid \mathbf{a}_j^T \mathbf{x}^* + b_j = 0 \right\} \tag{11.199}$$

We can be prove [27] that a local minimum point $\mathbf{x}^*$ of minimax problem (11.173) to (11.175) must satisfy the following set of equations:

$$\sum_{i\in A^*} \delta_i \nabla f_i(\mathbf{x}^*) - \sum_{j\in C^*} \lambda_j \mathbf{a}_j = 0$$

$$\sum_{i\in A^*} \delta_i - 1 = 0$$

$$f_{i_0}(\mathbf{x}) - f_i(\mathbf{x}) = 0 \tag{11.200}$$

$$\mathbf{a}_j^T \mathbf{x} + b_j = 0$$

where $A^* = A(\mathbf{x}^*)$, $C^* = C(\mathbf{x}^*)$, $\delta_i \geq 0$, $\lambda_j \geq 0$, and $i_0 \in A^*$ is fixed.

As for the constrained l_1 optimization problem, an approximate Newton method can be used to solve the system (11.200) in the variables $(\mathbf{x}, \boldsymbol{\delta}, \boldsymbol{\lambda})$.

The Lagrangian function corresponding to (11.173) to (11.175) is

$$L(\mathbf{x}, \boldsymbol{\delta}, \boldsymbol{\lambda}) = \sum_{i\in A} \delta_i f_i(\mathbf{x}) - \sum_{j\in C} \lambda_j \left(\mathbf{a}_j^T \mathbf{x} + b_j\right) \tag{11.201}$$

A Newton step for the nonlinear system (11.200) is the solution of the following set of linear equations:

$$\begin{bmatrix} \nabla_{xx} L(\mathbf{x}, \boldsymbol{\delta}, \boldsymbol{\lambda}) & \vdots & \mathbf{E} & \vdots & -\mathbf{F} \\ 0 \ 0 \ \cdots \ 0 & \vdots & 1 \ 1 \ \cdots \ 1 & \vdots & 0 \ 0 \ \cdots \ 0 \\ \mathbf{G}^T & \vdots & 0 & \vdots & 0 \\ \mathbf{F}^T & \vdots & 0 & \vdots & 0 \end{bmatrix} \begin{bmatrix} \Delta\mathbf{x} \\ \Delta\boldsymbol{\delta} \\ \Delta\boldsymbol{\lambda} \end{bmatrix} = \begin{bmatrix} \nabla_x L(\mathbf{x}, \boldsymbol{\delta}, \boldsymbol{\lambda}) \\ \sum \lambda_i - 1 \\ \mathbf{e} \\ \mathbf{c} \end{bmatrix}$$

$$\tag{11.202}$$

where $\mathbf{E}$ has the columns equal to $\nabla f_i(\mathbf{x})$, $i \in A(\mathbf{x})$:

$$\mathbf{E} = \left\{ \nabla f_i(\mathbf{x}) \mid i \in A(\mathbf{x}) \right\} \tag{11.203}$$

$\mathbf{F}$ has the columns equal to $\mathbf{a}_j$, $j \in C(\mathbf{x})$:

$$\mathbf{F} = \left\{ \mathbf{a}_j \mid j \in C(\mathbf{x}) \right\} \tag{11.204}$$

and $\mathbf{G}$ is the matrix whose columns are equal to $\nabla f_{i_0}(\mathbf{x}) - \nabla f_i(\mathbf{x})$, $i \in A \setminus \{i_0\}$:

$$\mathbf{G} = \left\{ \nabla f_{i_0}(\mathbf{x}) - \nabla f_i(\mathbf{x}) \mid i \in A \setminus \{i_0\} \right\} \tag{11.205}$$

($A \setminus B$ denotes the difference (or relative complement) of the sets A and B; that is, the set of those elements of A which do not belong to B.)

$\mathbf{e}$ = the vector whose elements are $f_{i_0}(\mathbf{x}) - f_i(\mathbf{x})$, $i \in A \setminus \{i_0\}$ and
$\mathbf{c}$ = the vector of active constraints $\mathbf{a}_j^T \mathbf{x} + b_j$, $j \in C(\mathbf{x})$.

The next estimate of the solution is found by (11.197).

An efficient implementation of this scheme in the form of a FORTRAN subroutine for linearly constrained minimax optimization is given in [20].

Example 11.7

Minimax formulation of an optimization problem is extremely attractive in computer-aided filter design. Low-pass and bandpass filters implemented on microstrip are good examples of such circuits. Formulation of a filter design problem as a minimax optimization problem avoids the complications normally associated with synthesis of high-degree networks and easily takes into account effects of all discontinuities and parasitics. Figure 11.18(a) illustrates microstrip realization of the filter of Figure 11.18(b). It is an elliptic filter with equal ripple insertion loss characteristic in both the passband and stopband.

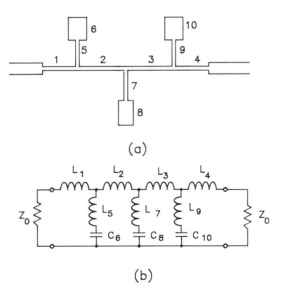

(a)

(b)

Figure 11.18 (a) Microstrip realization of a symmetric low-pass filter (b) for $n = 7$.

The value $Z_{0h}(2 Z_0 \leq Z_{0h} \leq 2.5 Z_0; \; Z_0 = 50 \; \Omega)$ of the high-impedance lines is chosen as the highest possible value corresponding to a reasonable strip width on the substrate material.

Let us assume that the circuit is symmetrical (i.e., $l_1 = l_4$, $l_2 = l_3$, $l_5 = l_9$, $l_6 = l_{10}$, $Z_{05} = Z_{09}$, $Z_{06} = Z_{010}$).

Design problems of the described low-pass microstrip filter may be formulated as a minimax optimization problem defined as

$$\min_{\mathbf{p} \in R^n} \max \{ f_i(\mathbf{p}, \omega_i) \}$$

where

$$f_i(\mathbf{p}, \omega_i) = \begin{cases} w_i \left[\left| \Gamma_i(\mathbf{p}, \omega_i) \right|^2 - r_p^2 \right], & \text{in the passband} \\ w_i \left[r_s^2 - \left| \Gamma_i(\mathbf{p}, \omega_i) \right|^2 \right], & \text{in the stopband} \end{cases}$$

$$\mathbf{p} = \left[l_1, l_2, l_5, l_6, l_7, l_8, Z_{05}, Z_{06}, Z_{08} \right]^T$$

r_p and r_s are the magnitudes of the input reflection coefficient corresponding to the assumed insertion loss, respectively, in the passband and in the stopband of the filter. $\Gamma_i(\mathbf{p}, \omega_i)$ is the reflection coefficient of the filter at the ith frequency point ω_i.

As constraints on the optimized parameters $\mathbf{p}$ of the circuit we introduce the following linear inequalities.

for all stripline lengths:

$$0 < l_i < \lambda_{fi}/4, \quad i = 1, 2, 5, 6, 7, 8$$

where λ_{fi} is the wavelength in the ith stripline, and

for all characteristic stripline impedances:

$$20 \; \Omega < Z_{0i} < Z_{0h}, \quad i = 5, 6, 8$$

Figure 11.19 presents design specifications for the low-pass filter with cut-off frequency f_c, a passband ripple, and minimum stopband attenuation.

Example 11.8

Optimization methods are widely used in modeling microwave semiconductor devices from measured scattering matrix parameters. A small-signal equivalent circuit of a microwave MESFET can be extracted from measured scattering matrix data in a number of ways. A very effective way among these is l_1 optimization.

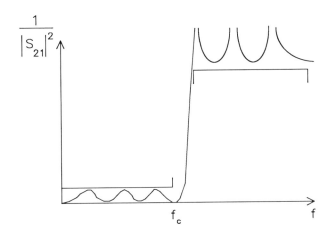

Figure 11.19 An example of design specifications for the minimax low-pass filter design problem.

For the MESFET modeling problem, we formulate an l_1 optimization problem as follows:

$$\min_{\mathbf{p} \in R^n} \sum_{i=1}^{l} \sum_{j=1}^{2} \sum_{k=1}^{2} \left\{ \left| \text{Re}\left[f_{jk}(\mathbf{p}, \omega_i) \right] \right| + \left| \text{Im}\left[f_{jk}(\mathbf{p}, \omega_i) \right] \right| \right\}$$

where

$$f_{jk}(\mathbf{p}, \omega_i) = S_{jk}^c(\mathbf{p}, \omega_i) - S_{jk}^m(\omega_i),$$

$S_{jk}^c(\mathbf{p}, \omega_i)$, $j, k = 1, 2$ are four scattering parameters calculated from a MESFET equivalent circuit, and

$S_{ik}^m(\omega_i)$, $j, k = 1, 2$ are measured scattering parameters of a MESFET.

The mathematical properties of l_1 optimization make it very attractive in application to equivalent circuit extraction from measurement data. As in Section 11.2.3, the l_1 norm ignores some large components of the sum forming the objective function. This means that a few large errors in measurement data will not drastically influence the quality of device modeling.

Figure 11.20 presents a MESFET small-signal equivalent circuit. The total number of its parameters is 13:

$$\mathbf{p} = \left[L_g, R_g, C_{gs}, R_i, C_{dg}, g_m, \tau, R_{ds}, C_{ds}, R_d, L_d, R_s, L_s \right]^T$$

This number is rather large. In conventional modeling procedures we may try to fit all

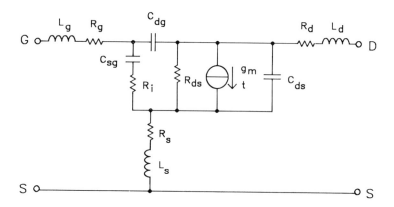

Figure 11.20 Small-signal equivalent circuit of a MESFET.

four scattering parameters simultaneously at all frequencies by using l_1 optimization. Because of the large number of circuit parameters and poor sensitivity of the objective function with respect to some circuit parameters, this procedure usually suffers from a so-called local minimum problem. To overcome the problem we can reformulate the approximation problem. For example, the modeling procedure may be divided into few consecutive l_1 optimization steps, in each of which only a group of selected circuit parameters is considered as a set of optimization variables $\mathbf{p}$. Simultaneously, at each step of optimization, different circuit functions may be considered over different

Table 11.1
Definition of Eight Objective Functions

Step	Optimization Variables	Objective Function	Frequency Range
1	R_{ds}, C_{ds}	S_{22}	Whole
2	C_{gs}	S_{11}	Whole
3	R_s, C_{dg}	S_{12}	Whole
4	g_m	S_{21}	Whole
5	R_d, L_d	S_{22}	Upper half
6	R_g, R_i, L_g	S_{11}	Upper half
7	L_s	S_{12}	Upper half
8	τ	S_{21}	Upper half

frequency bands. Such a procedure corresponds to many objective functions to be minimized consecutively, as opposed to a single objective function used in the conventional modeling procedure. In a MESFET modeling only a group of two or three circuit parameters may be considered as optimization variables to fit a specific scattering matrix parameter over a specific frequency range [54]. Table 11.1 presents eight combinations of selected circuit parameters, scattering parameters considered as parameters and fitted at selected frequency bands [54].

An entire modeling procedure requires eight l_1 optimizations performed consecutively. Such partitioning of the modeling procedure improves its accuracy [54].

REFERENCES

[1] P.E. Gill and W. Murray, eds., *Numerical Methods for Constrained Optimization*, Academic Press, New York, 1974.

[2] L.S. Lasdon, *Optimization Theory for Large Systems*, Macmillan, New York, 1970.

[3] J.W. Bandler, "Optimization Methods for Computer Aided Design," *IEEE Trans. Microwave Theory Tech.*, Vol. MTT-17, August 1969, pp. 533–552.

[4] J.W. Bandler, "Computer-Aided Circuit Optimization," in *Modern Filter Theory and Design*, G.C. Temes and S.K. Mitra, eds., John Wiley and Sons, New York, 1973, pp. 211–271.

[5] J.W. Bandler and S.H. Chen, "Circuit Optimization: The State of the Art," *IEEE Trans. Microwave Theory Tech.*, Vol. MTT-36, February 1988, pp. 424–443.

[6] H. Tromp, "The Generalized Tolerance Problem and Worst Case Search," *Proc. Conf. CAD Electronic and Microwave Circuits and Systems*, London, 1977, pp. 72–77.

[7] J.W. Bandler, P.C. Liu, and H. Tromp, "Integrated Approach to Microwave Design," *IEEE Trans. Microwave Theory Tech.*, Vol. MTT-24, September 1976, pp. 584–591.

[8] R. Fletcher and C. M. Reeves, "Function Minimization by Conjugate Gradients," *Computer J.*, Vol. 7, July 1964, pp. 149–154.

[9] J.W. Bandler and M.R.M. Rizk, "Optimization of Electrical Circuits," *Math. Programming Study*, Vol. 11, 1979, pp. 1–64.

[10] G.C. Temes and D.Y.F. Zai, "Least pth Approximation," *IEEE Trans. Circuit Theory*, Vol. CT-16, 1969, pp. 235–237.

[11] J.W. Bandler, W. Kellermann, and K. Madsen, "A Nonlinear l_1 Optimization Algorithm for Design, Modeling and Diagnosis of Networks," *IEEE Trans. Circuits and Systems*, Vol. CAS-34, 1987, pp. 174–181.

[12] J.W. Bandler, S.H. Chen, and S. Daijavad, "Microwave Device Modeling Using Efficient l_1 Optimization: A Novel Approach," *IEEE Trans. Microwave Theory Tech.*, Vol. MTT-34, December 1986, pp. 1282–1293.

[13] J. Hald and K. Madsen, "Combined LP and Quasi-Newton Methods for Nonlinear l_1 Optimization," *SIAM J. Numerical Analysis*, Vol. 22, 1985, pp. 68–80.

[14] J. Hald, "A Two-Stage Algorithm for Nonlinear l_1 Optimization," Report No. NI-81-01, Inst. for Numerical Analysis, Technical University of Denmark, Lyngby, 1981.

[15] M.R. Osborne and G.A. Watson, "On an Algorithm for Discrete Nonlinear l_1 Approximation," *Computer J.*, Vol. 14, 1971, pp. 184–188.

[16] W. Murray and M.L. Overton, "A Projected Lagrangian Algorithm for Nonlinear l_1 Optimization," *SIAM J. Science Statistical Computing*, Vol. 2, 1981, pp. 207–224.

[17] R.A. El-Attar, M. Vidyasagar, and S.R.K. Dutta, "An Algorithm for l_1-Norm Minimization with

Application to Nonlinear l_1-Approximation," *SIAM J. Numerical Analysis*, Vol. 16, 1979, pp. 70–86.

[18] D.D. Morrison, "Optimization by Least Squares," *SIAM J. Numerical Analysis*, Vol. 5, 1968, pp. 83–88.

[19] B.S. Perlman and V.G. Gelnovatch, "Computer Aided Design, Simulation and Optimization," in *Advances in Microwaves*, L. Young, ed., Academic Press, New York, 1974.

[20] J. Hald, "MMLA1Q, a FORTRAN Subroutine for Linearly Constrained Minimax Optimization," Report No. NI-81-01, Inst. for Numerical Analysis, Technical University of Denmark, Lyngby, 1981.

[21] R.H. Bartels, A.R. Conn, and J.W. Sinclair, "Minimization Techniques for Piecewise Differentiable Functions: The l_1 Solution to an Overdetermined Linear System," *SIAM J. Numerical Analysis*, Vol. 15, No. 2, April 1978, pp. 224–241.

[22] D. Agnew, "Improved Minimax Optimization for Circuit Design," *IEEE Trans. Circuits and Systems*, Vol. CAS-28, 1981, pp. 791–803.

[23] J.W. Bandler, S.H. Chen, S. Daijavad, and W. Kellermann, "Optimal Design of Multicavity Filters and Continuous-Band Multiplexers," *Proc. 14th European Microwave Conf.*, Liege, Belgium, 1984, pp. 863–868.

[24] J.W. Bandler, S.H. Chen, S. Daijavad, W. Kellermann, M. Renault, and Q.J. Zhang, "Large Scale Minimax Optimization of Microwave Multiplexers," *Proc. 16th European Microwave Conf.*, Dublin, 1986, pp. 435–440.

[25] J.W. Bandler, W. Kellermann, and K. Madsen, "A Superlinearly Convergent Minimax Algorithm for Microwave Circuit Design," *IEEE Trans. Microwave Theory Tech.*, Vol. MTT-33, 1985, pp. 1519–1530.

[26] C. Charalambous and A.R. Cohn, "Optimization of Microwave Networks," *IEEE Trans. Microwave Theory Tech.*, Vol. MTT-23, 1975, pp. 834–838.

[27] J. Hald and K. Madsen, "Combined LP and Quasi-Newton Methods for Minimax Optimization," *Math. Programming*, Vol. 20, 1981, pp. 49–62.

[28] K. Madsen, J. Schjaer-Jacobsen, and J. Voldby, "Automated Minimax Design of Networks," *IEEE Trans. Circuits and Systems*, Vol. CAS-22, 1975, pp. 791–796.

[29] M.R. Osborne and G.A. Watson, "An Algorithm for Minimax Optimization in the Nonlinear Case," *Computer J.*, Vol. 12, 1969, pp. 63–68.

[30] J.W. Bandler and C. Charalambous, "Theory of Generalized Least pth Approximation," *IEEE Trans. Circuit Theory*, Vol. CT-19, 1974, pp. 607–619.

[31] C. Charalambous, "Nonlinear least pth Optimization and Nonlinear Programming," *Math. Programming*, Vol. 12, 1977, pp. 195–225.

[32] R. Fletcher, "A New Approach for Variable Metric Algorithm," *Computer J.*, Vol. 13, 1970, pp. 317–322.

[33] R. Fletcher and M.J.D. Powell, "A Rapidly Convergent Descent Method for Minimization," *Computer J.*, Vol. 6, 1963, pp. 163–168.

[34] D.G. Luenberger, *Linear and Nonlinear Programming*, Addison Wesley, Reading, MA, 1984.

[35] J.E. Dennis and R.B. Schnabel, *Numerical Methods for Optimization and Nonlinear Equations*, Prentice-Hall, Englewood Cliffs, NJ, 1983.

[36] M.J.D. Powell, "A Fast Algorithm for Nonlinearly Constrained Optimization Calculations," *Proc Dundee Conf. Numerical Analysis*, A. Dold and B. Eckermann, eds., (Lecture Notes in Mathematics), Springer-Verlag, New York, 1977, p. 630.

[37] A.V. Fiacco and G.P. McCormick, *Nonlinear Programming: Sequential Unconstrained Minimization Techniques*, John Wiley and Sons, New York, 1968.

[38] D.M. Ryan, "Penalty and Barrier Functions," in *Numerical Methods for Constrained Optimization*, P.E. Gill and W. Murray, eds., Academic Press, New York, 1974, pp. 175–190.

[39] W.I. Zangwill, *Nonlinear Programming*, Prentice-Hall, Englewood Cliffs, NJ, 1966.

[40] M.C. Biggs, "Constrained Minimization Using Recursive Quadratic Programming: Some Alternative Subproblem Formulations," in *Towards Global Optimization*, L.C.W. Dixon and G.P. Szego, eds., North Holland, Amsterdam, 1975.

[41] S. P. Han, "Superlinearly Convergent Variable Metric Algorithms for General Nonlinear Programming Problems," *Math. Programming*, Vol. 11, 1976, pp. 263–282.

[42] M.J.D. Powell, "A New Algorithm for Unconstrained Optimization," in *Nonlinear Programming*, J.B. Rosen, O.L. Margasarian, and K. Ritter, eds., Academic Press, New York, 1970.

[43] C.G. Broyden, "A New Double-Rank Minimization Algorithm," *Notices American Math. Soc.*, Vol. 16, 1969, p. 670.

[44] D. Goldfarb, "A Family of Variable-Metric Methods Derived by Variational Means," *Math. Computing*, Vol. 24, 1970, pp. 23–26.

[45] D.F. Shanno, "Conditioning of Quasi-Newton Methods for Function Minimization," *Math Computing*, Vol. 24, 1964, pp. 647–656.

[46] D.P. Bertsekas, "Multiplier Methods: A Survey," *Automatica*, Vol. 2, 1976, pp. 133–145.

[47] M.J.D. Powell, "A Method for Nonlinear Constraints in Minimization Problems," in *Optimization*, R. Fletcher, ed., Academic Press, New York, 1969, pp. 283–298.

[48] D.P. Bertsekas, "On Penalty and Multiplier Methods for Constrained Minimization," *SIAM J. Control and Optimization*, Vol. 14, 1976, pp. 216–235.

[49] R. Fletcher, "Methods for Solving Nonlinearly Constrained Optimization Problems," in *The State of the Art in Numerical Analysis*, D. Jacobs, ed., Academic Press, New York, 1977, pp. 365–448.

[50] M.R. Lightner and S.W. Director, "Multiple Criterion Optimization for the Design of Electronic Circuits," *IEEE Trans. Circuits and Systems*, Vol. CAS-28, March 1981, pp. 169–179.

[51] J. Hald and K. Madsen, "Combined LP and Quasi-Newton Methods for Minimax Optimization," *Math. Programming*, Vol. 20, 1981, pp. 49–62.

[52] D. Marquard, "An Algorithm for Least-Squares Estimation of Nonlinear Parameters," *SIAM J. Applied Math.*, Vol. 11, 1963, pp. 431–441.

[53] R. Fletcher, "A Survey of Algorithms for Unconstrained Optimization," in *Numerical Methods for Unconstrained Optimization*, W. Murray, ed., Academic Press, London, 1972.

[54] H. Kondoh, "An Accurate FET Modeling from Measured S-Parameters," *1986 IEEE MTT Symp. Dig.*, pp. 377–380.

Appendix 1
Vector and Matrix Norms, Ranks

A norm, designated $\| \cdot \|$, is simply a measure of the size of the largest elements in the vector of matrix. The most common vector norms are

L_1
$$\|\mathbf{x}\|_1 = |x_1| + |x_2| + \cdots + |x_n| \tag{A1.1}$$

L_2
$$\|\mathbf{x}\|_2 = \sqrt{x_1^2 + x_2^2 + \cdots + x_n^2} \tag{A1.2}$$

L_∞
$$\|\mathbf{x}\|_\infty = \max_{1 \leq i \leq n} |x_i| \tag{A1.3}$$

The matrix norms are

L_1
$$\|\mathbf{A}\|_1 = \max_{1 \leq j \leq n} \sum_{i=1}^{n} |a_{ij}| \tag{A1.4}$$

L_2
$$\|\mathbf{A}\|_2 = \text{The square root of the largest eigenvalue of } \mathbf{A}^T\mathbf{A} \tag{A1.5}$$

L_∞
$$\|\mathbf{A}\|_\infty = \max_{1 \leq i \leq n} \sum_{j=1}^{n} |a_{ij}| \tag{A1.6}$$

All norms have the following properties.
For vector norms:

a: $\qquad \|\mathbf{x}\| \geq 0 \quad$ for all $\mathbf{x}$; $\|\mathbf{x}\| = 0$ if and only if $x = 0$.

b: $\qquad \|k\mathbf{x}\| = |k| \cdot \|\mathbf{x}\| \quad$ for any real number k.

c: $\qquad \|\mathbf{x} + \mathbf{y}\| \leq \|\mathbf{x}\| + \|\mathbf{y}\| \tag{A1.7}$

d: $\qquad \|\mathbf{x}^+\mathbf{y}\| \leq \|\mathbf{x}\| \cdot \|\mathbf{y}\|$

For matrix norms:

a: $\|\mathbf{A}\| \geq 0$ for all $\mathbf{A}$; $\|\mathbf{A}\| = 0$ if and only if $\mathbf{A} = 0$.

b: $\|k\mathbf{A}\| = |k| \cdot \|\mathbf{A}\|$ for any real number k.

c: $\|\mathbf{A} + \mathbf{B}\| \leq \|\mathbf{A}\| + \|\mathbf{B}\|$ (A1.8)

d: $\|\mathbf{A} \cdot \mathbf{B}\| \leq \|\mathbf{A}\| \cdot \|\mathbf{B}\|$

These conditions can be satisfied by suitably defining $\|\mathbf{A}\|$. We shall define the norm of $\mathbf{A}$ as

$$\|\mathbf{A}\| = \max_{\mathbf{x} \neq 0} \frac{\|\mathbf{A}\mathbf{x}\|}{\|\mathbf{x}\|} \tag{A1.9}$$

Let us consider now the Hermitian matrix $\mathbf{A}^+ = \mathbf{A}$. It is positive semidefinite and hence its eigenvalues are all nonnegative:

$$\sigma_1^2 \geq \sigma_2^2 \geq \cdots \geq \sigma_n^2 \geq 0$$

By definition the singular values of $\mathbf{A}$ are the positive square roots of $\sigma_1^2, \sigma_2^2, \ldots, \sigma_n^2$ namely, $|\sigma_1|, |\sigma_2|, \ldots, |\sigma_n|$.

We know from matrix theory, that we can always find a unitary matrix $\mathbf{U}$ $(\mathbf{U}^+ \cdot \mathbf{U} = \mathbf{I})$ such that

$$\mathbf{U}^+ \mathbf{A}^+ \mathbf{A} \mathbf{U} = \begin{bmatrix} \sigma_1^2 & 0 & \cdots & 0 \\ 0 & \sigma_2^2 & & \vdots \\ \vdots & & & 0 \\ 0 & & \cdots & \sigma_n^2 \end{bmatrix} \tag{A1.10}$$

and that any vector $\mathbf{x}$ can be expressed as a linear combination of the columns of $\mathbf{U}$

$$\mathbf{x} = \mathbf{U}\mathbf{c} = \mathbf{U}\begin{bmatrix} c_1 & c_2 & \cdots & c_n \end{bmatrix}^T \tag{A1.11}$$

It follows that

$$\frac{\mathbf{x}^+ \mathbf{A}^+ \mathbf{A} \mathbf{x}}{\mathbf{x}^+ \mathbf{x}} = \frac{(\mathbf{U}\mathbf{c})^+ \mathbf{A}^+ \mathbf{A} (\mathbf{U}\mathbf{c})}{(\mathbf{U}\mathbf{c})^+ (\mathbf{U}\mathbf{c})} = \frac{\sum\limits_{k=1}^{n} \sigma_k^2 |c_k|^2}{\mathbf{c}^+ \mathbf{c}} \leq \sigma_1^2 \tag{A1.12}$$

In this relation the equality holds for arbitrary σ_k^2 if and only if $c_2 = c_3 = \cdots = c_n = 0$.

Now using the definition of the norm of **A**, we have

$$\|\mathbf{A}\|^2 = \max_{\mathbf{x} \neq 0} \frac{\|\mathbf{A}\mathbf{x}\|^2}{\|\mathbf{x}\|^2} = \max \frac{\mathbf{x}^+\mathbf{A}^+\mathbf{A}\mathbf{x}}{\mathbf{x}^+\mathbf{x}} = \sigma_1^2 \qquad (A1.13)$$

From (A1.13) we see that the norm of **A** defined as (A1.9) equals maximum singular value of **A**:

$$\|\mathbf{A}\| = |\sigma_1| \qquad (A1.14)$$

We can prove that the norm of $\mathbf{A}^{-1}$ equals the reciprocal of the minimum singular value of **A**:

$$\|\mathbf{A}^{-1}\| = \frac{1}{|\sigma_n|} \qquad (A1.15)$$

We say that the vectors $\mathbf{x}_1, \mathbf{x}_2, \ldots, \mathbf{x}_k$ of $\mathbf{R}^n$ are linearly dependent if there exist complex numbers $c_1, c_2, \ldots, c_k$, which are not all zero, such that $c_1\mathbf{x}_1 + c_2\mathbf{x}_2 + \cdots + c_k\mathbf{a}_k = \mathbf{0}$.

If the vectors are not linearly dependent, we say that they are linearly independent.

We say that a system $\{\mathbf{x}_1, \mathbf{x}_2, \ldots, \mathbf{x}_k\}$ of vectors has the *rank r* if there are r linearly independent vectors among the vectors $\mathbf{x}_1, \mathbf{x}_2, \ldots, \mathbf{x}_k$ but any $r + 1$ vectors of $\mathbf{x}_1, \mathbf{x}_2, \ldots, \mathbf{x}_k$ are always linearly dependent. (Then the rank r is the maximum number of linearly independent vectors of given system.)

The *rank of a matrix* is the rank of the system of all vectors formed by the rows of the matrix. Thus, matrix **A** is of rank r if there are r linearly independent rows among its rows and every further row of the matrix is a linear combination of these r rows.

Appendix 2
Sparse Matrix Solver

The program presented in this appendix is designed to solve the linear sparse systems of complex equations generated by the frequency domain, sensitivity and noise analysis of linear microwave circuits described by the connection scattering matrix approach. It is based on the bifactorization method. The techniques incorporated into this program are fully described in Chapter 7. The solver contains four routines. The interface routine, SMCOMP, generates indexing, addressing, and ordering information to handle the nonzero terms of the connection scattering matrix stored as linked lists. The second subroutine, CMSOZ, performs symbolic ordering and reduction of the coefficient matrix. The subroutine CMRZ performs numeric reduction and the subroutine CMDSZ computes the solution vector.

The user interacts with the package through four subroutines whose functions and arguments are as follows.

1. SUBROUTINE SMCOMP(NP, NC, NS, IS, JS, ICON, JCON, LFMAX, LROW, ITAGR, LNXTR, NOZER, LCOL, ITAGC, LNXTC, NOZEC, NSEQ, LROWB, LCOLB, LF)

On input:

NP is the number of all ports of the analyzed circuit. NP equals the order of the **W** matrix of the analyzed circuit.

NC is the number of pairs of connected ports of the circuit; NC = NP/2.

NS is the number of scattering matrix terms of all circuit elements.

IS(K), JS(K), K = 1, 2, ..., NS all contain row-column indices of scattering matrix terms of all circuit elements. The ports of the circuit must be numbered from 1 to NP.

ICON(I), JCON(I), I = 1, 2, ..., NC all contain numbers of pairs of connected ports of the circuit.

LFMAX is the dimension parameter. It must be equal to or larger than NS.

On output:

LROW(NP), ITAGR(LFMAX), LNXTR(LFMAX), NOZER(NP),

LCOL(NP), ITAGC(LFMAX), LNXTC(LFMAX), NOZEC(NP), NSEQ(NP), LF are integer arrays and integer parameter containing indexing, addressing, and ordering information on the structure of the preordered **W** matrix of the circuit. These data should not be changed by the user.

LFMAX is set to -1 if the routine fails due to insufficient integer arrays (LFMAX < NS). Remedy is to set LFMAX $\geq$ NS and start execution again.

The subroutine generates the indexing, addressing and ordering information in the form of integer sequences describing the structure of the connection scattering matrix with rows interchanged so as to locate ones, the terms of Γ matrix, on the main diagonal (see Section 8.2). This subroutine is called once for a given microwave circuit (for a given **W** matrix structure).

2. SUBROUTINE CMSOZ(N, NV, LF, LROW, ITAGR, LNXTR, NOZER, LZEROR, LCOL, ITAGC, LNXTC, NOZEC, LZEROC, NSEQ)

On input:

N is the order of the matrix. N must be set to NP.

NV is the dimension of ITAGR, LNXTR, ITAGC, and LNXTC arrays. NV must be larger than the number of nonzero off-diagonal terms of the matrix. NV must be set to LFMAX (see SMCOMP description).

LROW(N), ITAGR(NV), LNXTR(NV), NOZER(N), LCOL(N), ITAGC(NV), LNXTC(NV), NOZEC(N), NSEQ(N), NF are the integer arrays and the integer pointer containing indexing, addressing, and ordering information on the structure of the matrix, received from SMCOMP routine.

On output:

LROW(N), ITAGR(NV), LNXTR(NV), NOZER(N), LZEROR(NV), LCOL(N), ITAGC(NV), LNXTC(NV), NOZEC(N), LZEROC(NV), NSEQ(N), NF all contain indexing, addressing, and ordering information on structures of left-hand and right-hand factor matrices and should not be changed by the user.

This subroutine performs symbolic reduction and orders the coefficient matrix. It is executed only once for a given matrix structure.

3. SUBROUTINE CMRZ(N, NV, LF, LROW, ITAGR, LNXTR, LZEROR, LCOL, ITAGC, LNXTC, LZEROC, NSEQ, RE, CE, DE)

On input:

N is the order of the matrix. N must be set to NP.

NV is the dimension of ITAGR, LNXTR, LZEROR, ITAGC, LNXTC, LZEROC arrays. NV must be set to LFMAX (see SMCOMP description).

LROW(N), ITAGR(NV), LNXTR(NV), LZEROR(NV), LCOL(N), ITAGC(NV), LNXTC(NV), LZEROC(NV), NSEQ(N) are integer arrays containing information on structures of left-hand and right-hand factor matrices. These data are supplied by CMSOZ subroutine and should not be changed by the user.

RE(K), K = 1, 2, ..., NV contain rowwise stored nonzero off-diagonal matrix terms.

CE(K), K = 1, 2, ..., NV contain columnwise stored nonzero off-diagonal matrix terms.

RE and CE must contain, the scattering matrix terms of all circuit elements, stored in both arrays in the same order. The ordering of the scattering matrix terms in RE and CE define IS(K) and JS(K), K = 1, 2, ..., NS (see SMCOMP description).

DE(I), I + 1, 2, ..., N all contain diagonal terms of the matrix. For the pre-ordered connection scattering matrix all DE entries must be set to ones (see Section 8.2).

On output:

RE(K), K = 1, 2, ..., NV all contain rowwise stored nonzero terms of the right-hand factor matrices.

CE(K), K = 1, 2, ..., NV all contain columnwise stored nonzero terms of the left-hand factor matrices.

DE(I), I = 1, 2, ..., N all contain diagonal terms of the left-hand factor matrices.

This subroutine performs the reduction of the matrix. It is executed at each change in the values of CE, RE, and DE (the values of the scattering matrix parameters).

4. SUBROUTINE CMDSZ(N, NV, LROW, ITAGR, LNXTR, LCOL, ITAGC, LNXTC, NSEQ, RE, CE, DE, V)

On input:

N is the order of the matrix. It must be set to NP.

NV is the dimension of ITAGR, LNXTR, ITAGC, LNXTC, RE and CE. NV must be set to LFMAX (see SMCOMP description).

LROW(N), ITAGR(NV), LNXTR(NV), LCOL(N), ITAGC(NV), LNXTC(NV), NSEQ(N) all contain information on structures of left-hand and right-hand factor matrices. These data are supplied by CMSOZ subroutine and should not be changed by the user.

RE(K), K = 1, 2, ..., NV all contain rowwise stored nonzero terms of the right-hand factor matrices.

CE(K), K = 1, 2, . . . , NV all contain columnwise stored nonzero terms of the left-hand factor matrices.

DE(I), I = 1, 2, . . . , N all contain diagonal terms of the left-hand factor matrices.

The entries of RE, CE, and DE arrays are supplied by the CMSOZ subroutine and should not be changed by the user.

On output:
V(I), I = 1, 2, . . . , N all contain the solution vector.

This subroutine computes the solution vector for a given right-hand vector. It is called at each change in the right-hand vector.

```
C**SMCOMP*********************************************************
C*
      SUBROUTINE SMCOMP(NP,NC,NS,IS,JS,ICON,JCON,LROW,ITAGR,
     1             LNXTR,NOZER,LCOL,ITAGC,LNXTC,NOZEC,NSEQ,
     2             LROWB,LCOLB,LF)
C*
C*THIS ROUTINE PERFORMS GENERATION OF INDEXING, ADDRESSING
C*AND ORDERING INFORMATION IN THE FORM OF INTEGER SEQUENCES.
C*THIS DATA DESCRIBES THE NONZERO STRUCTURE OF THE
C*PREORDERED CONNECTION SCATTERING MATRIX WITH ROWS
C*INTERCHANGED SO AS TO LOCATE 1'S, THE Γ MATRIX ENTRIES,
C*ON THE MAIN DIAGONAL
C*
      DIMENSION ITAGR(LFMAX),IS(NS),JS(NS),LROW(NP),LROWB(NP),
     1             LNXTR(LFMAX),NSEQ(NP),NOZER(NP),LCOL(NP),
     2             LCOLB(NP),LNXTC(LFMAX),ITAGC(LFMAX),NOZEC(NP),
     3             ICON(NC),JCON(NC)
      LFM = NS
      IF (LFM.GT.LFMAX) GO TO 990
      DO 100 M = 1,NS
      I = IS(M)
      J = JS(M)
      IF (I.NE.J) GO TO 100
      LROW(I) = M
      LROWB(I) = M
      ITAGR(M) = J
      NOZER(I) = 1
  100 CONTINUE
      DO 110 L = 1,NS
      I = IS(L)
```

```
      J = JS(L)
      IF (I.EQ.J) GO TO 110
      ITAGR(L) = J
      NOZER(I) = NOZER(I) + 1
      LR = LROW(I)
      LNXTR(LR) = L
      LROW(I) = L
110   CONTINUE
      LF = NS + 1
      DO 130 I = 1,NP
      LI = LROW(I)
      LNXTR(LI) = 0
      LROW(I) = LROWB(I)
      NSEQ(I) = I
130   CONTINUE
      IF (LF - LFMAX) 140,150,160
140   DO 170 I = LF,LFMAX
      LNXTR(I) = I + 1
170   CONTINUE
      GO TO 150
160   LF = 0
150   LNXTR(LFMAX) = 0
      DO 180 K = 1,NP
      LST = LROW(K)
      LANG = 0
190   IF (LST.LE.0) GO TO 180
      IMIN = ITAGR(LST)
      LMIN = LST
      LAMIN = 0
      LANU = LST
      L = LNXTR(LST)
200   IF (L.GT.0) GO TO 210
      IF (LAMIN.LE.0) GO TO 220
      LNXTR(LAMIN) = LNXTR(LMIN)
      GO TO 230
220   LST = LNXTR(LMIN)
230   IF (LANG.LE.0) GO TO 240
      LNXTR(LANG) = LMIN
      GO TO 250
240   LROW(K) = LMIN
250   LANG = LMIN
      GO TO 190
```

```
210 IF (ITAGR(L) - IMIN) 260,280,270
260 IMIN = ITAGR(L)
    LMIN = L
    LAMIN = LANU
270 LANU = L
    L = LNXTR(L)
    GO TO 200
280 LNXTR(LANU) = LNXTR(L)
    LNXTR(L) = LF
    LF = L
    L = LNXTR(LANU)
    GO TO 200
180 CONTINUE
    DO 510 K = 1,NC
    I = ICON(K)
    J = JCON(K)
    LI = LROW(I)
    NI = NOZER(I)
    LJ = LROW(J)
    NJ = NOZER(J)
    NOZER(I) = NJ
    LROW(I) = LJ
    LROW(J) = LI
    NOZER(J) = NI
510 CONTINUE
    DO 520 M = 1,NS
    I = IS(M)
    J = JS(M)
    IF (I.NE.J) GO TO 520
    LCOL(J) = M
    LCOLB(J) = M
    ITAGC(M) = I
    NOZEC(J) = 1
520 CONTINUE
    DO 320 L = 1,NS
    I = IS(L)
    J = JS(L)
    IF (I.EQ.J) GO TO 320
    ITAGC(L) = I
    NOZEC(J) = NOZEC(J) + 1
    LK = LCOL(J)
    LNXTC(LK) = L
    LCOL(J) = L
320 CONTINUE
```

```
      LF = NS + 1
      DO 340 I = 1,NP
      LI = LCOL(I)
      LNXTC(LI) = 0
      LCOL(I) = LCOLB(I)
340   CONTINUE
      IF (LF - LFMAX) 350,360,370
350   DO 380 I = LF,LFMAX
      LNXTC(I) = I + 1
380   CONTINUE
      GO TO 360
370   LF = 0
360   LNXTC(LFMAX) = 0
      DO 580 K = 1,NC
      I = ICON(K)
      J = JCON(K)
      DO 390 M = 1,NP
      LI = LCOL(M)
400   IR = ITAGC(LI)
      IF (IR.NE.I) GO TO 410
      ITAGC(LI) = J
      GO TO 420
410   IF (IR.NE.J) GO TO 420
      ITAGC(LI) = I
420   LI = LNXTC(LI)
      IF (LI.GT.0) GO TO 400
390   CONTINUE
580   CONTINUE
      DO 430 K = 1,NP
      LST = LCOL(K)
      LANG = 0
440   IF (LST.LE.0) GO TO 430
      IMIN = ITAGC(LST)
      LMIN = LST
      LAMIN = 0
      LANU = LST
      L = LNXTC(LST)
450   IF (L.GT.0) GO TO 460
      IF (LAMIN.LE.0) GO TO 470
      LNXTC(LAMIN) = LNXTC(LMIN)
      GO TO 480
470   LST = LNXTC(LMIN)
480   IF (LANG.LE.0) GO TO 490
      LNXTC(LANG) = LMIN
```

```
      GO TO 500
490  LCOL(K) = LMIN
500  LANG = LMIN
     GO TO 440
460  IF (ITAGC(L) - IMIN) 590,600,610
590  IMIN = ITAGC(L)
     LMIN = L
     LAMIN = LANU
610  LANU = L
     L = LNXTC(L)
     GO TO 450
600  LNXTC(LANU) = LNXTC(L)
     LNXTC(L) = LF
     LF = L
     L = LNXTC(LANU)
     GO TO 450
430  CONTINUE
     GO TO 900
990  LFMAX = -1
900  CONTINUE
     RETURN
     END


C**CMSOZ*********************************************************
C*
      SUBROUTINE CMSOZ(N,NV,LF,LROW,ITAGR,LNXTR,NOZER,
     1            LZEROR,LCOL,ITAGC,LNXTC,NOZEC,LZEROC,NSEQ)
      DIMENSION LROW(N),LCOL(N),ITAGR(NV),ITAGC(NV),LNXTR(NV),
     1            LNXTC(NV),NSEQ(N),NOZER(N),NOZEC(N),
     2            LZEROR(NV),LZEROC(NV)
C*THIS ROUTINE PERFORMS THE SYMBOLIC REDUCTION AND
C*ORDERING OF THE COEFFICIENT MATRIX
      LFR = LF
      LFC = LF
      LZR = 0
      LZC = 0
      NN = N - 1
      DO 10 J = 1,NN
      K = NSEQ(J)
      MIN = NOZER(K) + NOZEC(K)
      MINIL = NOZER(K)*NOZEC(K)
      M = J
```

```
      IF (MINIL.LE.0) GO TO 36
      JJ = J + 1
      DO 31 I = JJ,N
      K = NSEQ(I)
      NOZERC = NOZER(K) + NOZEC(K)
      ILNZRC = NOZER(K)*NOZEC(K)
      IF (ILNZRC.GT.0) GO TO 35
      M = I
      GO TO 36
   35 IF (ILNZRC - MINIL) 32,33,31
   33 IF (NOZERC - MIN) 37,31,31
   32 MINIL = ILNZRC
   37 MIN = NOZERC
      M = I
   31 CONTINUE
   36 KP = NSEQ(M)
      NSEQ(M) = NSEQ(J)
      NSEQ(J) = KP
      LR = LROW(KP)
      LK = LCOL(KP)
      IF (LR.LE.0) GO TO 11
      IF (LK.GT.0) GO TO 12
      LI = LCOL(KP)
      IP = N + 1
      GO TO 290
   11 IF (LK.LE.0) GO TO 10
      LI = LROW(KP)
      IP = N + 1
      GO TO 29
   12 LI = LROW(KP)
      IP = ITAGR(LI)
   29 K = ITAGC(LK)
      LA = 0
      L = LROW(K)
      I = ITAGR(L)
   13 IF (I - IP) 14,15,16
   14 IF (I.NE.KP) GO TO 17
      LN = LNXTR(L)
      IF (LA.LE.0) GO TO 18
      LNXTR(LA) = LN
      GO TO 19
   18 LROW(K) = LN
```

```
19 LNXTR(L) = LFR
   LFR = L
   LZR = LZR + 1
   LZEROR(LZR) = L
   NOZER(K) = NOZER(K) − 1
   L = LN
   GO TO 20
17 LA = L
   L = LNXTR(L)
20 IF (L.LE.0) GO TO 21
   I = ITAGR(L)
   GO TO 13
21 IF (LI.LE.0) GO TO 22
   I = N + 1
   GO TO 13
22 LK = LNXTC(LK)
   IF (LK.LE.0) GO TO 110
   IF (LR.LE.0) GO TO 29
   GO TO 12
15 LA = L
   L = LNXTR(L)
   IF (L.LE.0) GO TO 23
   I = ITAGR(L)
   GO TO 24
23 I = N + 1
   GO TO 24
16 IF (IP.EQ.K) GO TO 24
   IF (LFR.GT.0) GO TO 25
   LF = − 1
   GO TO 30
25 LN = LFR
   IF (LA.LE.0) GO TO 26
   LNXTR(LA) = LN
   GO TO 27
26 LROW(K) = LN
27 LFR = LNXTR(LN)
   LNXTR(LN) = L
   ITAGR(LN) = IP
   NOZER(K) = NOZER(K) + 1
   LA = LN
24 LI = LNXTR(LI)
   IF (LI.LE.0) GO TO 28
```

```
      IP = ITAGR(LI)
      GO TO 13
 28  IF (L.LE.0) GO TO 22
      IP = N + 1
      GO TO 13
110  IF (LR.LE.0) GO TO 10
120  LI = LCOL(KP)
      IP = ITAGC(LI)
290  K = ITAGR(LR)
      LA = 0
      L = LCOL(K)
      I = ITAGC(L)
130  IF (I - IP) 140,150,160
140  IF (I.NE.KP) GO TO 170
      LN = LNXTC(L)
      IF (LA.LE.0) GO TO 180
      LNXTC(LA) = LN
      GO TO 190
180  LCOL(K) = LN
190  LNXTC(L) = LFC
      LFC = L
      LZC = LZC + 1
      LZEROC(LZC) = L
      NOZEC(K) = NOZEC(K) - 1
      L = LN
      GO TO 200
170  LA = L
      L = LNXTC(L)
200  IF (L.LE.0) GO TO 210
      I = ITAGC(L)
      GO TO 130
210  IF (LI.LE.0) GO TO 220
      I = N + 1
      GO TO 130
220  LR = LNXTR(LR)
      IF (LR.LE.0) GO TO 10
      LK = LCOL(KP)
      IF (LK.GT.0) GO TO 120
      GO TO 290
150  LA = L
      L = LNXTC(L)
      IF (L.LE.0) GO TO 230
```

```
      I = ITAGC(L)
      GO TO 240
230 I = N + 1
      GO TO 240
160 IF (IP.EQ.K) GO TO 240
      IF (LFC.GT.0) GO TO 250
      LF = - 1
      GO TO 30
250 LN = LFC
      IF (LA.LE.0) GO TO 260
      LNXTC(LA) = LN
      GO TO 270
260 LCOL(K) = LN
270 LFC = LNXTC(LN)
      LNXTC(LN) = L
      ITAGC(LN) = IP
      NOZEC(K) = NOZEC(K) + 1
      LA = LN
240 LI = LNXTC(LI)
      IF (LI.LE.0) GO TO 280
      IP = ITAGC(LI)
      GO TO 130
280 IF (L.LE.0) GO TO 220
      IP = N + 1
      GO TO 130
 10 CONTINUE
      LZR = LZR + 1
      LZEROR(LZR) = 0
      LZC = LZC + 1
      LZEROC(LZC) = 0
 30 RETURN
      END


C**CMRZ*********************************************************
C*
      SUBROUTINE CMRZ(N,NV,LF,LROW,ITAGR,LNXTR,LZEROR,LCOL,
     1             ITAGC,LNXTC,LZEROC,NSEQ,RE,CE,DE)
      COMPLEX RE(NV),CE(NV),DE(N),D,RF,CF
      DIMENSION LROW(N),LCOL(N),ITAGR(NV),ITAGC(NV),LNXTR(NV),
     1             LNXTC(NV),NSEQ(N),LZEROR(NV),LZEROC(NV)
C*THIS ROUTINE PERFORMS THE NUMERIC REDUCTION OF THE
C*COEFFICIENT MATRIX
```

```
C*
    DO 5 J = LF,NVR
    RE(J) = (0.,0.)
    CE(J) = (0.,0.)
  5 CONTINUE
    J = 1
  6 LZR = LZEROR(J)
    IF (LZR.LE.0) GO TO 7
    RE(LZR) = (0.,0.)
    J = J + 1
    GO TO 6
  7 J = 1
  8 LZC = LZEROC(J)
    IF (LZC.LE.0) GO TO 9
    CE(LZC) = (0.,0.)
    J = J + 1
    GO TO 8
  9 CONTINUE
    DO 10 J = 1,N
    KP = NSEQ(J)
    D = (1.,0.)/DE(KP)
    DE(KP) = D
    LR = LROW(KP)
    LK = LCOL(KP)
    IF (LR.LE.0) GO TO 10
 11 RE(LR) = D*RE(LR)
    LR = LNXTR(LR)
    IF (LR.GT.0) GO TO 11
    IF (LK.LE.0) GO TO 10
 12 K = ITAGC(LK)
    RF = CE(LK)
    LI = LROW(KP)
    IP = ITAGR(LI)
    L = LROW(K)
 13 IF (L.GT.0) GO TO 14
    I = N + 1
    GO TO 15
 14 I = ITAGR(L)
 15 IF (I - IP) 16,17,20
 16 L = LNXTR(L)
    GO TO 13
 17 RE(L) = RE(L) - RF*RE(LI)
```

```
      L = LNXTR(L)
      IF (L.LE.0) GO TO 19
      I = ITAGR(L)
      GO TO 20
   19 I = N + 1
   20 LI = LNXTR(LI)
      IF (LI.LE.0) GO TO 21
      IP = ITAGR(LI)
      GO TO 15
   21 LK = LNXTC(LK)
      IF (LK.GT.0) GO TO 12
      LR = LROW(KP)
  120 K = ITAGR(LR)
      CF = RE(LR)
      LI = LCOL(KP)
      IP = ITAGC(LI)
      L = LCOL(K)
  130 IF (L.GT.0) GO TO 140
      I = N + 1
      GO TO 150
  140 I = ITAGC(L)
  150 IF (I - IP) 160,170,180
  160 L = LNXTC(L)
      GO TO 130
  170 CE(L) = CE(L) - CF*CE(LI)
      L = LNXTC(L)
      IF (L.LE.0) GO TO 190
      I = ITAGC(L)
      GO TO 200
  190 I = N + 1
      GO TO 200
  180 IF (IP.NE.K) GO TO 200
      DE(K) = DE(K) - CF*CE(LI)
  200 LI = LNXTC(LI)
      IF (LI.LE.0) GO TO 210
      IP = ITAGC(LI)
      GO TO 150
  210 LR = LNXTR(LR)
      IF (LR.GT.0) GO TO 120
   10 CONTINUE
      RETURN
      END
```

```
C**CMDSZ****************************************************
C*
      SUBROUTINE CMDSZ(N,NV,LROW,ITAGR,LNXTR,LCOL,ITAGC,
     1                 LNXTC,NSEQ,RE,CE,DE,V)
C*THIS ROUTINE COMPUTES THE SOLUTION VECTOR FOR THE GIVEN
C*RIGHT-HAND SIDE VECTOR
C*
      COMPLEX RE(NV),CE(NV),DE(N),V(N),CF,SUM
      DIMENSION LROW(N),LCOL(N),ITAGR(NV),ITAGC(NV),LNXTR(NV),
     1          LNXTC(NV),NSEQ(N)
      DO 10 J = 1,N
      K = NSEQ(J)
      CF = DE(K)*V(K)
      V(K) = CF
      L = LCOL(K)
   11 IF (L.LE.0) GO TO 10
      I = ITAGC(L)
      V(I) = V(I) - CE(L)*CF
      L = LNXTC(L)
      GO TO 11
   10 CONTINUE
      NN = N - 1
      DO 12 II = 1,NN
      J = N - II
      K = NSEQ(J)
      SUM = V(K)
      L = LROW(K)
   13 IF (L.LE.0) GO TO 14
      I = ITAGR(L)
      SUM = SUM - RE(L)*V(I)
      L = LNXTR(L)
      GO TO 13
   14 V(K) = SUM
   12 CONTINUE
      RETURN
      END


C**********************************************************
```

Appendix 3
Basics of Statistical Analysis

First we introduce a probability density function of a random variable. Let $[x_1, x_2]$ be an interval of the variable. We define now a function $h(x)$ over the interval $[x_1, x_2]$ such that the probability pr $(0 \leq \text{pr} \leq 1)$ of occurrence of the value x' $(x \leq x' \leq x + dx)$ is $h(x)\,dx$. So defined, function $h(x)$ is called a *probability density function* of a random variable, x. The pdf has the following properties:

(1)
$$h(x) \geq 0 \tag{A3.1}$$

(2)
$$\int_a^b h(x)\,dx = \text{pr}\{a \leq x \leq b\}; \quad x_1 \leq a \leq b \leq x_2 \tag{A3.2}$$

(3)
$$\int_{x_1}^{x_2} h(x)\,dx = 1 \tag{A3.3}$$

From (A3.2) we see that the probability that x lies between a and b is the area under the pdf $h(x)$ between a and b. Equation (A3.3) indicates that the total area under the pdf is equal to one.

Figure A3.1 presents two common forms of pdf. The first one is the uniform distribution function. The three general properties defined by (A3.1) to (A3.3) and applied to the uniform distribution of Figure A3.1(a) are

(1)
$$c > 0$$

(2)
$$\text{pr}\{a \leq x \leq b\} = c(b - a)$$

(3)
$$c(x_2 - x_1) = 1$$

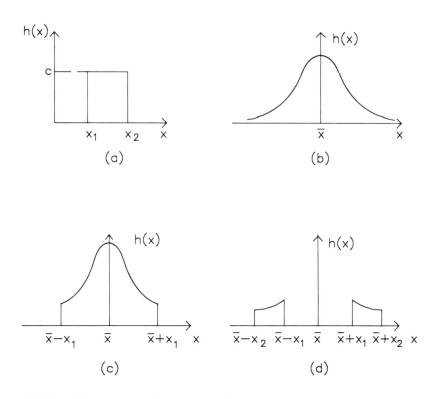

Figure A3.1 Probability distribution functions: (a) uniform distribution; (b) Gaussian (normal) distribution; (c) truncated Gaussian distribution; (d) binomial distribution.

Presented in Figure A3.1(b) is the Gaussian (normal) distribution given by

$$h(x) = \frac{1}{\sigma\sqrt{2\pi}} e^{-(x-\bar{x})^2/2\sigma^2} \tag{A3.4}$$

If two or more variables vary in a correlated way, then a joint probability density function is specified. For example,

$$\int_c^d \int_a^b h(x_1, x_2)\, dx_1\, dx_2 = \mathrm{pr}\{a \le x_1 \le b \text{ and } c \le x_2 \le d\} \tag{A3.5}$$

If individual pdf values are considered for each random variable, that is, the variables are independent, then the joint pdf is the product of the individual pdf values (two random variables x_1 and x_2 are said to be independent if and only if $\mathrm{pr}\{a \le x_1 \le$

b and $c \leq x_2 \leq d\} = \mathrm{pr}\{a \leq x_1 \leq b\} \cdot \mathrm{pr}\{c \leq x_2 \leq d\})$. All forms of pdf have the property (A3.3) that says that their integrals over the entire space are equal to one.

A set of conditions similar to (A3.1) to (A3.3) can be given for a pdf of a random variable taking only a discrete set of values.

The shape of a pdf is characterized by its mean (expectation) and variance. The expected value is a measure of the location of the center of $h(x)$; the variance is an indication of the deviation of $h(x)$ from this center. The definitions of these parameters are the following.

Mean (expected value, average, first central moment):
Discrete case:

$$E[x] = \bar{x} \triangleq \frac{1}{n} \sum_{i=1}^{n} x_i h_i \qquad (A3.6)$$

Continuous case:

$$E[x] = \bar{x} \triangleq \int_{x_1}^{x_2} x h(x)\, dx \qquad (A3.7)$$

Standard deviation (σ), variance (σ^2):
Discrete case:

$$\mathrm{var}[h(x)] = \sigma^2 = E[x - E[x]]^2$$

$$= \frac{1}{n} \sum_{i=1}^{n} (x_i - \bar{x})^2 h_i = \frac{\sum_{i=1}^{n} x_i^2 h_i}{n} - \bar{x}^2 \qquad (A3.8)$$

Continuous case:

$$\mathrm{var}[h(x)] = \sigma^2 = E[x - E(x)]^2 = \int_{x_1}^{x_2} (x - \bar{x})^2 h(x)\, dx \qquad (A3.9)$$

The variance has the following property. If the variance of x is σ^2, then the variance of ax is $a^2\sigma^2$. This property shows the effect of scaling.

When two or more random variables are considered simultaneously, the coupling between variables is important. The basic parameters describing n-dimensional random variable $\mathbf{X} = [x_1, x_2, \ldots, x_n]^T$ are as follows.

The means (expected values):

$$E[x_i] = \bar{x}_i \triangleq \int_{-\infty}^{+\infty} \cdots \int_{-\infty}^{+\infty} x_i h(x_1, x_2, \ldots, x_n)\, dx_1\, dx_2 \ldots dx_n$$

(A3.10)

Variances:

$$\text{var}[x_i] = k_{ii} = \sigma^2 \triangleq E[x_i - E[x_i]]^2 = E[x_i - \bar{x}_i]^2 \quad (A3.11)$$

Covariances:

$$\text{cov}[x_i, x_j] = k_{ij} \triangleq E[(x_i - \bar{x}_i)(x_j - \bar{x}_j)] \quad (A3.12)$$

Variances and covariances (also called the *second central moments*) form a quadratic $n \times n$ covariance matrix $\mathbf{K}$:

$$\mathbf{K} = E[(\mathbf{X} - E(\mathbf{X}))(\mathbf{X} - E(\mathbf{X}))^T] = \begin{bmatrix} k_{11} & k_{12} & \cdots & k_{1n} \\ k_{21} & k_{22} & & \\ \vdots & & \ddots & \vdots \\ k_{n1} & \cdots & & k_{nn} \end{bmatrix} \quad (A3.13)$$

The correlation coefficients r_{ij}, relating variables in pairs, are defined as

$$r_{ij} \triangleq \frac{k_{ij}}{\sigma_i \sigma_j} \quad (A3.14)$$

where σ_i and σ_j are standard deviations of the variables x_i and x_j, respectively. The correlation coefficients form the correlation matrix $\mathbf{R} = [r_{ij}]$.

For a finite number of pairs of values of x_i and x_j the correlation coefficient k_{ij} may be estimated using the expression

$$\hat{k}_{ij} = \frac{1}{n} \sum_{l=1}^{n} (x_{il} - E[x_i])(x_{il} - E[x_j]) \quad (A3.15)$$

If the random variables are uncorrelated, the correlation matrix $\mathbf{R}$ is the unity matrix, $\mathbf{R} = \mathbf{I}$. If random variables $x_1, x_2, \ldots, x_n$ are described by their variances $\sigma_1^2, \sigma_2^2, \ldots, \sigma_n^2$ and the correlation matrix $\mathbf{R}$, then the variance of the random variable

$$x = \sum_{i=1}^{n} x_i \quad (A3.16)$$

equals

$$\sigma_x^2 = \sigma \mathbf{R} \sigma^T \quad (A3.17)$$

For uncorrelated random variables $\mathbf{R} = \mathbf{I}$, (A3.17) becomes

$$\sigma_x^2 = \sum_{i=1}^{n} \sigma_i^2 \quad (A3.18)$$

The Strong Law of Large Numbers

If a random variable f_i is a function of another random variable $x_1, x_2, \ldots, x_N$ of the arbitrary pdf $h(x)$, and the integral

$$\bar{f} = \int_{-\infty}^{+\infty} f(x)h(x)\,dx \tag{A3.19}$$

exists, then

$$\hat{f} = \frac{1}{N}\sum_{i=1}^{N} f(x_i) \tag{A3.20}$$

The central limit theorem states that, for large N,

$$(\hat{f} - \bar{f})/\sqrt{V(\bar{f})/N} \tag{A3.21}$$

is approximately distributed as Gaussian with zero mean and unit variance. In (A3.21),

$$V(\bar{f}) = \overline{f^2} - (\bar{f})^2 \tag{A3.22}$$

$$\overline{f^2} = \int_{-\infty}^{+\infty} f^2(x)h(x)\,dx \tag{A3.23}$$

This result permits the construction of confidence intervals for $\bar{f}$ based on $\hat{f}$. If $K_{\alpha/2}$ is such a point on the x axis that the integral of the Gaussian pdf from $K_{\alpha/2}$ to infinity is $\alpha/2$, then

$$\mathrm{pr}\left\{ |\hat{q} - \bar{q}| \le K_{\alpha/2}\sqrt{\frac{V(\bar{f})}{N}} \right\} = 1 - \alpha \tag{A3.24}$$

which means that

$$|\hat{q} - \bar{q}| \le K_{\alpha/2}\sqrt{\frac{V(\bar{f})}{N}} \tag{A3.25}$$

with probability $1 - \alpha$.

The variance $V(\bar{f})$ is not known and must be estimated as

$$V(\bar{f}) \approx s^2 = \frac{1}{N-1}\sum_{i=1}^{N} \left[f(x_i) - \hat{f} \right]^2$$

$$= \frac{1}{N-1}\left[\left\{ \sum_{i=1}^{N} f^2(x_i) \right\} - N\hat{f}^2 \right] \tag{A3.26}$$

INDEX

The Artech House Microwave Library

Algorithms for Computer-Aided Design of Linear Microwave Circuits, Stanislaw Rosloniec

Analysis, Design, and Applications of Fin Lines, Bharathi Bhat and Shiban K. Koul

Analysis Methods for Electromagnetic Wave Problems, Eikishi Yamashita, ed.

Automated Smith Chart Software and User's Manual, Leonard M. Schwab

Capacitance, Inductance, and Crosstalk Analysis, Charles S. Walker

C/NL: Linear and Nonlinear Microwave Circuit Analysis and Optimization Software and User's Manual, Stephen A. Maas

Computer-Aided Design of Microwave Circuits, K.C. Gupta, R. Garg, and R. Chadha

Design of Impedance-Matching Networks for RF and Microwave Amplifiers, Pieter L.D. Abrie

Digital Microwave Receivers, James B. Tsui

Electric Filters, Martin Hasler and Jacques Neirynck

E-Plane Integrated Circuits, P. Bhartia and P. Pramanick, eds.

Evanescent Mode Microwave Components, George Craven and Richard Skedd

Feedback Maximization, Boris J. Lurie

Filters with Helical and Folded Helical Resonators, Peter Vizmuller

GaAs FET Principles and Technology, J.V. DiLorenzo and D.D. Khandelwal, eds.

GaAs MESFET Circuit Design, Robert A. Soares, ed.

GASMAP: Gallium Arsenide Model Analysis Program, J. Michael Golio, et al.

Handbook of Microwave Integrated Circuits, Reinmut K. Hoffmann

Handbook for the Mechanical Tolerancing of Waveguide Components, W.B.W. Alison

Handbook of Microwave Testing, Thomas S. Laverghetta

HEMTs and HBTs: Devices, Fabrication, and Circuits, Fazal Ali, Aditya Gupta, and Inder Bahl, eds.

High Power Microwave Sources, Victor Granatstein and Igor Alexeff, eds.

Introduction to Microwaves, Fred E. Gardiol

Introduction to Computer Methods for Microwave Circuit Analysis and Design, Janusz A. Dobrowolski

Introduction to the Uniform Geometrical Theory of Diffraction, D.A. McNamara, C.W.I. Pistorius, and J.A.G. Malherbe

Monolithic Microwave Integrated Circuits: Technology and Design, Ravender Goyal, *et al.*

Nonlinear Microwave Circuits, Stephen A. Maas

Optical Control of Microwave Devices, Rainee N. Simons

PLL: Linear Phase-Locked Loop Control System Analysis Software and User's Manual, Eric L. Unruh

Receiving Systems Design, Stephen J. Erst

Scattering Parameters of Microwave Networks with Multiconductor Transmission Lines: Software and User's Manual, A.R. Djordjevic, et al.

Solid-State Microwave Devices, Thomas S. Laverghetta

Stripline Circuit Design, Harlan Howe, Jr.

Terrestrial Digital Microwave Communications, Ferdo Ivanek, *et al.*

Time-Domain Response of Multiconductor Transmission Lines: Software and User's Manual, A.R. Djordjevic, et al.